Environmental Science and Engineering

The ultimate goal of this series is to contribute to the protection of our environment, which calls for both profound research and the ongoing development of solutions and measurements by experts in the field. Accordingly, the series promotes not only a deeper understanding of environmental processes and the evaluation of management strategies, but also design and technology aimed at improving environmental quality. Books focusing on the former are published in the subseries Environmental Science, those focusing on the latter in the subseries Environmental Engineering.

Sughosh Madhav · Himanshu Gupta ·
Arun Lal Srivastav · Sirajuddin Ahmed
Editors

Municipal Sewage and Sludge: Treatment and Disposal Strategies

Management of Sewage and Sludge

Springer

Editors
Sughosh Madhav
Department of Civil Engineering
Jamia Millia Islamia University
New Delhi, Delhi, India

Himanshu Gupta
Department of Chemistry
School of Sciences, IFTM University
Moradabad, Uttar Pradesh, India

Arun Lal Srivastav
Chitkara University
Baddi, Himachal Pradesh, India

Sirajuddin Ahmed
Department of Environmental Sciences
Jamia Millia Islamia University
New Delhi, Delhi, India

ISSN 1863-5520 ISSN 1863-5539 (electronic)
Environmental Science and Engineering
ISBN 978-3-032-07596-3 ISBN 978-3-032-07597-0 (eBook)
https://doi.org/10.1007/978-3-032-07597-0

This Springer imprint is published by the registered company Springer Nature Switzerland AG
The registered company address is: Gewerbestrasse 11, 6330 Cham, Switzerland

Contents

Chapter 1
Municipal Sewage: Impact and Treatment Methods

Saurabh Kumar, Naveen Patel, Brijesh Kumar Yadav, Aditi Baranwal, Akansha Patel, Vinod Kumar Chaudhary, Arun Lal Srivasatav, Sughosh Madhav, and Himanshu Gupta

Abstract Municipal Sewage is highly contaminated wastewater that contains several types of pollutants that including heavy metals, total organic carbon, grease and oil, nutrients, total suspended solids, chemical oxygen demand (COD), biochemical oxygen demand (BOD) and microbial contaminants. The chemical composition of sewage is very complex and variable, depending on different sources. Nowadays, microplastics and other emerging contaminants are also found in sewage. The improper disposal of untreated or partially treated sewage may cause soil and water pollution. The toxic contaminants present in sewage affect aquatic life and human health. The conventional sewage treatment methods include filtration, coagulation, flocculation, adsorption and disinfection. Membrane technology, advanced oxidation processes (AOP), activated carbon method, membrane bioreactor (MBR) and activated sludge process (ASP) are the advanced treatment methods. This chapter includes characterisation of sewage, various treatment methods as well as its impact on human health, and environment.

Keywords Municipal sewage · Disposal · Water pollution · Treatment methods · Human health

---

S. Kumar · B. K. Yadav · A. Baranwal · A. Patel · V. K. Chaudhary
Department of Environmental Sciences, Dr. Rammanohar Lohia Avadh University, Ayodhya, Uttar Pradesh, India

N. Patel (✉)
Department of Civil Engineering, Dr. Rammanohar Lohia Avadh University, Ayodhya, Uttar Pradesh, India
e-mail: naveenrbverma30@gmail.com

A. L. Srivasatav
Chitkara University School of Engineering and Technology, Chitkara University, Solan – 174103, Himachal Pradesh, India

S. Madhav
Department of Civil Engineering, Jamia Millia Islamia, New Delhi, India

H. Gupta
Department of Chemistry, School of Sciences, IFTM University, Moradabad, Uttar Pradesh, India

S. Madhav et al. (eds.), *Municipal Sewage and Sludge: Treatment and Disposal Strategies*, Environmental Science and Engineering,
https://doi.org/10.1007/978-3-032-07597-0_1

1.1 Introduction

By 2030, the world's population is predicted to increase to up to 8.5 billion. Numerous toxins are continuously released into the environment due to rapid growth, urbanisation, and industrialisation, posing a negative impact on both humans and the environment (Patel et al. 2020; Liu et al. 2022). Over the last few decades, domestic waste, agricultural practices, and municipal and industrial effluents have significantly contaminated the world's surface and groundwater (Chauhan et al. 2019; Patel et al. 2021; Khan et al. 2023). Megacities are the result of population density in and around metropolitan areas due to global megatrends like population growth and demographic shift to cities. The high demand for food, energy, and water resources in large urban regions is a result of the dense population, which leads to waste generation, including chemicals used by consumers. Water use will continue to rise in many areas experiencing water scarcity due to increased extraction and climate change (He et al. 2021). Presently, 380 trillion litres of sewage wastewater are produced annually worldwide. It is projected that the amount of wastewater produced worldwide would increase by 24% by 2030 and 51% by 2050. There is a serious freshwater crisis occurring all over the world as a result of the direct discharge of wastewater effluents into surface water (Chaudhary et al. 2022). In India, 72,368 million litres of sewage per day (MLD) were generated by various cities in 2021, and sewage treatment plants (STP) could treat 36,668 MLD of sewage (Vaidya et al., 2024). Sewage is typically the result of wastewater discharged from residential areas. Although the term sewage primarily refers to materials originating after human use, it is a general term for animal fecal matter (Topare et al. 2011). In addition to intact living entities (viruses and bacteria), sewage is predominantly mixture of organic and inorganic materials that combine to form an extremely complicated cocktail. Sewage water's composition and properties change depending on its use, source, and location. Various contaminants like endocrine disruptors, heavy metals, nutrients, suspended particles, medications, and pathogens, are present in sewage discharge along with feces and fecal viruses. These substances are excellent indicators of sewage pollution, much like fecal viruses. Sewage contamination is also indicated by a number of standard physicochemical measures, including pH, temperature, electrical conductivity, chemical oxygen demand (COD), biochemical oxygen demand (BOD), dissolve oxygen (DO) etc. (Barrett et al. 2025). Sewage pollution of surface waters also poses negative impact on animals and natural habitats along with health risk (Prüss-Ustün et al. 2019). Because there are insufficient sewage pipes for disposing of wastewater in lower and middle income countries and leaks in on-site sanitation facilities, significant amount of wastewaters seep into the soil, contaminating aquifers and groundwater (Ahmed et al. 2017). It can also affect water quality, making it unsafe to drink and increasing the risk of diseases like cholera, typhoid, and dysentery. Considering the severity of water pollution, sewage water management must be done properly and sustainably. Different nations use various approaches for sewage management. This approach typically consists of stabilization ponds, wetland, and underground wastewater infiltration systems (Chen et al. 2022). Recent decades have seen the adoption of advanced sewage treatment plant (WWTP) techniques due to the rapid growth in economy and increase in awareness among government and residents regarding environmental protection. Wastewater treatment typically involves primary, secondary, and occasionally advanced treatment processes physical, biological and chemical technologies, such as oxidation ditches, sequencing batch reactors (SBR), anaerobic anoxic–oxic, and conventional activated sludge treatment (Jin et al. 2014). In addition to these approaches, there are other innovative methods like membrane technology, advanced oxidation processes, gasification, incineration, and activated carbon treatment through pyrolysis, among others, that are crucial in resolving sewage management issues in both the surrounding environment and human settlement areas (Raheem et al. 2018). This book chapter emphasizes the current status of sewage pollution in the global and Indian scenario, characteristics and composition of sewage, impact of sewage on the environment (such as land area, water bodies) and human health, as well as conventional and non-conventional treatment approaches that are applied in wastewater treatment plants.

1.2 Characteristics of Sewage

For sewage water to be controlled and treated sustainably, assessment of its characteristics is highly important (Ren et al. 2020). Sewage water characterisation is particularly challenging due to higher cost and lack of resources required to analyse all parameters in a single laboratory. On the other hand, sewage usually contains a complex blend of inorganic and organic materials. The composition of sewage has been discussed in a number of studies to monitor bulk characteristics such as nutrients (phosphorus and nitrogen), total suspended solids (TSS), total organic carbon (TOC), chemical oxygen demand (COD), and biochemical oxygen demand (BOD). The pollution of the surrounding environment is caused by heavy metals like nickel, mercury, chromium, and lead etc. that are present in sewage water along with micropollutants and other organic pollutants (Moretti et al. 2024). One of the contaminants in sewage effluent that persists for a longer duration is heavy metals. They accumulate rapidly throughout the food chain and, unlike organic pollutants, cannot be metabolised, which may threaten human health and cause ecological disturbances (Etsuyankpa et al. 2024). Hydrophilic acids and aquatic humic compounds make up the majority of the dissolved organic matter (DOM) found in canal water. Nutrients and human waste in the water lead to microbial blooms that are majorly harmful to human health. Ecologically induced processes like eutrophication and a drop in dissolved oxygen (DO) are caused by the introduction of such waste products into freshwater bodies. The most commonly used method for determining the oxygen demand for microorganisms' breakdown of organic pollutants in wastewater is the 5-day biological oxygen demand (BOD_5) (Maddah 2022). COD is always higher than BOD in sewage water. The COD measures chemicals that can be oxidised both chemically and physiologically. The COD: BOD ratio is a helpful indicator of the amount of organic material found in wastewaters. All of these parameters get reduced when sewage water is properly treated using treatment processes (Zheng et al. 2013). The resulting effluent is then either recycled for application in agriculture or sent to natural water bodies. Furthermore, bacteria, viruses, algae, and protozoans are examples of microbiological contaminants that act as water quality indicators. The detection, isolation, and identification of various forms of microbiological contaminants in wastewater are usually challenging, costly, and time-consuming (Finnerty et al. 2023). Nowadays, emerging pollutants like microplastics, personal care products, and pharmaceutical products are also identified in sewage water, posing a threat to human health and responsible for issues in organisms, as well as degrading soil quality (Table 1.1).

1.3 Impact of Sewage

Due to the increasing demand of land and water resources caused by industrialization and population growth, the amount of sewage generated has increased immensely. Sewage, sometimes referred to as wastewater, is a mixture of solid waste and water that is dumped into pipelines called sewers from residential and commercial properties and buildings. Numerous studies demonstrate the effect of sewage on the ecosystem, including the land, water, and human health as follows:

1.3.1 Studies on Environment

Wastewater is a complicated resource that has certain positive and negative aspects. The farming communities and society at large can profit greatly from the use of wastewater and its nitrogen contents for crop production. However, the use of wastewater can potentially have detrimental effects on ecosystems and human beings (Mishra et al. 2023). The extensive usage of wastewater that contains toxic wastes and a lack of financing for treatment are likely to accelerate environmental degradation and are responsible for wastewater-borne illnesses. The loss of dissolved oxygen is a result of eutrophication driven by an excessive amount of nutrients (Ansari et al. 2011). Fish survival is impacted by low dissolved oxygen levels in several ways, including increased susceptibility to diseases or quick mortality. The species composition may alter as a result of continuous decreases in dissolved oxygen levels. All aquatic species have distinct tolerance limits and temperature preferences. A rise in average water body temperature causes ecological effects (Owhonka et al. 2021). Water with significant levels of sewage pollution, are responsible for high level of endocrine disruptors in freshwater creatures including fish, mussels, and insects. Recent research has shown that sewage pollution and urban runoff, rather than pesticides, are the main causes of skewed sex ratios in amphibians (Wear et al. 2021). Sewage pollution also affects terrestrial ecosystems, animals and people that live in the nearby areas. Around the world, sewage effluents are frequently used as fertiliser

Table 1.1 Common characteristics of raw sewage water extracted from urban areas

S. No.	Physicochemical parameter	values	References
1.	Electrical Conductivity (mS/m)	70-120	Valdés et al., (2021); Ogunbiyi et al.,(2021); Finnerty et al., (2023)
2.	pH	7-8.5	
3.	Density (kg/m^3)	~1,000	
4.	Temperature (^0C)	15-30	
5.	Surface tension (Dyn/cm^2)	50-60	
6.	Total suspended solids (TSS) mg/L	3,000-8,000	
7.	Volatile suspended solids (mg/L)	200-480	
8.	Absolute viscosity (kg/m.s)	0.001	
9.	BOD (mg/L)	230-560	
10.	COD (mg/L)	500-1200	
11.	Volatile fatty acids (as acetate) (mg/L)	10-80	
12.	Total Organic Carbon (TOC) (mg/L)	15-400	
13.	Cyanide CN^- (mg/L)	0.05	
14.	Chloride Cl^- (mg/L)	600	
15.	Sodium Na^+ (mg/L)	300-3000	
16.	Potassium K^+ (mg/L)	20-200	
17.	Magnesium Mg^{2+} (mg/L)	20-150	
18.	Calcium Ca^{2+} (mg/L)	20-200	
19.	Cadmium Cd (μg/L)	0.1-10	Xu et al., (2022); Finnerty et al., (2023)
20.	Sulphate $SO^{2-}{}_4$ (mg/L)	20-800	
21.	Bromide Br^- (mg/L)	0.5-50	
22.	Copper Cu (μg/L)	10-200	
23.	Nickel Ni (μg/L)	10-2500	
24.	Lead Pb (μg/L)	0.5-160	
25.	Zinc Zn (μg/L)	10-1600	
26.	Boran B (mg/L)	0.5-20	
27.	Total nitrogen (mg/L)	0.1-17	
28.	Total phosphorus (mg/L)	0.1-30	

(continued)

Table 1.1 (continued)

S. No.	Physicochemical parameter	values	References
29.	Chromium Cr (μg/L)	0.5-120	
Microbial component (number of organisms per 100 mL)			
1.	*Escherichia coli*	5 x 10^8	García-Aljaro et al., (2019)
2.	*Salmonella*	1-10^4	
3	*Coliforms*	~ 10^{13}	
4.	*Rotavirus*	~ 10^7	
5.	*Campylobacter*	100-10^5	
6.	*Fecal Streptococci*	10^8	
7.	*Enterovirus*	1-10^6	
Emerging contaminants (mg/L)			
1.	Poly aromatic hydrocarbons (PAH)	0.5-2.5	Vaz-Ramos et al., (2025)
2.	Phenol	less than 10	Fseha et al.,(2023)
3.	Methylene chloride	1- 10	Wu et al., (2024)
4.	Chloroform	0.02-0.15	Sharma et al., (2021)
5.	Nonylphenol ethoxylates	0.01-0.08	Klontza et al., (2010)
6.	Microplastics	–	–
7.	Disinfection byproducts (DBPs)	–	–

and to water croplands. The presence of heavy metals, medications, endocrine disruptors, hydrocarbons, and other substances in these effluents are usually left untreated, poses risk to human health and the environment.

According to few recent studies, the placement of treated or untreated sewage waste on the surface of wetlands and bottomland forests reduces biodiversity, among other effects on terrestrial ecosystems. Concentrated sewage effluent, sludge and other sewage waste are especially loaded with pollutants, chemicals, and toxins. Groundwater contamination and the food chain are the two possible ways through which heavy metals as well as other pollutants may reach neighbouring ecosystems (Manzetti et al. 2015). The soil, freshwater, and marine ecosystems are negatively impacted by numerous newly discovered pollutants. In terrestrial ecosystems, endocrine-disrupting chemicals (EDCs), which are frequently found in household sewage, cause severe issues. The risk of exposure through tuber vegetables is higher for pharmaceuticals that primarily get collected in the roots, such as triclocarban, triclosan, etc., whereas the risk of exposure through leafy vegetables is higher for compounds that accumulate in the leaves, like carbamazepine, diclofenac, dilantin, etc. The health risks of pathogen exposure are undeniable, and the presence of a variety of pathogens in sewage water should not be neglected. The majority of the microorganisms found in sewage come from faeces. Sewage water is where enteric bacteria are most frequently discovered. The pathogenic oocytes can also be directly exposed through the mouth or through animal carriers (Khatri et al. 2021). Furthermore, sewage water has certain beneficial effects. For example, it is usual in agriculture to apply fertilisers containing nitrogen (N), potassium (K) and phosphorus (P), to the soil because crops require the highest levels of these elements and the soil contains fewer of them. One of the main reasons sewage water is recommended for irrigation is that it contains higher quantities of N, P, and K. Long-term irrigation is the only way for these nutrients to concentrate in the soil, which eventually reduces the demand for fertilisers and the related expenses. Sewage water also contains other vital elements like calcium (Ca), magnesium (Mg), and sulphur (S) in their salt forms that help in increasing the soil fertility (Kominko et al. 2024).

1.3.2 Studies on Human Health

The most common diseases linked to untreated drinking as well as recreational waters are those brought on by bacteria, viruses, and protozoa. Human and animal wastes are the primary sources of these microbiological pollutants in wastewater. These include a diverse range of bacteria, viruses, and protozoa that can reach into receiving water bodies or drinking water supplies (Naidoo and Olaniran 2014). Numerous microbial infections found in wastewater can result in long-term physiological consequences that are expensive, including degenerative cardiovascular disease and gastrointestinal tract conditions, including ulcer in stomach. More than 100 distinct kinds of enteric viruses that might cause illness or infection are expelled by humans. Human liver cancer is occasionally believed to be linked to cyanobacterial toxin exposure through the drinking water line; typically, this exposure occurs through tainted drinking water or recreational water contact. Numerous nations, including Zimbabwe, United States of America, United Kingdom, China, Canada, Brazil, and Australia, have reported health issues linked to cyanotoxins (Owhonka et al. 2021). In freshwater environments, sewage pollution is responsible for a rise in the concentration of pollutants and spread of human diseases. These heavy metals are also present in sewage, such as chromium, arsenic, mercury, and cadmium, can seep into groundwater and enhancing the concentrations that could be harmful to human health (Prüss-Ustün et al. 2019). Additionally, pollutants can also cause birth abnormalities and infertility, along with other reproductive problems. Exposure to such pollution increases the likelihood of negative health impacts for vulnerable groups, including children of developing age and the elderly with compromised immune systems. Targeted interventions, more stringent laws, and public health campaigns can help in lowering pollution levels and limit exposure to dangerous pollutants are necessary to protect these vulnerable groups and preserve the well-being and health of all people.

1.4 Treatment Methods of Sewage

Water or wastewater treatment plants utilise several techniques to remove contaminants such as chemicals, microbes, and turbidity that could be harmful to human health (Saravanan et al. 2021). Conventional and non-conventional techniques are the two types of water treatment techniques that are accessible (Yazdani et al. 2024). Figure 1.1 show the different treatment techniques available for the treatment of sewage. These methods are typically used in wastewater treatment plants (WWTP) and sewage treatment plants (STP).

1.4.1 Conventional Method for Sewage Treatment

Physical, chemical, and biological treatments are the techniques of the traditional water treatment process. The traditional method of treating water entails first screening the raw water to remove large suspended particles (primary (mechanical) treatment), then aerating the water by adding air (oxidation), which encourages secondary treatment by microbial growth to biodegrade pollutants. Tertiary water treatment includes a variety of physical techniques like filtration and adsorption as well as chemical treatments like disinfection and coagulation (Sangamnere et al. 2023).

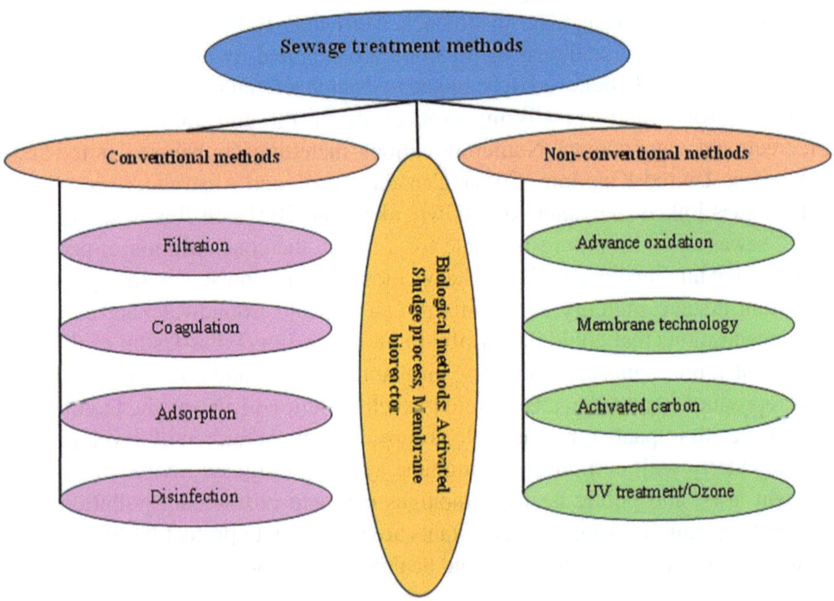

Fig. 1.1 Treatment technique of sewage treatment

1.4.1.1 Filtration

In filtration technique, the water passing through filters (porous media) like charcoal, gravel, sand etc., eliminates microbes, fine suspended particles, and other contaminants from sewage water. The foundation of filtration is based on adsorption process, which involves liquid molecules adhering to surface of solid materials with a large surface area, called adsorbents. Activated carbon, because of its large surface area and extremely porous structure, is considered as most widely used adsorbent. Activated carbon filtration, because of its cost-effectiveness in removing the micropollutants, has recently attracted more interest (Kumar et al. 2024).

1.4.1.2 Coagulation

The basic but essential process for treating water and wastewater is coagulation. When positive charge coagulants, balance negative charge on colloidal particles, floc is produced during the coagulation process (Bratby 2016). Flocculation accelerates the settling of coagulated particles in a solution or wastewater by dramatically increasing their size and density. Coagulants may come from natural or chemical sources. Dolomite (calcium magnesium carbonate) or limestone (calcium carbonate) is the two materials used for creating lime. Both coagulants are effective in purifying water of a variety of chemical and physical impurities. However, chemical coagulants produce sludge and residues in treated water that are detrimental (Sukmana et al. 2021). Algal, bacterial, and fungal substances are examples of natural coagulants, as are components derived from plants and animals, such as fruit waste and plant extracts. As they are sustainable, biodegradable, plant-based coagulants are gaining acceptance as alternatives to traditional inorganic coagulants (Benalia et al. 2024).

1.4.1.3 Adsorption

Adsorption is a widely used, affordable, easy, and sustainable wastewater treatment process that removes harmful substances from contaminated water, including heavy metals like cadmium and mercury, organic chemicals like organic colors, and pharmaceuticals (Smith et al. 2009). Ion-exchange resins, inorganic materials and activated carbon are examples of conventional adsorbents. Biosorbents, activated carbon from solid wastes, and adsorbents made of natural materials are examples of non-conventional adsorbents. Walnut-based adsorbent (Juglans regia) formed from walnut shell has been found to be a promising biosorbent to remove different metal ions from water (Crini et al. 2019).

1.4.1.4 Disinfection

Disinfection refers to the killing, eliminating, or destroying of harmful bacteria present in water. A variety of physical and chemical disinfection procedures were utilised for the disinfection of water. Most used extensively around the globe is chlorination, which is affordable, simple, and efficient (Gelete et al. 2020). Chlorine is used to prevent and eliminate odours and iron, as well as to disinfect water and sewage wastewater. However, the main limitation of the method is the generation of carcinogenic byproducts as a result of chemical disinfectants' reaction (Jin et al. 2014; Mazhar et al. 2024).

1.4.2 *Non-conventional Methods for Sewage Treatment*

Non-conventional methods for treatment of sewage include: reverse osmosis, ozonation, ultraviolet (UV) light, biological activated carbon treatment, photocatalysis, bioaugmentation, membrane technology (ultrafiltration), and improved oxidation processes, etc. (Yazdani et al. 2024). In comparison with the standard approach, the non-conventional method uses more advanced tools and technologies. Non-conventional approaches utilise more sophisticated technology and have less of an adverse environmental impact than conventional treatments (Sangamnere et al. 2023).

1.4.2.1 Membrane Technology

Membrane technology is mostly used to enhance the efficiency of the filtration method. Membrane filtration is a popular advanced wastewater treatment method that utilises a semipermeable membrane to filter dissolved pollutants, particles, and microorganisms. Ultrafiltration (UF), nanofiltration (NF), and Reverse osmosis (RO) are a few membrane filtration techniques (Aziz et al. 2024). When wastewater is subjected to high pressure in an RO system, it is forced through a membrane that only permits water molecules to flow through it, leaving behind organic compounds, dissolved salts, and other impurities. Because UF uses membranes with bigger pores than NF and RO, macromolecules and colloidal particles are retained, while larger particles like germs, viruses, and suspended solids can pass through it (Lin et al. 2024).

1.4.2.2 Ozonation

Ozonation is modern treatment method which eliminates bacteria, organic contaminants, and substances that cause odours. Ozone gas (O_3) is created and utilised in the wastewater, where it undergoes an oxidation reaction with the pollutants.

Ozone molecules break down during ozonation to produce very reactive oxygen radicals, namely hydroxyl radicals ($OH^{\bullet}$), that target and oxidize organic components in wastewater. Complex chemical molecules are broken down by this oxidation process into simpler, less dangerous compounds like water, CO_2, and other byproducts (Lim et al. 2022).

1.4.2.3 Advanced Oxidation Processes (AOP)

In AOP, highly reactive oxygen species (ROS) such as hydroxyl radicals ($OH^{\bullet}$) are produced by using strong oxidizing agents. Other radicals, including superoxide or sulphate radicals, may also be present in AOP. Ozone-based, UV-based, catalytic, Fenton's reagent, photo-Fenton, electrochemical, and non-$OH^{\bullet}$ producing AOPs are some of the types of AOP. ROS from these activities interact with organic contaminants; complex molecules are broken down into less dangerous byproducts like carbon dioxide, water, and minerals. AOP are efficient in breaking down various pollutants, like industrial chemicals, dyes, pesticides, and pharmaceuticals. Additionally, it can eliminate pathogens and enhance the treated wastewater quality (Lupu et al. 2023).

1.4.2.4 Activated Carbon Treatment

In recent years, activated carbon made from industrial and agricultural waste has been frequently used to remove pollutant from contaminated water by adsorption. Carbon is activated (modified) in order to boost the adsorption capabilities of the adsorbent. Any carbonaceous material (wood, coal, plant parts, etc.) can be used to make activated carbon (Ahmad et al. 2024). The process of carbonisation means the thermal breakdown (pyrolysis) of biomass at high temperatures (400–800 °C) in an oxygen-free environment, producing biochar, with a low surface area, that further requires activation to create a more porosity and higher surface area. The carbon is activated by oxidising biochar at temperatures between 800 and 1100 °C. Carbon activation is also accomplished via magnetic, microwave, and physicochemical methods. The main application of activated carbon technology is removal of inorganic (metals, non-metals) and organic (dyes, pesticides, etc.) contaminants from contaminated water (Garg et al. 2023).

1.4.3 Biological Treatment

1.4.3.1 Activated Sludge Process (ASP)

ASP is a primary method for treating sewage in biological treatment. This technique is used for the elimination of many kinds of organic waste, including phosphorus and nitrogen. Ardern and Lockett developed this procedure in 1912–1914. An aeration tank serves as a bioreactor in this process, while a settling tank separates wastewater and activated sludge solid from treated water by acting as a fine clarifier. The settled activated sludge is then transported from the clarifier to the influent of the aeration tank using a return activated sludge (Wanner and Torregrossa 2017). *Achromobacter*, *Escherichia coli*, *Sphaerotilus*, *Bacillus*, *Citromonas*, and *Nocardia* are among the bacteria typically employed in the activated sludge process to produce a biological floc. The temperature of the activated sludge determines the type of microbial species to be selected. Temperature, COD, BOD, food for microbes, and oxygen all affect the efficiency of biological degradation (Katare et al. 2023).

1.4.3.2 Membrane Bio-Reactor (MBR)

MBR is an innovative treatment method that combines membrane filtration with biological treatment processes. Membrane filtration used along with MBR to separate bacteria and materials that are broken down by them in the reactor (Iorhemen et al. 2016). Because of the greatly increased membrane efficiency and selectivity at lower costs, it demonstrates great promise in producing high-quality effluent. MBRs are more economical than conventional bioreactors because they can eliminate both difficult and simple biodegradable organic pollutants. The porous membranes of MBR systems frequently contain polymeric materials to keep bacteria, sediments, and other tiny particles out of the treated wastewater (Jijingi et al. 2024).

1.5 Conclusion

Sewage water management is critical for public health as well as the environment. Globally, it can be observed that sewage water generation, characterisation of sewage and treatment depend on the economy of the country. Developing countries such as India, Brazil, Colombia, Egypt, Sri Lanka, Bangladesh and Mexico are using confined anaerobic and advanced treatment systems in wetlands or synthetic lagoon systems to treat sewage water from domestic sources. Moreover, this book chapter highlights the composition and characteristics of sewage that depend on the source and location. In sewage water, some emerging pollutants such as personal care products, pharmaceutical contaminants, disinfection byproduct and microplastics are also reported to be hazardous for the nearby environment (such as water bodies and soil ecosystems) and human population. Treatment approaches i.e. conventional (filtration, adsorption, flocculation, coagulation, disinfection, etc.) and non-conventional (membrane technology, AOP, ozonation, as well as biological methods like ASP, MBR), are discussed in this book chapter.

Acknowledgements The authors would like to thank Dr. Rammanohar Lohia Avadh University, Ayodhya for their support and help.

References

Ahma, AA, Alwahbi A, Al Khatib LA, Dammag H (2024) Biomass-based activated carbon. In: From biomass to biobased products. IntechOpen

Ahmed MB, Zhou JL, Ngo HH, Guo W, Thomaidis NS, Xu J (2017) Progress in the biological and chemical treatment technologies for emerging contaminant removal from wastewater: a critical review. J Hazard Mater 323:274–298

Ansari AA, Gill SS, Khan FA (2011) Eutrophication: threat to aquatic ecosystems. Eutrophication: causes, consequences and control, pp 143–170

Aziz S, Mazhar AR, Ubaid A, Shah SMH, Riaz Y, Talha T, Jung DW (2024) A comprehensive review of membrane-based water filtration techniques. Appl Water Sci 14(8):169

Barrett LR, Beasy P, Palacios Delgado YM, Boyce JD, Leder K, McCarthy DT, Henry R (2025) Beyond borders: a systematic review and meta-analysis of human-specific faecal markers across geographical settings. Critical Rev Environ Sci Technol 1–18

Bratby J (2016) Coagulation and flocculation in water and wastewater treatment. IWA publishing

Benalia A, Derbal K, Amrouci Z, Baatache O, Khalfaoui A, Pizzi A (2024) Application of plant-based coagulants and their mechanisms in water treatment: a review. J Renewable Mater 12(4):667–698

Chauhan D, Rai D, Khan MZA, Patel N (2019) A statistical analysis of consumer's willingness to pay for improved water supply system: a study of some villages of Uttar Pradesh India. Inter J Res Adv Techno 10(6):15–23

Chaudhary VK, Srivastav AL, Rinku NP, Singh RB, Bhardwaj AK (2022) Water quality index of Yamuna River in Delhi region due to community lockdown amid COVID-19 pandemic

Chen P, Zhao W, Chen D, Huang Z, Zhang C, Zheng X (2022) Research progress on integrated treatment technologies of rural domestic sewage: a review. Water 14(15):2439

Crini G, Lichtfouse E, Wilson LD, Morin-Crini N (2019) Conventional and non-conventional adsorbents for wastewater treatment. Environ Chem Lett 17:195–213

Etsuyankpa MB, Augustine AU, Musa ST, Mathew JT, Ismail H, Salihu AM, Mamman A (2024) An overview of wastewater characteristics, treatment and disposal: a review. J Appl Sci Environ Manag 28(5):1553–1572

Finnerty CT, Childress AE, Hardy KM, Hoek EM, Mauter MS, Plumlee MH, Elimelech M et al (2023) The future of municipal wastewater reuse concentrate management: drivers, challenges, and opportunities. Environ Sci Technol 58(1):3–16

Fseha YH, Shaheen J, Sizirici B (2023) Phenol contaminated municipal wastewater treatment using date palm frond biochar: optimization using response surface methodology. Emerg Contam 9(1):100202

García-Aljaro C, Blanch AR, Campos C, Jofre J, Lucena F (2019) Pathogens, faecal indicators and human-specific microbial source-tracking markers in sewage. J Appil Microbiol 126(3):701–717

Garg R, Garg R, Sillanpää M, Alimuddin Khan MA, Mubarak NM, Tan YH (2023) Rapid adsorptive removal of chromium from wastewater using walnut-derived biosorbents. Sci Rep 13(1):6859

Gelete G, Gokcekus H, Ozsahin DU, Uzun B, Gichamo T (2020) Evaluating disinfection techniques of water treatment. Desalin Water Treat 177:408–415

He C, Liu Z, Wu J, Pan X, Fang Z, Li J, Bryan BA (2021) Future global urban water scarcity and potential solutions. Nat Commun 12(1):4667

Iorhemen OT, Hamza RA, Tay JH (2016) Membrane bioreactor (MBR) technology for wastewater treatment and reclamation: membrane fouling. Membranes 6(2):33

Jijingi HE, Yazdia SK, Abakr YA, Etim E (2024) Evaluation of membrane bioreactor (MBR) technology for industrial wastewater treatment and its application in developing countries: a review. Case Stud Chem Environ Eng 100886

Jin L, Zhang G, Tian H (2014) Current state of sewage treatment in China. Water Res 66:85–98

Katare AK, Tabassum A, Sharma AK, Sharma S (2023) Treatment of pharmaceutical wastewater through activated sludge process—a critical review. Environ Monit Assess 195(12):1466

Khatri N, Tyagi S, Tharmavaram M, Rawtani D (2021) Sewage water: from waste to resource–a review. Environ Claims J 33(2):108–135

Khan WU, Ahmed S, Dhoble Y, Madhav S (2023) A critical review of hazardous waste generation from textile industries and associated ecological impacts. J Indian Chem Soc 100(1):100829

Klontza EE, Koukouraki EE, Diamadopoulos E (2010) Determination of nonylphenol ethoxylates in wastewater samples with SPME/GC-MS. Desalination and Water Treatment, 23(1-3):80–88

Kominko H, Gorazda K, Wzorek Z (2024) Sewage sludge: a review of its risks and circular raw material potential. J Water Proc Eng 63:105522

Kumar J, Choudhary M, Dikshit PK, Kumar S (2024) Recent advancements in utilizing plant-based approaches for water and wastewater treatment technologies. Cleaner Water, 100030

Lim S, Shi JL, von Gunten U, McCurry DL (2022) Ozonation of organic compounds in water and wastewater: a critical review. Water Res 213:118053

Lin H, Wang Y, Dong Y (2024) A review of methods, influencing factors and mechanisms for phosphorus recovery from sewage and sludge from municipal wastewater treatment plants. J Environ Chem Eng 12(1):111657

Liu K, Chen J, Sun F, Liu Y, Tang M, Yang Y (2022) Historical development and prospect of intimately coupling photocatalysis and biological technology for pollutant treatment in sewage: a review. Sci Total Environ 835:155482

Lupu GI, Orbeci C, Bobirică L, Bobirică C, Pascu LF (2023) Key principles of advanced oxidation processes: a systematic analysis of current and future perspectives of the removal of antibiotics from wastewater. Catalysts 13(9):1280

Maddah HA (2022) Predicting optimum dilution factors for BOD sampling and desired dissolved oxygen for controlling organic contamination in various wastewaters. Int J Chem Eng 2022(1):8637064

Manzetti S, van der Spoel D (2015) Impact of sludge deposition on biodiversity. Ecotoxicology 24:1799–1814

Mazhar MA, Madhav S, Ahmed S, Kumar P, Springer C (2024) Drinking water chlorination and disinfection by-products: formation, history, and regulations. In: Drinking water disinfection by-products: sources, fate and remediation. Springer Nature Switzerland, Cham, pp 21–34

Mishra S, Kumar R, Kumar M (2023) Use of treated sewage or wastewater as an irrigation water for agricultural purposes-Environmental, health, and economic impacts. Total Env Res Themes 6:100051

Moretti A, Ivan HL, Skvaril J (2024) A review of the state-of-the-art wastewater quality characterization and measurement technologies. Is the shift to real-time monitoring nowadays feasible? J Water Proc Eng 60:105061

Naidoo S, Olaniran AO (2014) Treated wastewater effluent as a source of microbial pollution of surface water resources. Int J Environ Res Public Health 11(1):249–270

Ogunbiyi O, Saththasivam J, Al-Masri D, Manawi Y, Lawler J, Zhang X, Liu Z (2021) Sustainable brine management from the perspectives of water, energy and mineral recovery: a comprehensive review. Desalination 513:115055

Owhonka A, Fubara EF, Justice OB (2021) Wastewater quality-It's impact on the environment and human physiology: a review. Int J Adv Res Innov 9:43–58

Patel N, Khan MD, Shahane S, Rai D, Chauhan D, Kant C, Chaudhary VK (2020) Emerging pollutants in aquatic environment: source, effect, and challenges in biomonitoring and bioremediation-a review. Pollution 6(1):99–113

Patel A, Chaudhary VK, Singh A, Rai D, Patel N (2021) Pollution in river Ganga due to heavy metal toxicity and various mitigation plans-a review. Ecol Environ Conserv 27(1):382–393

Prüss-Ustün A, Wolf J, Bartram J, Clasen T, Cumming O, Freeman MC, Johnston R et al (2019) Burden of disease from inadequate water, sanitation and hygiene for selected adverse health outcomes: an updated analysis with a focus on low-and middle-income countries. Int J Hyg Environ Health 222(5):765–777

Raheem A, Sikarwar VS, He J, Dastyar W, Dionysiou DD, Wang W, Zhao M (2018) Opportunities and challenges in sustainable treatment and resource reuse of sewage sludge: a review. Chem Eng J 337:616–641

Ren J, Ren X, Liu Y, Man Y, Toniolo S (2020) Sustainability assessment framework for the prioritization of urban sewage treatment technologies. In: Waste-to-Energy. Academic Press, pp 153–176

Saravanan A, Kumar PS, Jeevanantham S, Karishma S, Tajsabreen B, Yaashikaa PR, Reshma B (2021) Effective water/wastewater treatment methodologies for toxic pollutants removal: Processes and applications towards sustainable development. Chemosphere 280:130595

Sangamnere R, Misra T, Bherwani H, Kapley A, Kumar R (2023) A critical review of conventional and emerging wastewater treatment technologies. Sustainable Water Res Manage 9(2):58

Sharma N, Mohapatra S, Padhye LP, Mukherji S (2021) Role of precursors in the formation of trihalomethanes during chlorination of drinking water and wastewater effluents from a metropolitan region in western India. J water process eng 40:101928

Smith KM, Fowler GD, Pullket S, Graham ND (2009) Sewage sludge-based adsorbents: a review of their production, properties and use in water treatment applications. Water Res 43(10):2569–2594

Sukmana H, Bellahsen N, Pantoja F, Hodur C (2021) Adsorption and coagulation in wastewater treatment–review. Prog Agric Eng Sci 17(1):49–68

Topare NS, Attar SJ, Manfe MM (2011) Sewage/wastewater treatment technologies: a review. Sci Revs Chem Commun 1(1):18–24

Vaidya R, Verma K, Kumar M, Hoysall C, Rao L (2024) Assessing wastewater management challenges in developing countries: a case study of India, current status and future scope. Environ Dev Sustain 26(8):19369–19396

Valdés H, Saavedra A, Flores M, Vera-Puerto I, Aviña H, Belmonte M (2021) Reverse osmosis concentrate: physicochemical characteristics, environmental impact, and technologies. Membranes 11(10):753

Vaz-Ramos J, Le Calve S, Begin S (2025) Polycyclic aromatic hydrocarbons in water environments: impact, legislation, depollution processes and challenges, and magnetic iron oxide/graphene-based nanocomposites as promising adsorbent solutions. J Hazardous Mater 490:137726

Wanner J, Torregrossa M (2017) Aeration tank and secondary clarifier as one system. Activated Sludge Separation Probl Theory, Control Measures, Pract Exper 67

Wear SL, Acuña V, McDonald R, Font C (2021) Sewage pollution, declining ecosystem health, and cross-sector collaboration. Biol Cons 255:109010

Wu M, Zhao D, Gu B, Wang Z, Hu J, Yu Z, Yu J (2024) Efficient degradation of aqueous dichloromethane by an enhanced microbial electrolysis cell: degradation kinetics, microbial community and metabolic mechanisms. J Environ Sci 139:150–159

Xu Z, Zhang Q, Li X, Huang X (2022) A critical review on chemical analysis of heavy metal complexes in water/wastewater and the mechanism of treatment methods. Chem Eng J 429:131688

Yazdani R, Harirchi S, Lak M, Mirshafiei M, Nojoumi SA, Ramezani M Yazdian F et al (2024) Recent advances in conventional and modern wastewater treatment approaches. Microbial Nexus for Sustainable Wastewater Treatment, 83–115

Zheng C, Zhao L, Zhou X, Fu Z, Li A (2013) Treatment technologies for organic wastewater. In: Water treatment. IntechOpen

Raheem A, Sikarwar VS, He J, Dastyar W, Dionysiou DD, Wang W, Zhao M (2018) Opportunities [illegible]

Chapter 2
Characterization of Sewage and Sludge

Pawan Kumar Shukla, Amrita Kaushik, Amanpreet Kaur, and Monika Chauhan

Abstract Sewage and sludge characterisation is essential to comprehending their effects on the environment and improving treatment procedures. The physical, chemical, and biological characteristics of sewage and the accompanying sludge produced during wastewater treatment are the main subjects of this research. A number of parameters were examined, including pH, temperature, total suspended solids (TSS), chemical and biological oxygen demands (BOD and COD), nutrients (phosphorus and nitrogen), heavy metals, and microbial composition. The findings show that the composition varies significantly according on whether the sewage comes from a mixed, industrial, or residential source. The high organic content, nutritional richness, and possible pollutants, such as harmful metals and pathogenic bacteria, were noted during sludge characterisation. These results are essential for choosing the best treatment methods, guaranteeing secure disposal or reuse, and lowering environmental and public health hazards. The research emphasises how crucial thorough characterisation is to promoting resource recovery and sustainable wastewater treatment techniques.

Keywords Domestic sewage · Characterization · Mass spectrometry · Nutritional composition

P. K. Shukla (✉) · A. Kaur
Department of Chemistry, School of Sciences, IFTM University, Moradabad, Uttar Pradesh, India
e-mail: pks3030@rediffmail.com

A. Kaushik
School of Environmental and Civil Engineering, Seneca Polytechnic, Toronto, Canada

M. Chauhan
School of Basic Sciences, Galgotia University, Greater Noida, Gautam Budh Nagar, Uttar Pradesh, India

S. Madhav et al. (eds.), *Municipal Sewage and Sludge: Treatment and Disposal Strategies*, Environmental Science and Engineering,
https://doi.org/10.1007/978-3-032-07597-0_2

2.1 Introduction

Characterization of solid waste and sewage is an important step in understanding the public health and environmental hazards associated with them. The composition and type of wastes generated are changing with the change in our lifestyles (Gusiatin et al. 2024). This topic is very important for reducing public health risks, ensuring regulatory compliance, and improving recovery processes for sustainable waste treatment practices (Sava et al. 2024).

With this scenario, it is evident that the microplastics, hazardous chemicals, and carcinogenic elements in water, wastewater, sludges and sewage are growing day by day (Kumar and Gupta 2020; Gupta and Gupta 2016; Singh et al. 2018). Characterization of sewage and sludge types develops proper management strategies (Messaoudi et al. 2024). There is a wide range of pollutants found in various types of sewage, including organic matter, nutrients such as nitrogen and phosphorus, heavy metals, and emerging contaminants, along with pathogens (Brisolara and Ochoa 2016). Table 2.1 highlights different key components in sewage and sludge.

By systematically analyzing these components, researchers and wastewater treatment professionals can gain a clearer understanding of the potential risks linked to discharging treated effluent into natural water bodies. This understanding enables them to devise more effective treatment strategies tailored to specific waste streams (Abrahão Cezarini et al. 2024).

The sludge and sewage characterization process leads to the optimization of treatment processes (Mladenovič et al. 2017). There are several technologies available for contaminant removal from sewage and sludge, using physical, chemical, and biological methods. Proper understanding of the chemical and physical properties of these materials includes pH, biochemical oxygen demand (BOD), total suspended solids (TSS), total dissolved solids (TDS), and chemical oxygen demand (COD) that enable operators to select the correct treatment process in order to enhance

Table 2.1 Key components in sewage and sludge

Components	Potential risks	Examples	References
Organic matter	Increases BOD, eutrophication	Carbs, proteins, fats	Cezarini et al. (2024)
Heavy metals	Toxicity to human and as well as aquatic life	Lead, mercury, cadmium	Gusiatin et al. (2024)
Pathogens	Public health hazards	Bacteria, viruses, protozoa	Łomińska-Płatek et al. (2024)
Emerging components	Heavily persistent in the environment	Pharmaceuticals, microplastics, hormones	Raveendravarrier et al. (2025)
Hazardous chemicals	Carcinogenic and mutagenic effects	Pesticides, industrial chemicals	Yáñez-Hernández et al. (2024)

Table 2.2 Treatments for contamination removal

Treatments	Techniques involved	Target contaminants	Advantages	References
Physical treatment	Filtration, sedimentation	Suspended solids, large particles	Primary cost-effective method, can be easily operated	Soni et al. (2024)
Chemical treatment	Coagulation, precipitation, oxidation	Nutrients, toxins	Effective for specific contaminants	Jin et al. (2025),
Biological treatment	Activated sludge, Biofilm reactors	Pathogens and organic matter	Sustainable to reduce pollutants	Li (2025)

the desired contaminant removal while minimizing toxic sludge production, hence cost-effectiveness with energy-saving curated treatment mechanisms for optimum outcomes. Besides, sludge characterization is also an excellent approach in terms of sustainable downstream management and possible reuse as biosolids. These biosolids will be appropriate for repurposing in agriculture, landscaping, or even as a resource for energy generation (Li and Yuan 2024, Ghazali et al. 2024).

Another critical aspect of sewage and sludge characterization is regulatory compliance. Many environmental regulations put strong limits on the discharge of pollutants and the quality of biosolids (Kujawiak et al. 2024). The same can be made sure that the wastewater treatment facilities abide by these legal requirements through effective characterization studies and thus escape penalties while helping to contribute to national and local goals on the environment (Brisolara and Ochoa 2016; Kominko et al. 2024). Table 2.2 represents various methods for contamination removal.

The importance of characterizing sewage and sludge extends to public health. Pathogens and toxic substances in wastewater can pose risks to communities and ecosystems (Chen et al. 2024a, b) Characterization efforts focus on the identification and quantification of harmful agents to address them with appropriate treatment and disposal measures (Rezig et al. 2024). This becomes of particular importance with emerging contaminants, which may not be understood in their entirety but might have very critical health implications (Paffrath et al. 2024; Khalil et al. 2024). Figure 2.1 represents different stages of Sludge passing through the sewage treatment plant.

2.1.1 Types of Sewage and Sludge

Sewage and sludge may be classified according to origin, treatment processes, pathogen content, which define further uses and methods of disposal (Gusiatin et al. 2024).

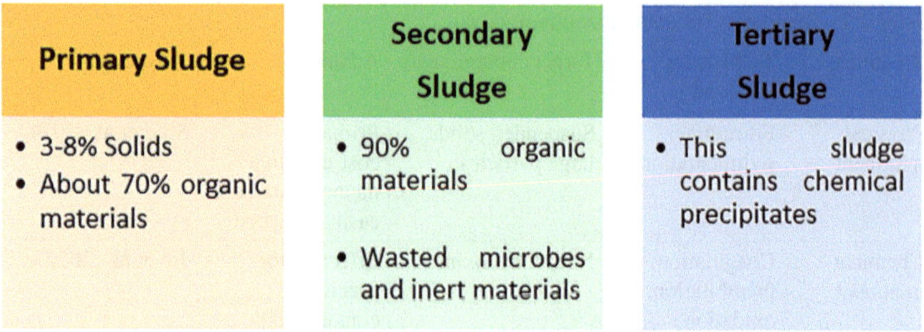

Fig. 2.1 Different stage of the sludge and their composition

2.1.1.1 Domestic Sewage

Domestic sewage is primarily wastewater originating from households, including human excreta, food residues, and many other trace chemicals. This type of sewage normally undergoes treatment in municipal wastewater treatment facilities to eliminate pathogens and pollutants before being discharged into the environment or being used again (Raveendravarrier et al. 2025).

2.1.1.2 Industrial Wastewater

Industrial sewage originates from manufacturing and industrial processes; it can contain a wide variety of pollutants such as heavy metals, micro-pollutants, and chemicals not normally present in domestic sewage. The treatment of this type of sewage often presents special challenges, as the normal methods used in municipal treatment may not effectively remove the pollutants (Li et al. 2023).

2.1.1.3 Class A Sludge

Class A sludge has been treated to kill specified pathogens, and therefore, it can be used without any restriction in agricultural uses and also freely sold to the public. Treatments to achieve Class A status may include pasteurization, heat drying, and thermophilic composting. This class of sludge poses the least risk to public health because of the undetectable presence of pathogens (Chang et al. 2023).

2.1.1.4 Class B Sludge

Class B sludge, on the other hand, may still contain some pathogen content, and thus its use is subject to more restrictions. This type of sludge is land-applied in such a manner that it does not present great risks either to human health or to the environment under controlled management (Madala 2025).

2.2 Overview of the Methods Used for Characterization of Sewage and Sludge

The characterization of sewage and sludge, in general, is a complex array of processes using diversified methodologies to assess their chemical, physical, and biological characteristics. Every characterization technique gives important information necessary for the management of wastewater and assessment of the environmental consequences apart from resource recovery. A summary of the main methods used in each of these categories is presented here (Alonso et al. 2023).

2.2.1 Chemical Characterization Techniques

Chemical characterization involves the analysis of the chemical composition of sewage and sludge with a view to determining the presence and concentrations of various substances, including both organic and inorganic contaminants (Facchini et al. 2021). Table 2.3 highlights the major chemical composition of sewage and sludge.

The main methodologies include:

Table 2.3 Different chemical components present in sewage and sludge

Components	Typical concentration range	Function/significance
Carbohydrates	10–20%	Energy source for microbial activity
Proteins	20–30%	Contribute to nutrient content (N Source)
Lipids	5–15%	Potential for biofuel production
Volatile fatty acids (VFAs)	1–5%	Precursors for bioplastics and bioenergy
Heavy metals	Variable concentrations	Environmental concerns, requires monitoring
Pathogens	Variable	Public health risk, necessitates treatment

2.2.1.1 Spectroscopic Techniques

Ultraviolet–visible spectroscopy is used to analyze the chromatic substances and organic components contained in wastewater. This method measures absorbance at specific wavelengths to assess dissolved organic material concentrations (Płonka et al. 2025). Infrared spectroscopy furnishes information regarding molecular structure and functional groups in organic materials by way of their characteristic absorption bands.

2.2.1.2 Chromatographic and Diffraction Methodologies

Gas Chromatography (GC) is applied for the volatile compounds. GC separates and quantifies organic pollutants like hydrocarbons, solvents, and pesticides (Ashekuzzaman et al. 2019). High-Performance Liquid Chromatography (HPLC) is in use for non-volatile compounds and polar, HPLC is quite effective for the detection of polycyclic organic substances in wastewater (Gupta 2018). X-ray diffraction (XRD) is the method of determining crystalline structures by the analysis of X-ray scattering.

2.2.1.3 Mass Spectrometry (MS)

Coupled with chromatography, mass spectrometry has been applied to the detailed molecular weight information and structural elucidation of organic compounds, thereby allowing for the identification of complicated mixtures present in sewage and sludge (Salva et al. 2025).

2.2.1.4 Elemental Analysis

ICP-MS and AAS are the techniques applied in the quantification of heavy metals and trace elements, which have a very significant role in the determination of toxicity levels and compliance with environmental regulations (Septiariva et al. 2022).

2.2.1.5 Chemical Oxygen Demand (COD) and Biochemical Oxygen Demand (BOD)

COD measures the total amount of oxygen that can oxidize organic matter present in wastewater, while BOD measures the amount of oxygen that microorganisms in the water will use to degrade organic matter biologically. These parameters are very relevant for the determination of organic load and treatment efficiency (Wiercik et al. 2022).

Table 2.4 Different techniques for chemical characterization

Techniques	Method	Purpose/applications
Spectroscopic techniques	UV–Vis spectrophotometry	Analyzes chromatic substances and dissolved organic materials
	IR spectroscopy	Determines molecular structures and specified functional groups in organic/ inorganic constituents
	Mass spectrometry	Hyphenated technique, specifically coupled with different Gas Chromatographic techniques such as LC–MS, GC–MS, HPLC–MS, etc The techniques specifically identify molecular weight and charge to mass ratio of different analytes
	ICP-MS and AAS	Quantitation of heavy metals and trace elements to assess toxicity
Chromatographic techniques	Gas chromatography	Separates and quantifies volatile compounds (e.g., hydrocarbons)
	High-performance liquid chromatography	Detects non-volatile polar compounds (e.g., pharmaceutical dyes)
X-ray based techniques	X-Ray diffraction	Determines crystalline structures via X-ray scattering

Sewage treatment plants involve various stages and these stages thereby followed by different stages. The chemical characterisation techniques involve. An objective summary of different characterization techniques is presented in Table 2.4.

2.2.2 *Physical Characterization Techniques*

Physical characterization of sewage and sludge refers to the measurable and observable properties that may impact treatment processes. Key methods involved include:

2.2.2.1 Particle Size Distribution Analysis

Techniques like laser diffraction and sieving are used in the determination of the size distribution of the solid particles in sludge. The information is critical in optimizing processes like sedimentation and filtration (Domini et al. 2022). Figure 2.2 highlights different techniques associated to particle size distribution analysis.

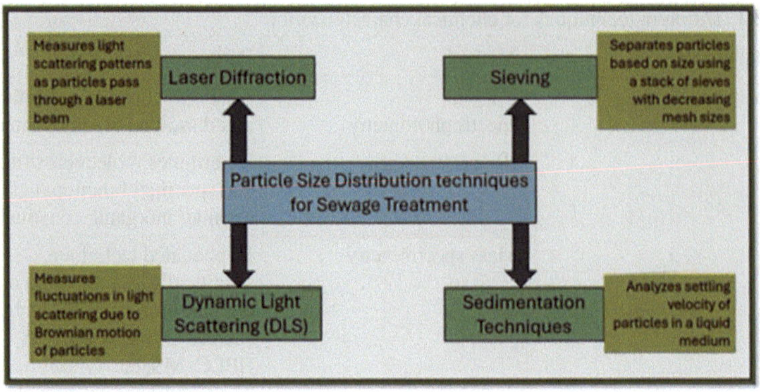

Fig. 2.2 Different particle size distribution techniques and their principles

2.2.2.2 TSS and VSS Measurement

Measurements of the Total Suspended Solids (TSS) and Volatile Suspended Solids (VSS) are useful for establishing the solids concentration of wastewater. The VSS, because it measures the organic material present, is particularly important to evaluate the efficiency of biological treatment operations (El Hayany et al. 2022).

2.2.2.3 Test for Density and Specific Gravity

Gravimetric procedures test and density gradient centrifugation types tests measure the density of sewage and sludge, parameters influencing settling behaviors and treatment systems design (van den Berg et al. 2020).

2.2.2.4 pH and Electrical Conductivity

The pH level is very important in establishing the level of acidity or alkalinity of the wastewater for microbial activity and chemical reactions. Electrical conductivity indicates ionic content, which affects the efficiency of the treatment process (Iticescu et al. 2018).

2.2.3 *Biological Characterization Techniques*

Biological characterization methodologies assess the presence and functionality of microorganisms in sewage and sludge, which are very essential in the processes of biological treatment (Yu et al. 2021). Main techniques include:

2.2.3.1 Microbial Community Analysis

Techniques such as Polymerase Chain Reaction (PCR) and next-generation sequencing (NGS) are used in the detection and quantification of microbial populations in sewage and sludge. This information is indispensable in understanding the processes occurring during biodegradation and assesses the effectiveness of biological treatment systems (El Houari et al. 2020).

2.2.3.2 Viable Cell Counts

Methods such as membrane filtration and culturing techniques allow for the quantification of viable bacteria and pathogens in wastewater, a crucial step in assessing public health risks (Michalska et al. 2022).

2.2.3.3 Enzyme Activity Assays

Measurements of the activity of specific enzymes, for example, dehydrogenase and cellulase, give very important information on the metabolic potential of microbial communities and their ability to decompose organic matter (Galintin et al. 2021).

2.2.3.4 Bioassays Used to Evaluate Toxicity

Bioassays use living organisms—such as algae and daphnia to determine the toxic impact of sewage and sludge on aquatic life. This helps determine the potential environmental impact of discharging treated effluent (Terekhin et al. 2024).

The examination of sewage and sludge using chemical, physical, and biological methods is very important for the effective management of wastewater and protection of the environment. Each of the adopted techniques provides unique information that is relevant in optimizing treatment processes, meeting regulatory standards, and safeguarding public health. By adopting these different techniques, wastewater treatment plants can obtain an overall understanding of their influent and effluent characteristics, thereby improve the outcome of treatment and create sustainable operating systems (Rangabhashiyam et al. 2022).

2.3 Environmental Implications

Sewage sludge, a residual byproduct of the treatment of wastewater, is a complex mixture of both organic and inorganic materials. Its composition varies depending on the characteristics of the influent and the treatment processes used. Its high content of nutrients, mainly nitrogen, phosphorus, and potassium, has resulted in its application as a soil amendment to increase the productivity of agriculture fields and soils. However, the application of sewage sludge raises serious concerns regarding soil quality and fertility, and environmental health because of the possible existence of harmful contaminants that include heavy metals, pathogens, and new pollutants like microplastics and pharmaceuticals (Brisolara and Ochoa 2016).

2.3.1 Composition of Sewage Sludge

Sewage sludge is a complex mixture of organic and inorganic materials that accumulates during the treatment of wastewater. The composition of the sludge varies considerably with the type of wastewater treated, the treatment processes used, and the peculiar characteristics of the influent.

2.3.1.1 Nutritional Composition

Sewage sludge contains essential plant nutrients, especially nitrogen (N), phosphorus (P), and potassium (K). The types and amounts of these nutrients determine their availability to crops. In sludge, nitrogen is found in many forms: organic nitrogen, ammonium ($NH_4 +$), and nitrate (NO_3). While the plants can directly utilize ammonium and nitrate, the organic nitrogen must undergo mineralization, which slowly releases its nutrients. Phosphorus, usually found at lower concentrations in biosolids than nitrogen, is of special significance because of its importance in agriculture and the eventual depletion of traditional sources of phosphorus (Campo et al. 2022).

2.3.1.2 Pollutants and Contaminants

In addition to nutrients, sewage sludge is expected to carry a variety of pollutants, including heavy metals, pathogens, and organic micro-pollutants. The EPA of the United States has established pollutant limits for various constituents in sludge because these may pose risks to both human health and the environment.

Micro-pollutants, such as pharmaceuticals and endocrine-disrupting compounds, can be contained in dried sludge at levels up to hundreds of mg/kg, which calls for much concern about their possible ecological and health impacts once the sludge is used in agricultural practices (Jothinathan et al. 2023). Different sources of pollutants and contaminants are presented in Fig. 2.3.

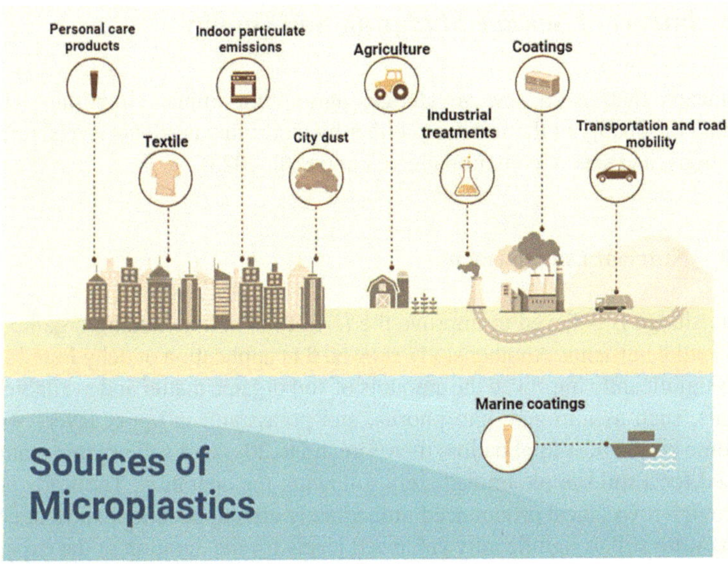

Fig. 2.3 Sources of microplastics as pollutants and contaminants

2.3.1.3 Pathogens

Pathogenic microorganisms are one of the major components making up sewage sludge. Pathogens such as *Salmonella*, as well as species of the genera *Cryptosporidium* and *Giardia*, can survive in untreated or improperly treated sludge and may pose a health risk unless properly inactivated. Various treatment methodologies, which include anaerobic digestion and alkaline stabilization, are used in an effort to reduce pathogen concentrations, hence improving the safety of the sludge for agricultural use (Appels et al. 2008).

2.3.2 Physical Characterisation of Sewage

The physical characteristics of sewage sludge, such as its water content, specific gravity, and size distribution, are important in handling, transportation, and agricultural application. In general, the sludge can be classified into primary sludge from physical treatment processes and secondary sludge generated during biological treatment processes. Good management of these characteristics is very vital in maximizing the beneficial uses of sewage sludge and minimizing all potential risks that may be associated with the application in the soil (Dramé et al. 2023).

2.3.3 *Impact of Sewage Sludge on Soil Quality*

This means that using sewage sludge, agriculture implies important changes regarding the quality of the soil itself, either by modifying nutrients levels, microbial populations, and sets of contaminants (Chang et al. 2023).

2.3.3.1 Nutrient Enhancement

Sewage sludge is believed to improve the fertility of soil by adding organic matter and essential nutrients. Another study showed that application of dehydrated sewage sludge significantly improved the amounts of soil organic matter and available major nutrients, such as nitrogen, phosphorus, and potassium, in upper layers within a short time period post-application; therefore, up to 20.6% of soil organic matter was obtained for a soil sample immediately following the operation. The study showed that, though it was most pronounced immediately after application, nutrient concentrations were still at significantly enhanced levels for the duration of the experiment a clear testimony to the ability of sewage sludge to enrich soil fertility with vital nutrients (Eid et al. 2020).

2.3.3.2 Soil Type and Toxicity

The toxicity associated with sewage sludge is contingent upon the type of soil, which profoundly influences its overall implications for soil quality. Empirical studies have demonstrated that diverse soil types, including sandy, loamy, and OECD artificial soil (OECD 207, 1984), display differing degrees of toxicity when subjected to sewage sludge treatment, particularly impacting the growth of plant roots and the viability of invertebrate species.

The EC(50) values for the inhibition of root growth in the different soils showed that OECD artificial soil was particularly sensitive to the application of sewage sludge, further underlining the need for consideration of the soil characteristics when assessing risks associated with sludge application (Domene et al. 2010a, b).

2.3.3.3 Pollutants and Monitoring

Notwithstanding its advantages, the use of insufficiently processed sewage sludge presents significant risks related to soil contamination, notably concerning heavy metals and pathogenic microorganisms. Ongoing assessment of sludge composition is crucial to minimize environmental hazards, especially considering the possible existence of toxic agents and heavy metals. The amalgamation of domestic, surface, and industrial wastewater can lead to the introduction of pollutants into the sludge, thereby requiring a comprehensive evaluation of its quality before application on land (Dramé et al. 2023).

2.3.3.4 Microbial Community and Soil Health

It has also been determined that sewage sludge may have a positive impact on the structure of the microbial community in soils, which will promote biological activity and improve soil health. Greater microbial biomass, resulting from the addition of biological solids, leads to better retention of moisture in the soil and reduced erosion, promoting better soil fertility over time. Long-term ecological effects due to heavy metals in sludge may alter microbial functions and overall soil health (Curci et al. 2020).

2.3.3.5 Fertility Enhancement

Using sewage sludge as fertilizer is known to offer great potential in enhancing the fertility of soils and further improving crop yield. Research shows that sewage sludge contributes organic matter to the soil, which is important in maintaining its health and fertility (Muter et al. 2022). In the following, the effects of different fertilization methods nitrogen and sewage sludge applications on soil organic matter quality and crop productivity were compared in five treatments established in a long-term field experiment initiated in 1993.

2.3.3.6 Nutrient Contribution and Organic Matter

Sewage sludge applies considerable amounts of organic matter to the soil. Annual carbon additions have been determined at 879 kg /ha year for the S120 treatment and 1758 kg /ha year for the S240 treatment. Despite these contributions, sewage sludge generally has a lower quality of organic matter compared to farmyard manure and thus may be less effective in improving CSOM (coarse-resolution soil organic matter) content than traditional organic fertilizers. For example, the S treatments did not exhibit a significant elevation in CSOM compared to the Con treatments, which might be explained by the relatively low C/N ratio of sewage sludge, stimulating rapid mineralization of organic matter. It is an end product of wastewater treatment, sewage sludge: that organic and inorganic mixture has a complex nature and varies with the influent characteristics and the treatment methods applied. Its high nutrient content, mainly nitrogen, phosphorus and potassium, has been used as an amendment to soil for increased agricultural productivity. There are concerns about soil quality and fertility, and environmental health that arise from using this material, as it is likely to contain hazardous contaminants such as heavy metals, pathogens, and emerging contaminants like microplastics and pharmaceuticals. The impact of

Table 2.5 Nutrient contribution and organic matter that contributes soil fertility

Aspects	Descriptions
Organic matter	879 kg/ha/year (S120), 1758 kg/ha/year (S240); lower quality than manure
CSOM impact	No Significant increase due to low C/N ratio causing rapid mineralization
Nutrient content	Rich in nitrogen, phosphorus, potassium; boosts crop productivity

sewage sludge on crop yield is considerable. In the mentioned study, the control treatment averaged 9.76 t DM/ha year while the other fertilization treatments, including those using sewage sludge, yielded an average of 12.3 t DM/ha year (Eyser et al. 2015). Table 2.5 represents different nutrient contribution and organic constituents establish soil fertility.

While there was no significant difference in yield from the fertilization treatments, the application of sewage sludge was associated with an increase in glomalin content, which is shown to correlate positively with levels of soil organic carbon. This would indicate that, though the yield improvement may not always be significantly greater in the short term, the future application of sewage sludge would well be worthwhile for soil health and productivity.

2.3.3.7 Quality of Soil Organic Matter

Soil organic matter quality parameters were evaluated using hot water and $CaCl_2$ extractions. Results showed that the mineral nitrogen treatment (N240) decreased soil organic matter quality as evidenced by lower mineralizable carbon and extractable carbon levels. S240 treatment has further presented significant increases in both hot water extractable organic matter and carbon content, which compares to those found in the control treatment, emphasizing the potential of sewage sludge for stabilizing organic matter in soil over time (Balík et al. 2022).

2.3.4 Evaluation of Toxicity from Sewage Sludge

The evaluation of sewage sludge toxicity is vital in understanding its environment-related ramifications, particularly in terms of soil quality and fertility. Sewage sludge, the solid residual produced by the wastewater treatment process, contains various toxic and hazardous materials that affect plant as well as animal life when applied to land as fertilizer.

Table 2.6 Toxicity of sewage sludge based on soil type

Type of soil	Test species	Toxicity indicator	EC (50)/LC (50) range	Observations
OCED artificial soil	*Lepidium sativum, Sorghum saccharatum*	EC (50) shows root growth	0.1–6.4%	Highest negative effects on root growth
Sandy soil	*Lepidium sativum, Sorghum saccharatum*	EC (50) shows root growth	0.03–9.4%	More tolerant compared to artificial soil
Loamy soil	*Lepidium sativum, Sorghum saccharatum*	EC (50) shows root growth	6.6–22.1%	Highest tolerance to sludge toxicity
Other soils	*Heterocypris incongruens*	LC (50) reprents survival rate	0.26–11.5%	Toxicity varies with sludge and soil

2.3.4.1 Influence of Soil Type on Toxicity of the Sludge

Some studies indicate that the toxicity level of sewage sludge is usually dependent on the soil type to which it is applied. In one such study, the effect of sandy, loamy, and OECD artificial soil on toxic sewage sludge was tested on three plant species, *Lepidium sativum*, *Sorghum saccharatum*, and *Sinapis alba*, and the invertebrate species *Heterocypris incongruens*. The OECD artificial soil proved to have the greatest negative effects on root growth, with EC(50) values ranging from 0.1 to 6.4%. Sandy soil had a more tolerant EC(50) value of 0.03–9.4%, while loamy soil also exhibited higher levels of tolerance, with EC(50) values of 6.6–22.1% aforementioned for OECD artificial soil. The highest LC(50) values for H. incongruencies range from 0.26–11.5%, depending on the sludge tested thus maintaining the idea of soil property in mediating sludge toxicity (Domene et al. 2010a, b).

Yet, research now points out that the toxic qualities of sewage sludge are sharply altered by the specific type of soil it is used in. Table 2.6 presents the toxicity of sewage sludge based on soil type.

2.3.4.2 Indicate the Unique Ways of Soil Type to Cause Toxicity

Some studies have shown that the soil application provides the necessary conditions for varying toxicity xenobiotics concerning sludge. One such study concerning sandy-loamy soil type and OECD artificial soil emphasizes the different toxic effects of sewage sludge across many plant species. According to the composition of contaminants, sewage sludge contains various contaminants, including heavy metals, organic compounds, and bacteria, which greatly vary in their toxic effects. According to the

EPA's 2009 Targeted National Sewage Sludge Survey, numerous toxic substances, including endocrine disrupters and metals, were found in sewage sludge- posing harmful effects for human and ecosystem health. According to the survey, 27 metals will end up in the mixture, depending on the concentration of semi volatile organics and polycyclic aromatic hydrocarbons. Because of the varying concentrations of bacteria, for instance, Salmonella, the health risks during the application of sludge on farmland grow exponentially (Domene et al. 2010a, b).

2.3.4.3 Emerging Pollutants and Microplastics

The emergence of new pollutants such as microplastics and per- and polyfluoroalkyl substances (PFAS) threatens to further complicate sewage sludge management. These contaminants pose an added risk due to their ability to build up in soils, leading to subsequent uptake by crops and raising the risk of inhalation and ingestion. Compounding the risk assessment for sludge application to agriculture is the variety of compounds and their synergistic interactions that may cause cumulative toxicological effects (Alvarenga et al. 2015).

2.3.5 Monitoring and Control

Sewage sludge toxicity risks will need to be effectively controlled and reduced. Wastewater treatment plants should put in place intermediate measures of control towards at-sources entry of toxic substances so as to improve and manage sludge quality before use. This is where the urgent need for new analytical methods to detect those contaminants emerges, since traditional approaches have failed to offer adequate data regarding the presence of hazardous substances, like microplastics. Continuous research on the impacts of sludge on soil ecosystems and food safety should be prioritized to guarantee everyone is protected from contracting diseases (Kulikowska et al. 2016).

2.3.6 Environmental Risks

The application of sewage sludge, generally called biosolids, to agricultural lands raises important issues about environmental health, mainly due to the presence of emerging pollutants and heavy metals that persist in the environment and impact soil quality and fertility. Land application is one practice adopted in the waste management of biosolids; approximately 40% of the six million dry metric tons of sewage sludge produced in a year in the United States are treated and applied on land, a practice that presents some dangers to the environment (Kirchmann et al. 2017).

2.3.7 Emerging Pollutants

Biosolids may act as a sink for a variety of emerging pollutants such as microplastics and other toxic substances. These pollutants impact not only the soil but also human health through bioaccumulation in the food chain. Evidence also suggests that the land application of biosolids may facilitate microplastic transfer to agricultural soils and ultimately into the greater environment. In addition, as systematic monitoring programs and standards pertaining to biosolid application are currently not in place, the hazards presented by biosolid application have amplified the need for more rigorous regulatory frameworks necessary to protect human health and the environment (Karlen et al. 2021).

2.3.7.1 Heavy Metal Contamination

Heavy metals, including lead, cadmium, and mercury, constitute a complex mixture of contaminants found in biosolids due to their prevalence in sewage. Some of these metals are regarded as essential at low doses, e.g., iron and nickel; yet others remain toxic and may cause metabolic defects in organisms that ingest crops cultivated in contaminated soils. Most heavy metals, once introduced into the soil, do not have quite the same shelf life biological or chemical degradation and remain on an environmental level as a persistent pollutant with a risk for bioaccumulation in the food chain. This is mainly alarming in developing countries, where the presence of insufficient control on landfill leachate stimulated the appearance of groundwater contamination that led to various health hazards threatening the community around (Saha et al. 2017).

2.3.8 Recommendations for Mitigation

To tackle these emerging environmental health threats, investments should ensue for continuous research on infrastructural changes toward wastewater treatment and proper land application of biosolids, as well as programs that raise public awareness about the dangers of the overuse of pharmaceuticals and plastics. Supportive regulations that consider the health of the environment and dispose of threats should be floated. The promotion of environmentally friendly products with better quality control in biosolids could minimize those risks (Nunes et al. 2021).

2.3.8.1 Mitigation Strategies

Treatment Processes

Effective treatment methods for municipal sewage sludge are critical to mitigating the potential health and environmental hazards associated with raw sewage sludge, which typically contains 1 to 4% solids and high levels of pathogens. Various treatment processes aim to stabilize sewage sludge, reduce pathogen content, and enhance solids concentration. These methods are often complex and require specific conditions such as holding time, temperature, pH, and solids content to be effective (Fusi et al. 2017).

Regulatory Framework

In addition to treatment processes, stringent regulations play a vital role in managing the risks associated with the land application of biosolids. In Pennsylvania, for instance, the biosolids regulations are more restrictive than federal guidelines, requiring adherence to additional risk-management measures. These include prohibitions on applying biosolids in environmentally sensitive areas, on steep slopes, and near water sources, as well as implementing soil conservation plans. Furthermore, mandatory training for personnel involved in land application programs ensures that practices comply with safety and environmental standards.

Public Communication and Awareness

Effective communication concerns the implications regarding the ingestion of pharmaceutical substances, or the environmental issues caused by agricultural runoff and their potential to start changing the entire scene. Raising public awareness will lead to a realization of the risk of such misuse of pharmaceuticals or plastics and may start exerting adequate pressure on decision-makers for the implementation of greener alternatives, finally finding ways to lessen the volumes of contaminants in the environment.

Innovative Technologies

Research continues into innovative degradation processes for persistent contaminants, such as perfluorinated compounds. Low-temperature mineralization methods, for example, are introduced as potential devices for breaking down perfluoro carboxylic acids through advanced chemical pathways. In building new technologies, it will also be crucial for future technologies to ensure dual benefits-they must be effective and environmentally friendly, catering to a wide range of emerging pollutants while minimizing energy consumption and capital expenditures (Ćwiertniewicz-Wojciechowska et al. 2023).

2.4 Potential of Circularity and Other Forms of Usage of Sewage and Sludge

Increased global concentration on sustainability and the circular economy means a change in the paradigm of sewage and sludge management. Once considered wastes, sewage and sludge are now gaining acknowledgment as alternative valuable resources for applications in many sectors. This review aims to summarize the current understanding of sewage and sludge composition, emphasizing their potential for circular uses. Chemical, physical, and biological characteristics of the two materials have been discussed with applications in energy, agriculture, and construction. Thereafter, the barriers and opportunities for the application of circular economy principles in sewage and sludge context are analyzed (Karlen et al. 2021).

The world is facing unprecedented environmental challenges, which include climate change, water scarcity, and waste management. The concept of circular economy has emerged as a promising solution, aiming to reduce waste and the consumption of resources by promoting the reuse and recycling of materials. Sewage and sludge, traditionally considered waste, are now seen as vital resources applicable in several sectors. The chemical characterization of sewage and sludge has led to the observation of the presence of different compounds, such as carbohydrates, proteins, lipids, and volatile fatty acids (VFAs). The details of the chemical composition of sewage and sludge can be found in various amid each other (Alvarenga et al. 2015).

2.4.1 *Energy Production*

Energy recovery is a process through which sewage and sludge can be converted into a useful product, including anaerobic digestion, gasification, and pyrolysis. Anaerobic digestion is a known technology for biogas production, which is a mixture of methane and carbon dioxide that can be used as a renewable energy source. Gasification and pyrolysis are thermal processes that convert organic matter into syngas and bio-oil, respectively, for use as fuels (Hidaka et al. 2016; Gahlot et al. 2022).

2.4.2 *Agricultural Applications*

Sewage and sludge can thus be seen as fertilizers and soil amendments in agriculture, providing nutrients to foster plant growth. The application of sewage sludge as fertilizers has been shown to improve soil fertility and promote crop yields sustainably. Nevertheless, the pollutants and pathogens in sewage and sludge introduce risks to human health and the natural environment; hence, effective treatment and management will be warranted to mitigate the risks (Zhang et al. 2022).

2.4.3 Construction Materials

Sewage and sludge can serve as raw materials in the production of construction materials, such as bricks, concrete, and cement. The incorporation of sewage sludge ash as an addition to cement is proposed to improve the mechanical properties of concrete (Chang et al. 2020).

2.4.4 Challenges and Opportunities

So many challenges, such as the presence of pollutants and pathogens, public perception, and the existing regulatory frameworks, face the implementation of circular economy principles into sewage and sludge management. The potential benefits of circular usage, including energy production, agricultural applications, and construction materials, present opportunities for sustainable development and economic growth (Shafii et al. 2019).

2.5 Conclusion

An essential part of efficiently treating and managing wastewater is characterising sewage and sludge. The complexity of sewage and sludge, which differ greatly based on their source and the procedures they go through, is highlighted by this research. Important details on their composition and possible effects on the environment are provided by key factors such nutrients, heavy metals, microbial content, chemical oxygen demand (COD), and biological oxygen demand (BOD). Although there are options for resource recovery, such as composting and biogas generation, because to the high organic matter and nutritional content, the presence of hazardous pollutants, such as heavy metals and pathogens, need rigorous monitoring and management to avoid environmental contamination and health hazards. All things considered, this research emphasises how crucial comprehensive sewage and sludge analysis is to satisfying environmental regulations, streamlining treatment procedures, and encouraging sustainable practices. In order to turn wastewater from a disposal issue into a useful resource, thorough characterisation will remain essential given the growing emphasis on waste reduction and resource recovery. Enhancing management plans and environmental results requires ongoing study and the use of cutting-edge analytical methods.

References

Abrahão Cezarini EC, Peterli Z, Castillo Santiago Y, José Venturini O (2024) Sewage sludge energy potential assessment: characterization approach. CET J-Chem Eng Trans 109

Alonso JM et al (2023) Chemical characteristics and valuation of sewage sludge from four different wastewater treatment plants. Environ Monitor Assessm 196(1). https://doi.org/10.1007/s10661-023-12211-8

Alvarenga P, Mourinha C, Farto M, Santos T, Palma P, Sengo J, Morais MC, Cunha-Queda C (2015) Sewage sludge, compost and other representative organic wastes as agricultural soil amendments: benefits versus limiting factors. Waste Manage 40:44–52

Appels L et al (2008) Principles and potential of the anaerobic digestion of waste-activated sludge. Prog Energy Combust Sci 34(6):755–781. https://doi.org/10.1016/j.pecs.2008.06.002

Ashekuzzaman SM, Forrestal P, Richards K, Fenton O (2019) Dairy industry derived wastewater treatment sludge: generation, type and characterization of nutrients and metals for agricultural reuse. J Clean Prod 230:1266–1275

Balík J, Kulhánek M, Černý J, Sedlář O, Suran P, Procházková S, Asrade DA (2022) The impact of the long-term application of mineral nitrogen and sewage sludge fertilizers on the quality of soil organic matter. Chem Biol Technol Agricul 9(1):86

Brisolara KF, Ochoa H (2016) Biosolids and sludge management. Water Environ Res 88(10):1230–1248. https://doi.org/10.2175/106143016x14696400494731

Campo R, Carretti E, Lubello C, Lotti T (2022) Recovery of structural extracellular polymeric substances (sEPS) from aerobic granular sludge: Insights on biopolymers characterization and hydrogel properties for potential applications. J Environ Manage 324:116247

Cezarini ECA, Peterli Z, Santiago YC, Venturini OJ (2024) Sewage sludge energy potential assessment: characterization approach. Chem Eng Trans 109:247–252

Chang Z, Long G, Zhou JL, Ma C (2020) Valorization of sewage sludge in the fabrication of construction and building materials: a review. Resour Conserv Recycl 154:104606

Chang H et al (2023) Mini-review of inventory data for the dewatering and drying of sewage sludge. Waste Managem Res J Sustain Circular Econ 41(6):1081–1088. https://doi.org/10.1177/0734242x221139170

Chen Q et al (2024a) Characteristics of advanced anaerobic digestion sludge-based biochar and its application for sewage sludge conditioning. Biores Technol 402:130833. https://doi.org/10.1016/j.biortech.2024.130833

Chen Q, Fu X, Wang J, Zhao Y, Wang W, Memon AG, Zhang H, Wen Y (2024b) Characteristics of advanced anaerobic digestion sludge-based biochar and its application for sewage sludge conditioning. Biores Technol 402:130833

Curci M, Lavecchia A, Cucci G, Lacolla G, De Corato U, Crecchio C (2020) Short-term effects of sewage sludge compost amendment on semiarid soil. Soil Syst 4(3):48

Ćwiertniewicz-Wojciechowska M, Cema G, Ziembińska-Buczyńska A (2023) Sewage sludge pretreatment: current status and future prospects. Environ Sci Pollut Res 30(38):88313–88330. https://doi.org/10.1007/s11356-023-28613-7

Domene X et al (2010a) Role of soil properties in sewage sludge toxicity to soil collembolans. Soil Biol Biochem 42(11):1982–1990. https://doi.org/10.1016/j.soilbio.2010.07.019

Domene X, Colón J, Uras MV, Izquierdo R, Àvila A, Alcañiz JM (2010b) Role of soil properties in sewage sludge toxicity to soil collembolans. Soil Biol Biochem 42(11):1982–1990

Domini M, Bertanza G, Vahidzadeh R, Pedrazzani R (2022) Sewage sludge quality and management for circular economy opportunities in Lombardy. Appl Sci 12(20):10391

Dramé SI, Oba OA, Aydinlik NP (2023) Environmental and ecological risk assessment of municipal sewage sludge management using a sustainable solar drying system. Water Sci Technol 88(1):151–168. https://doi.org/10.2166/wst.2023.199

Eid EM, Alamri SA, Shaltout KH, Galal TM, Ahmed MT, Brima EI, Sewelam N (2020) A sustainable food security approach: controlled land application of sewage sludge recirculates nutrients to agricultural soils and enhances crop productivity. Food Energy Secur 9(2):e197

El Hayany B, Rumpel C, Hafidi M, El Fels L (2022) Occurrence, analysis of microplastics in sewage sludge and their fate during composting: a literature review. J Environ Manage 317:115364

El Houari A, Ranchou-Peyruse M, Ranchou-Peyruse A, Bennisse R, Bouterfas R, Goni Urriza MS, Qatibi AI, Guyoneaud R (2020) Microbial communities and sulfate-reducing microorganisms abundance and diversity in municipal anaerobic sewage sludge digesters from a wastewater treatment plant (marrakech, morocco). Processes 8(10):1284

Eyser CV et al (2015) Pharmaceutical load in sewage sludge and biochar produced by hydrothermal carbonization. The Sci Total Environ 537:180–186. https://doi.org/10.1016/j.scitotenv.2015.08.021

Facchini F, Mummolo G, Vitti M (2021) Scenario analysis for selecting sewage sludge-to-energy/matter recovery processes. Energies 14(2):276

Fusi A, González-García S, Moreira MT, Fiala M, Bacenetti J (2017) Rice fertilised with urban sewage sludge and possible mitigation strategies: an environmental assessment. J Clean Prod 140:914–923

Gahlot P et al (2022) Principles and potential of thermal hydrolysis of sewage sludge to enhance anaerobic digestion. Environ Res 214:113856. https://doi.org/10.1016/j.envres.2022.113856

Galintin O, Rasit N, Hamzah S (2021) Production and characterization of eco enzyme produced from fruit and vegetable wastes and its influence on the aquaculture sludge. Biointerface Res Appl Chem 11(3):10205–10214

Ghazali MSM, Syed-Hassan SSA (2024) Characterization of sewage sludge for sustainable urban environments: assessing heavy metal enrichment, thermal decomposition, and pyrolysis behavior. Malaysian J Sustain Environ 11(2):119–134

Gupta H, Gupta B (2016) Adsorption of polycyclic aromatic hydrocarbons on banana peel activated carbon. Desalinat Water Treatm 57(20):9498–9509

Gupta H (2018) PAH determination in effluent and sludge samples of paper industry. Environ Technol Innov 9:115–121. https://doi.org/10.1016/j.eti.2017.11.009

Gusiatin MZ, Kulikowska D, Bernat K (2024) Municipal sewage sludge as a resource in the circular economy. Energies 17(11):2474

Hidaka T et al (2016) Anaerobic codigestion of grass and sewage sludge: laboratory experiments and feasibility analysis. Water Environ Res 88(11):2070–2081. https://doi.org/10.2175/106143016x14733681695320

Iticescu C, Georgescu LP, Murariu G, Circiumaru A, Timofti M (2018) The characteristics of sewage sludge used on agricultural lands. In: AIP conference proceedings, November, vol 2022(1). AIP Publishing

Jin F, Lu J, Sun F, Yang F, Li Z (2025) Application and development of sludge-based materials for environmental pollution remediation: a bibliometric review from 2004 to 2024. RSC Adv 15(10):8072–8087

Jothinathan H, Singh AP (2023) Fecal sludge characterization, treatment, and resource recovery options: a state-of-the-art review on fecal sludge management. Environ Sci Pollut Res 30(57):119549–119567

Karlen DL, Stott DE, Mikha MM (2021) Soil health series: Laboratory methods for soil health analysis

Khalil NA, Lajulliadi AF, Abedin FNJ, Fizal ANS, Safie SI, Zulkifli M, Taweepreda W, Hossain MS, Ahmad Yahaya AN (2024) Multifaceted impact of lipid extraction on the characteristics of polymer-based sewage sludge towards sustainable sludge management. Polymers 16(18):2646

Kirchmann H, Börjesson G, Kätterer T, Cohen Y (2017) From agricultural use of sewage sludge to nutrient extraction: a soil science outlook. Ambio 46:143–154

Kominko H, Gorazda K, Wzorek Z (2024) Sewage sludge: a review of its risks and circular raw material potential. J Water Process Eng 63:105522

Kujawiak S, Makowska M, Janczak D, Czekała W, Krzesiński W, Antonowicz A, Kupryaniuk K (2024) An analysis of the physicochemical and energy parameters of briquettes manufactured from sewage sludge mixtures and selected organic additives. Energies 17(18):4573

Kulikowska D (2016) Kinetics of organic matter removal and humification progress during sewage sludge composting. Waste Manage 49:196–203

Kumar A, Gupta H (2020) Activated carbon from sawdust for naphthalene removal from contaminated water. Environ Technol Innov 20:101080

Li Z, Yuan D (2024) Global performance and trends of research on emerging contaminants in sewage sludge: a bibliometric analysis from 1990 to 2023. Ecotoxicol Environ Saf 281:116597

Li L, Hua Y, Zhao S, Yang D, Chen S, Song Q, Gao J, Dai X (2023) Worldwide research progress and trend in sludge treatment and disposal: a bibliometric analysis. ACS ES&T Eng 3(8):1083–1097

Li Z (2025) Global research progress on land application of sewage sludge: a comprehensive bibliometric review. Waste Biomass Valorizat 1–21

Liao BQ et al (2001) Surface properties of sludge and their role in bioflocculation and settleability. Water Res 35(2):339–350. https://doi.org/10.1016/s0043-1354(00)00277-3

Liew C-S et al (2021) Stabilization of heavy metals loaded sewage sludge: reviewing conventional to state-of-the-art thermal treatments in achieving energy sustainability. Chemosphere 277:130310. https://doi.org/10.1016/j.chemosphere.2021.130310

Liu S et al (2023) Effects of microplastics on the properties of different types of sewage sludge and strategies to overcome the inhibition: a review. The Sci Total Environ 902:166033. https://doi.org/10.1016/j.scitotenv.2023.166033

Liu X et al (2024) Pre-drying limitedly affected the yield, fuel properties, pyrolysis and combusion behavior of sewage sludge hydrochar. Waste Manage 184:63–71. https://doi.org/10.1016/j.wasman.2024.05.032

Łomińska-Płatek D, Figura Ł (2024) Analysis of the possibilities of using sewage sludge for agricultural use. Gaz, Woda i Technika Sanitarna

Madala P (2025) The role of sludge dewatering technologies in modern wastewater management. J Mech Civil Indus Eng 6(1):11–18

Mailler R et al (2014) Priority and emerging pollutants in sewage sludge and fate during sludge treatment. Waste Manage 34(7):1217–1226. https://doi.org/10.1016/j.wasman.2014.03.028

Merdun H et al (2022) Effects of biological pre-digestion of sewage sludge processed by fast pyrolysis on bio-oil yield and biochar toxicity. Waste Manage 157:149–158. https://doi.org/10.1016/j.wasman.2022.12.016

Messaoudi H, Koukouch A, Bakhattar I, Asbik M, Bonnamy S, Bennouna EG, Boushaki T, Sarh B, Rouboa A (2024) Physicochemical characterization, thermal behavior, and pyrolysis kinetics of sewage sludge. Energies 17(3):582

Michalska J, Turek-Szytow J, Dudło A, Surmacz-Górska J (2022) Characterization of humic substances recovered from the sewage sludge and validity of their removal from this waste. EFB Bioecon J 2:100026

Mladenovič A, Hamler S, Zupančič N (2017) Environmental characterisation of sewage sludge/paper ash-based composites in relation to their possible use in civil engineering. Environ Sci Pollut Res 24:1030–1041

Muter O, Dubova L, Kassien O, Cakane J (2022) Application of the sewage sludge in agriculture: soil fertility, technoeconomic, and life-cycle. Hazardous waste management, pp 125

Nilsson C et al (2021) Phosphorus speciation in sewage sludge from three municipal wastewater treatment plants in Sweden and their ashes after incineration. Waste Managem Res J Sustain Circular Econ 40(8):1267–1276. https://doi.org/10.1177/0734242x211065231

Nunes N, Ragonezi C, Gouveia CS, Pinheiro de Carvalho MÂ (2021) Review of sewage sludge as a soil amendment in relation to current international guidelines: a heavy metal perspective. Sustainability 13(4):2317

Oleszczuk P, Hollert H (2011) Comparison of sewage sludge toxicity to plants and invertebrates in three different soils. Chemosphere 83(4):502–509. https://doi.org/10.1016/j.chemosphere.2010.12.061

Paffrath SF, Carissimi E, Schner JC, Ferrari KF, Etchepare RG (2024) Understanding the interplay of sludge characteristics and dewatering in water treatment plants: a review. Environ Technol Rev 13(1):794–813

Płonka I, Kudlek E, Pieczykolan B (2025) Municipal sewage sludge disposal in the Republic of Poland. Appl Sci 15(6):3375

Raksasat R et al (2022) Enriched sewage sludge from anaerobic pre-treatment in spurring valorization potential of black soldier fly larvae. Environ Res 212:113447. https://doi.org/10.1016/j.envres.2022.113447

Rangabhashiyam S, dos Santos Lins PV, de Magalhães Oliveira LM, Sepulveda P, Ighalo JO, Rajapaksha AU, Meili L (2022) Sewage sludge-derived biochar for the adsorptive removal of wastewater pollutants: a critical review. Environ Pollut 293:118581

Raveendravarrier B, Palatel A, Chandrasekharan M (2025) Physico-chemical characterization of sewage sludge for thermochemical conversion processes. Energy Sources Part a: Recovery Utilization Environ Effects 47(1):2641–2665

Rezig W, Haoui KK, Hadjel M (2024) Structural analysis of heat-calcined modified sewage sludge from WWTP in gas complex for environmental risk assessment. The J Solid Waste Technol Managem 50(1):479–496

Saha S, Saha BN, Pati S, Pal B, Hazra GC (2017) Agricultural use of sewage sludge in India: benefits and potential risk of heavy metals contamination and possible remediation options–a review. Int J Environ Technol Manage 20(3–4):183–199

Salva J, Sečkár M, Schwarz M, Samešová D, Mordáčová M, Poništ J, Veverková D (2025) Analysis of the current state of sewage sludge treatment from the perspective of current European directives. Environ Sci Eur 37(1):1–27

Samara E et al (2019) Use of clay minerals for sewage sludge stabilization and a preliminary assessment of the treated sludge's fertilization capacity. Environ Sci Pollut Res 26(35):35387–35398. https://doi.org/10.1007/s11356-019-05132-y

Santos T, Sommaggio LRD, Marin-Morales MA (2022) Phyto-genotoxicity assessment of different associations between sludges from water and sewage treatment plants, before and after the bioremediation process. Environ Sci Pollut Res 29(26):40029–40040. https://doi.org/10.1007/s11356-022-18820-z

Sava C, Iluţiu-Varvara DA, Mare R, Roman MD, Rada S, Pică EM, Jäntschi L (2024) Physico-chemical characterization and possible uses of sludge processed from an urban sewage treatment plant. Heliyon 10(8)

Septiariva IY, Suryawan IWK, Zahra NL, Putri YNK, Sarwono A, Qonitan FD, Lim JW (2022) Characterization sludge from drying area and sludge drying bed in sludge treatment plant Surabaya city for waste to energy approach. J Ecol Eng 23(7):268–275

Shafii MA, Ling ECM, Shaffie E (2019) The use of sewage sludge and its ash in construction and agriculture industry: a review. Jurnal Teknologi (Sci Eng) 81(5)

Singh S, Parveen N, Gupta H (2018) Adsorptive decontamination of rhodamine-B from water using banana peel powder: a biosorbent. Environ Technol Innov 12:189–195. https://doi.org/10.1016/j.eti.2018.09.001

Singh RP et al (2011) Land application of sewage sludge: physicochemical and microbial response. In: Reviews of environmental contamination and toxicology, pp 41–61. https://doi.org/10.1007/978-1-4614-0668-6_3

Soni NK, Bhale PV (2024) Chemical insight into thermal degradation, chemical kinetics, and thermodynamics behavior of sewage sludge through thermogravimetry and Fourier transform infrared analysis. Thermal Sci Eng Progress 51:102617

Steele JC et al (2022) Comparative meta-analysis of organic contaminants in sewage sludge from the United States and China. The Sci Total Environ 821:153423. https://doi.org/10.1016/j.scitotenv.2022.153423

Terekhin AP, Maryandyshev PA, Brillard A (2024) Investigation of the morphology, composition, and structure of dehydrated sewage sludge of a pulp and paper enterprise. Solid Fuel Chem 58(6):459–464

Tytła M, Widziewicz K, Zielewicz E (2015) Heavy metals and its chemical speciation in sewage sludge at different stages of processing. Environ Technol 37(7):899–908. https://doi.org/10.1080/09593330.2015.1090482

Van den Berg P, Huerta-Lwanga E, Corradini F, Geissen V (2020) Sewage sludge application as a vehicle for microplastics in eastern Spanish agricultural soils. Environ Pollut 261:114198

Vítková M et al (2024a) Sustainable use of composted sewage sludge: metal(loid) leaching behaviour and material suitability for application on degraded soils. The Sci Total Environ 929:172588. https://doi.org/10.1016/j.scitotenv.2024.172588

Vítková M, Zarzsevszkij S, Šillerová H, Karlova A, Šimek P, Wimmerová L, Martincová M, Urbánek B, Komárek M (2024b) Sustainable use of composted sewage sludge: metal(loid) leaching behaviour and material suitability for application on degraded soils. The Sci Total Environ 929:172588. https://doi.org/10.1016/j.scitotenv.2024.172588

Wiercik P, Kuśnierz M, Kabsch-Korbutowicz M, Plucińska A, Chrobot P (2022) Evaluation of changes in activated sludge and sewage sludge quality by FTIR analysis and laser diffraction. Desalin Water Treat 273:114–125

Yáñez-Hernández O, Ríos-Lira AJ, Pantoja-Pacheco YV, Ruelas-Santoyo EA, Asato-España ML, Valdovinos-García EM (2024) Characterization of wastewater in an activated sludge treatment plant of the food sector. Water 16(18):2647

Yu B, Luo J, Xie H, Yang H, Chen S, Liu J, Zhang R, Li YY (2021) Species, fractions, and characterization of phosphorus in sewage sludge: a critical review from the perspective of recovery. Sci Total Environ 786:147437

Zhang H, Qi HY, Zhang YL, Ran DD, Wu LQ, Wang HF, Zeng RJ (2022) Effects of sewage sludge pretreatment methods on its use in agricultural applications. J Hazard Mater 428:128213

Chapter 3
Impact of Sewage Water on Properties of Soil, Water Resources and Environment

Rashmi Verma, Richa Saxena, Shreya Kotnala, Shalini Tiwari, B. S. Rawat, and P. Negi

Abstract Human activities are the main cause of wastewater that imitates from houses, industries, and agriculture. The wastewater or sewage water thus pollutes soil, drinking water, and also negatively impacts on environment. A huge amount of it is also disposed of into groundwater, which adversely affects human health. Water is a necessary source for living. Presently, agricultural land is exposed to wastewater. This wastewater is used for irrigation and provides nutrients to plants. Contamination occurs by sewage water through irrigation to food crops via the soil. The book chapter gives a comprehensive understanding of the composition, sources, and characteristics of sewage water. It provides a detailed understanding of the effect of sewage water on the physical, chemical, and biological properties of soil and also describes the effect of sewage water on various water resources, the environment, and human health. The chapter also highlights the health risks associated with sewage wastewater and provides mitigation and control measures. The chapter will help researchers and scientists to confiscate the problems of sewage wastewater improvement and control measures from future perspectives.

R. Verma
Department of Biotechnology, School of Basic and Applied Sciences, Shri Guru Ram Rai University, Dehradun, India

R. Saxena (✉)
Department of Physics, School of Basic and Applied Sciences, Shri Guru Ram Rai University, Dehradun, India
e-mail: saxena.richa23@gmail.com

S. Kotnala · S. Tiwari
Department of Chemistry, School of Basic and Applied Sciences, Shri Guru Ram Rai University, Dehradun, India

B. S. Rawat
Department of Physics, FUIT, Uttaranchal University, Dehradun, Uttarakhand, India

P. Negi
Department of Chemistry, GRD Institute of Management and Technology, Dehradun, Uttarakhand, India

S. Madhav et al. (eds.), *Municipal Sewage and Sludge: Treatment and Disposal Strategies*, Environmental Science and Engineering,
https://doi.org/10.1007/978-3-032-07597-0_3

Keywords Sewage water · Soil · Water · Environment · Human health

3.1 Introduction

The cogent utilisation and protection of water resources, as well as preventing their pollution, are the critical aspects of environmental protection. Effective management of wastewater, particularly the collection and disposal of municipal wastewater, is undeniably pivotal in this framework. The resources are getting depleted day by day because of the development of industrial activity and rapid growth in population. Rather than polluting resources by releasing waste into the air, water, and land, strategies for treatment, recycling and reusing resources should be adopted (Singh et al. 2018, 2022; Kumar et al. 2023; Gupta 2016). The quantity of water for utilisation can be reduced by reusing wastewater for irrigation purposes in agriculture (Gregory 2000; Kirkham 1974). It is a possible way to lessen the need for freshwater in order to release zero water and prevent contamination from entering the receiving sources. The infrastructure for disposing of urban wastewater, wastewater agriculture methods, water quality, health effects, and the amount of pollution all need to be considered in the modern period. The type and volume of industrial effluent released into sewage systems show the formation and composition of sewage waters (Larsen 1975; Arora 1984; Anderson 1973). This results in different compositions of soluble salts, oils, grease, and fat, as well as dangerous metals and plant nutrients. The sewage water, if used in agricultural land and its formation affects the plants, animals, soil, and people's health (Soda et al. 2017). In addition to having large evapotranspiration requirements, the Mediterranean basin has a shortage of water (Bedbabis et al. 2014). The acceptable yield, irrigation was very delicate to achieve in agriculture in dry and semi-arid areas (Sdiri et al. 2020). In this way, a huge amount of conventional water is being used in this scenario by increasing populations and various agricultural activities. Large amounts of effluent were produced because of this, and they were released into the environment (Abegunrin et al. 2016). Most of the wastewater released into the environment is either not treated at all or only partially treated, which poses a risk to the environment due to the introduction of heavy metals and other hazardous substances into the ecosystem (Thapliyal 2011; Metcalf and Eddy 2007). Large salinity conditions are treated as the main cause of land abandonment and resulting in less crop production. On the other hand, wastewater is a good source of nutrients and organic matter for plants for increasing crop production. Another possible degradation of the environment is in the form of air, land surface, and soil, etc. that is caused by undertreated effluents (Samaraweera et al. 2024). Irrigating crops by using industrial effluent off can enters the food chain and effects the quality of the crop. Waterborne illnesses such as diarrhoea, giardiasis, typhoid, cholera, hepatitis, jaundice, and cancer are caused by pollution of the water supply. The sewage water can absorb large amount of chemicals, pesticides and medicines (Obayomi et al. 2024).

The potential for harmful emissions such as dioxins, heavy metals, and particulate matter when burning sewage sludge can have a serious impact on human health and ecosystems if not properly managed through effective pollution control technologies and practices. This is why air pollution from municipal sewage sludge incineration is important globally. This is especially important as urbanization increases the generation of sewage sludge globally. For minimising the growing contaminants some remedies must be introduced so that sewage water can be reused, that can reduce the use of fresh water consumption (Khalid et al. 2018).

3.2 Composition and Characteristics of Sewage Water

The sewage water composition depends on the utilization of water per capita, and it also changes with place and season. Sewage can be caused because of waste from homes, industries, surface debris, and domestic waste. The wastes from homes are primarily removed by residents of residential areas and institutions. Commercial waste is the form of trash that is considered domestic if its nature of flow is of household origin. The majority of commercial garbage is emitted from stores and other service providers that prioritize helping the local community. Industrial wastewater is a waste that results from manufacturing or product-producing activities in the manufacturing sector. Many waste management businesses offer the service of collecting commercial garbage. This includes the daily routine of taking and disposing of commercial wastes that include waste from the office, waste from the retail industry, industry waste, and waste caused by construction (Boudjabi et al. 2022). Commercial waste gathering services include regular taking up, on-site commercial recycling services, and unique choices of disposal for hazardous materials. These services are usually customized to meet the unique demands of a business is of Two types of Commercial waste are biodegradable and non-biodegradable. Surface runoff is also a waste of precipitation, moving from a surface into a designated channel. They carry all the chemicals from the atmosphere while dissolving and leaching compounds from soil and vegetation. All these pollutants are collected, and after this, they run to a particular connecting point. Due to excessive garbage discharge and improper disposal, home trash is one of the big factors that seriously harm the rural eco-environment (Boudjabi et al. 2022).

3.2.1 Total Dissolved Solids

The term total dissolved solids (TDS) refer to a mixture of larger-than-2-micron organic and inorganic components, such as minerals and ions, that stay dissolved in water after standard filtering (Adjovu et al. 2023). The usage of water softeners is causing an increase in TDS. The range of TDS level in Municipal waste water effluent is 150–380 mg L^{-1} (Namkung and Rittmann 1986). Major ions such as sodium, sulphate, or chloride have a limited effect on hardness but can raise the TDS value by up to 75, 44, or 74 mg L^{-1}, respectively. They also have an impact on the acceptability and flavour of drinking water. While the Bureau of Indian Standards (BIS) states that TDS levels may increase, the World Health Organization recommends a TDS threshold of 300 ppm (Abascal et al. 2022).

3.2.2 Dissolved Organic Carbon (DOC)

Microorganisms in water are supported in their growth by dissolved carbon, often known as dissolved organic carbon, or DOC. It is a representation of the carbon compounds present in water that can have past filters having particles with diameters between 0.22 and 0.7 μm (Bradai et al. 2016). Microorganisms can grow by getting nutrients from DOC. In MWW effluents, soluble microbial products (SMP) and natural and manufactured organic matter are present as DOC (Gleisberg et al. 1976; Thomas et al. 2017). Water is contaminated by a variety of chemicals that come from factories, homes, and other sources. These include individual care substances, insecticides, surfactants, and steroidal hormones, which are expelled by humans and add to the DOC. LAS, or linear alkylbenzene sulfonates, are anionic detergent components that are frequently found in raw wastewater together with other (Spietz et al. 2015; Verma and Singh 2013).

3.2.3 Dissolved Oxygen Demand

The dissolved oxygen means the amount of oxygen present in the sample of water in a condition of dissolving. DO indicates the quality of water, and it is vital to ensure that the DO concentration of the effluent is at least 4 mg/L following its discharge via the treatment plant. Aquatic creatures are negatively impacted below this range (Aniyikaiye et al. 2019). The efficacy of the biological treatment phase is shown by the DO concentration of the effluents following wastewater treatment operations.

3.2.4 Biochemical Oxygen Demand

The term biochemical oxygen demand (BOD) means the amount of oxygen required or utilised by microbes in order to breakdown the biodegradable materials in a water sample. The amount of pollution in domestic and industrial wastewater can be examined, as well as the receiving water bodies ability to be purified, by making use of the BOD test. The BOD is measured using a 5-day incubation time at 20 °C; this measurement is known as the BOD. Any water sample, including wastewater, that has a high BOD indicates a significant concentration of organic matter. BOD levels in drinking water, clean water, and contaminated water are 1–2, 3–5, and 6–9 ppm, respectively (Huang et al. 2010a).

3.2.5 Chemical Oxygen Demand

Chemical oxygen demand (COD) shows the amount of oxygen (mg L^{-1} or g m^{-3}) required to break down the organic matter in a water sample. The dissolved oxygen (DO) value is inversely proportional to the amount of oxidizable organic matter (POM) in the sample of water, and the two are related to each other. Water supplies created because of food waste from cans, bottles, emulsified oils, etc., have raised the COD levels. Water-soluble pollutants raise COD levels and are mostly carried from stormwater to streams. Within a few hours, COD assays yield data (Liu et al. 2021). However, COD does not provide information about the relative amount of biodegradable organic matter. In comparison to other fibres, domestic wastewater consists of a high portion of protein and sugar. Of the total organic carbon (TOC), protein, sugar, and fibre make up 12.38%, 10.65%, and 20.64%, respectively. Thirty percent of the COD of DWW consists of the soluble fraction of degradable bioproducts. Other nutrients found in DWW are phosphorus and nitrogen compounds that also contribute to raising the level of COD (Huang et al. 2010b).

3.2.6 Characteristics of Sewage Water

A program for the efficient and cost-effective management of waste must include waste characterization. It aids in the selection of treatment techniques, determining the degree of treatment, evaluating the advantageous applications of wastes, and making use of the natural processes ability to purify waste.

3.2.7 Physical Characteristics

The naked eye can typically detect the colour of sewage, which indicates its freshness. If the colour is yellowish, grey, or light brown, it confirms fresh sewage; if it is black or dark brown, it shows stale and septic sewage. Industrial wastewater contributes colour to sewage, and the colour depends on the chemical processes in industries. Sewage temperature data can be used to determine the solubility of oxygen, which influences the rate of biological activity and the capacity of transfer of aeration equipment in aerobic systems. Sedimentation and biological treatment system effectiveness are both negatively impacted by extremely low temperatures. The pH scale, which represents the concentration of hydrogen ions, is an important metric for biological unit operation (Popa et al. 2012). The fresh sewage has a pH that is somewhat higher than the community's water supply. On the other hand, the pH can fluctuate greatly in the presence of industrial wastewater and decrease due to the breakdown of organic materials. Raw sewage typically has a pH between 5.5 and 8.0 (Sossou et al. 2024). Although sewage only includes around 0.1% particles, the remainder being water, the annoyance that these solids produce cannot be disregarded because they are highly putrescible and must be disposed of properly. The main nitrogenous substances found in household sewage include urea, proteins, amines, and amino acids. These organic components break down bacterially, which leads to the nitrogenous ammonia seen in sewage. Since nitrogen is a necessary component of biological protoplasm, its concentration is important for the efficient operation of biological treatment systems and land disposal. The amount of nitrogen concentration in untreated sewage water is 20–50 mg/L (Kiziloglu et al. 2008; Mohammad and Mazahreh 2003). Constant application of untreated or partially treated sewage can compact soil, which impairs drainage and aeration. Although the presence of organic content in sewage may initially aid in the aggregation of soil particles, prolonged exposure to excess waste can weaken the structure of the soil. Organic debris and suspended solids in sewage water can block soil pores, decreasing water-holding capacity and infiltration rate, leading to waterlogging and poor drainage (Chahal et al. 2016).

3.2.8 Chemical Characteristics

The pH level of sewage water is from 6.5 to 8.0. Its value varies depending on industrial waste. Heavy metals like lead, mercury, and cadmium, etc., are also constituents of sewage wastewater from industries. Toxic chemicals like pesticides, pharmaceuticals, and solvents are also constituents of sewage water. The primary metrics used to assess the concentration of organic contaminants in water bodies are BOD and COD. Overly high BOD and COD concentrations point to high DO consumption and reductive organics contents, making the water environment more vulnerable to issues like odorous and black water. Rural home sewage has greater organic contents (up to 500 mg/L) and better biodegradability (BOD5/COD, 0.45–0.55) than municipal sewage treatment plant intake. The organics can be efficiently eliminated through biochemical treatment (Qiang et al. 2022). Table 1 depicts the properties of sewage water (Hashem and Qi 2021).

3.2.9 Biological Properties of Sewage Water

Microorganisms and organic material are the biological properties. The activities of human beings and different biological processes result in biological components. Microorganisms include bacteria, protozoa, viruses, fungi, and helminths. Sewage water consists of pathogenic and non-pathogenic bacteria like *E. Coli*, Salmonella. Biological properties play a significant role in the usage of sewage wastewater (Xie

Table 1 Properties of sewage wastewater (published in an open-access journal, available via license: (CC BY 4.0)

Physical properties	Sources
Odour	Domestic and industrial wastes and decay of natural materials
Colour	Decomposing waste water and industrial waste
Temperature-high	Domestic and industrial wastes
Solids	Domestic water supply, domestic and industrial wastes, soil erosion
Chemical properties	
Organic	
Carbohydrates	Domestic, commercial, and industrial wastes
Fats, oils and grease	Domestic, commercial, and industrial waste
Pesticides	Agricultural wastes
Phenol	Industrial wastes
Proteins	Domestic, commercial and industrial wasters
Biological constituents	
Animals	Open watercourses, and treatment plants
Plants	Open watercourses, and treatment plants
Bacteria	Domestic wastes, surface water infiltration, treatment plants
Virus	Domestic wastes
Protista	Domestic wastes, treatment plants

et al. 2022). Viruses include norovirus, hepatitis in sewage water. Unicellular organisms like Giardia, cryptosporidium are protozoa, cause various diseases, and are found in wastewater. Some fungi are also found in wastewater. Ascaris, hookworms are the eggs and larvae of worms that are found in sewage water. Organic matter consists of biochemical oxygen demand and chemical oxygen demand. The amount of chemical oxygen demand is higher than biological oxygen demand (Tuholske et al. 2021). Sewage water consists of nutrients in the form of nitrogen compounds and phosphorus compounds. Ammonia, nitrates, etc., are the different nitrogen compounds found in sewage water. Sewage water consists of phosphates that can increase the growth of algae. Pathogens pose a risk to human health.

Suspended solids contain organic matter and microbial biomass. Microorganisms break down these solid waste materials during the treatment process. Temperature increases the biological activities as well degradation of biological activities. The pH level of sewage water also plays an important role in the growth of microorganisms (Rout et al. 2021).

3.3 Impact of Sewage Water on the Properties of Soil

Sewage wastewater contains organic, inorganic, chemical, and pathogen mixtures that affect soil properties both in a positive and negative manner. Soil properties are affected by the treatment of wastewater.

3.3.1 *Physical Properties*

The continuous consumption of sewage water causes compression in the soil texture by reducing its porosity and permeability. This results in an increase in susceptibility to waterlogging. Wastewater and groundwater were treated using sprinkler and surface irrigation systems for sugar beet, corn, and sunflower crops. The average concentration of lead in the soil is found to be 1.87 and 1.90 mg/l for the beginning and last of the starting stages, respectively. The concentration of lead in soil increases with depth. As compared with the starting stage, the use of wastewater with 0.016 mg/l lead has no effect on the gathering of soil (Abedi-Koupai et al. 2006). This gathering increases with the sewage water treatment. The use of wastewater in soil increases the infiltration rate. The emergence of biological mat or crust (Balks et al. 1997; Kristiansen 1981) biological clogging, with microbial extracellular particles (McAuliffe et al. 1982; Thomas et al. 1966), collapse in the structure of soil because of dissolution (Nevo and Mitchell 1967) of organic matter and inter soil space blockage because of suspended particles like colloidal clay and particles of algal cell (Berend 1967). Several studies indicate that using wastewater in irrigation for a long time enriches soils with heavy metals to such an extent that it could be hazardous to people's health, the environment, and soil properties (Kwiatkowska-Malina 2018). Cao et al. examined that the vertical distribution of heavy metal elements varies with soil texture and that they were primarily concentrated in the soil's surface layer within a depth of 50 cm of the soil surface (Cao 2004). In another study, the usage of sewage water is studied, Sewage water may have lower concentrations of heavy metals, but long-term usage of the sewage water may cause the soil to accumulate large levels of heavy metals. Also, in comparison to fields irrigated with fresh water, clay soil that was irrigated with sewage effluent had higher deposition of accessible Cu, Cd, Pb, Cr, Ni, and Zn (Dotaniya et al. 2018).

3.3.2 *Chemical Properties*

The acidity or alkalinity of the soil can be impacted by the fluctuating pH level of sewage water. Sewage with an acidic or alkaline pH can alter the soil's capacity to hold nutrients and promote plant development. The nutrients nitrogen, phosphorus, and potassium found in sewage water have the potential to temporarily improve soil fertility. But overuse can cause toxicity, salt buildup, and nutritional imbalances. Heavy metals such as arsenic, mercury, cadmium, and lead are frequently found in untreated sewage. These metals can poison humans, animals, and plants when they build up in the soil, contaminating the food chain. The presence of sewage water, particularly which is saline, can raise the salinity and solidity of the soil. This can influence the osmotic potential of the soil and the plants' ability to absorb water, which can lower soil fertility and impede plant growth. Rattan et al. examined that the continuous utilization of sewage increased the amount of water that was accessible in the soils (Rattan et al. 2005). As compared to irrigated soils, the pH level of soil watered using sewage effluents was not considerably affected. The salt concentration of home sewage contributed to the increased electrical conductivity in sewage-irrigated soils. That was, nevertheless, below the cutoff point that would have put the soil at risk of salinity. Soils watered with sewage, a mean organic carbon (OC) level that was higher, at 1.05%. Sewage-water is used to irrigate soils that have a high organic carbon content because sewage is applied over a long period of time (Begmatov et al. 2022).

The continued wastewater irrigation times resulted in higher levels of organic carbon, total nitrogen, microbial biomass of Carbon and Nitrogen, and activity of microbial (Ramirez-Fuentes et al. 2002). Under wastewater irrigation, soil concentrations of accessible concentrations of Pb, Cd, and Cr, as well as total concentrations of Mg, Hg, Mo, Ca, Cu, and Cr, rise dramatically while staying below the dangerous thresholds (Ramirez-Fuentes et al. 2002).

3.3.3 Biological Properties of Soil

The various biological properties help in the growth of plants, the nutrient cycle, and the sustainability of the ecosystem. Microbial organisms, including bacteria, fungi, protozoa, algae, and actinomycetes, contribute in the biological properties of soil. Bacteria found in the soil decompose the organic matter and fix the nitrogen. This helps in nutrient cycling. Fungi break down the organic compounds and which helps in enriching nutrients. The filamentous bacteria break down the complex organic compounds and provide an earthy smell to the soil. The microbial growth is regulated by protozoa, which in turn helps in enhancing soil availability. Nitrogen fixation is also contributed by algae (Hussain et al. 2021).

Earthworm enhances the soil texture by digesting organic matter. Nematodes play both roles, some are good for soil nutrients, while others are soil pathogens. Insects and mites help in the decomposition of organic matter. Humus enriches the soil texture. The soil texture is also indicated by the content of soil organic carbon. The decomposition of organic matter helps in soil growth by plant growth. Nitrogen fixation also contributes to the plant biological properties of the soil (Ahkola et al. 2021; Nawaz et al. 2024). Figure 1 depicts the different sources of sewage wastewater.

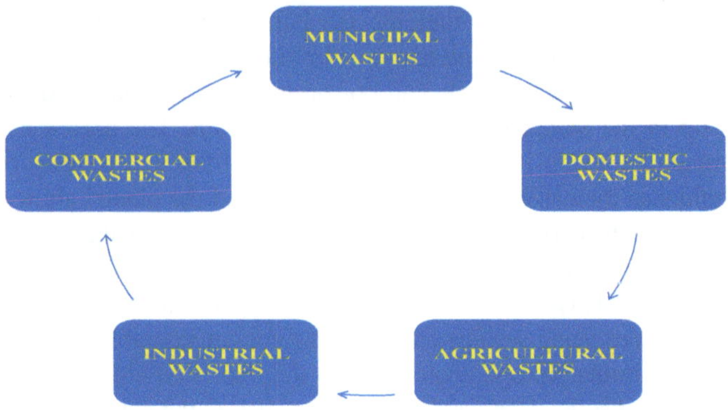

Fig. 1 Sources of sewage wastewater

3.4 Impact of Sewage Water on Water Resources

Water resources are unspoiled natural water supplies that are beneficial to people, such as irrigation or drinking water. Water is a natural resource present in sufficient quantities. It is essential for life on Earth. It is widely used in different processes, including drinking, cooking, cleaning, bathing, washing, and irrigation, among other commercial and residential uses. Rain, groundwater, ice, rivers, lakes, streams, and natural reservoirs are the primary sources of water (Lautze 2014; Paul et al. 2012). These are the sources from which we get water for daily use. Surface water, submerged river flow, groundwater, and frozen water are examples of natural freshwater sources. Desalinated saltwater and treated sewage (reclaimed water) are two examples of artificial freshwater sources. Nearly every process that makes out life on Earth can be split into chemical reactions, and a majority of these processes need water to break down chemicals so they can flow and interact freely. Because it is a solvent, liquid water is a necessary condition for life on Earth (Leverenz and Asano 2011; Warsinger et al. 2018).

3.4.1 Types of Various Pollutants of Water

In general, both point and non-point sources can contaminate water. Point source pollution refers to contamination that originates from one source. One example of this kind of pollution is the discharge of wastewater into bodies. On the other hand, diffuse sources, such as agricultural runoff, are the source of non-point sources of pollution (Ahamad et al. 2020).

However, the following categories classify water contamination on Earth as chemical pollution, surface water pollution, ground water pollution, thermal Pollution, oil Spillages, microbial pollution, oxygen-depletion Pollution, and suspended

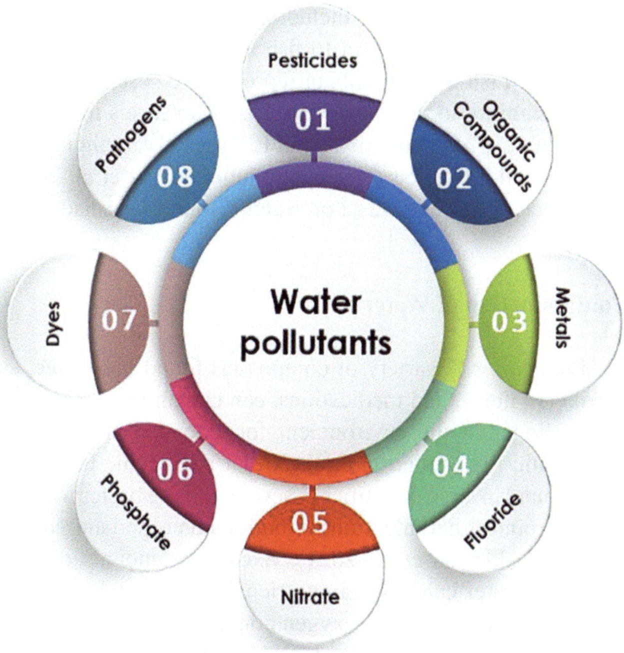

Fig. 2 Various types of water pollution (published in open access https://creativecommons.org/licenses/by-nc-nd/4.0/).

Matter Pollution, as shown in Fig. 2 (Naseem and Durrani 2021). The main issues with contaminated sewage water are biological contaminants, primarily caused by pathogens such as bacteria, viruses, and parasites. These lead to watery illnesses like cholera, dysentery, and hepatitis, which pose a major risk to public health. *Escherichia coli* (*E. coli*) and faecal coliforms are occasionally utilized as markers of sewage contamination in water. Growing numbers of antimicrobial-resistant bacteria found in sewage discharges make treatment more difficult and pose new health risks to the world (Wasimuddin et al. 2020; Rout et al. 2021).

3.4.2 Sewage and Water Pollution

One of the most important issues for the environment and public health is the effect of sewage water on water resources. Sewage that has been improperly or untreated can have a significant impact on the sustainability and quality of water bodies. Marine life, surface waters, and groundwater can all be contaminated by sewage. Pathogens, chemicals, and minerals may be introduced by this pollution, rendering the water unfit for human and animal consumption. Algal blooms brought on by sewage can decrease the amount of oxygen present in water. This may result in dead zones and the destruction of aquatic life, including fish and shellfish. The presence of heavy metals in the water is harmful to fish and humans and can rise as a result of sewage pollution. Fish and amphibian reproductive systems may be harmed by sewage pollution's ability to raise endocrine disruptor levels in the water. Diseases include salmonella, *E. Coli,* hepatitis A and E, dysentery, typhoid, and cholera can infect humans and wildlife due to sewage pollution (Wear et al. 2021a). There are many adverse effects of sewage waste on water resources, as discussed below.

3.4.2.1 Contamination of Water

Chemical Contaminants: A variety of compounds found in sewage, such as detergents, personal care items, and medications, can contaminate water sources. These substances may linger in the environment for extended periods of time and be hazardous to aquatic life (Alderton et al. 2021). Pathogenic bacteria, viruses, and protozoa can be found in large quantities in sewage. These bacteria can cause cholera, diarrhea, and gastrointestinal infections in both humans and animals when they get into water bodies (Alderton et al. 2021). Excessive amounts of phosphorus and nitrogen in sewage water can result in nutrient pollution, which in turn can cause algal blooms that lower the water's oxygen content. As a result, there may be dead zones where aquatic life is exterminated (Olds et al. 2018).

3.4.2.2 Degradation of the Aquatic Ecosystem

Aquatic habitats' natural balance may be upset by sewage contamination. It may result in a change in the species makeup of aquatic plants and animals, which could lower biodiversity and modify food webs. The quality of the environment is lowered when sewage sludge accumulates at the bottom of water bodies, suffocating benthic organisms, which live at the bottom (Bashir et al. 2020).

3.4.2.3 Drinking Water Source Contamination

Drinking Water Source Contamination: If surface and groundwater resources contaminated with sewage are used as sources of drinking water, they can provide serious health problems. For pathogens and contaminants to be removed from contaminated water, costly and intricate treatment procedures may be needed (Sharma and Bhattacharya 2017).

3.4.2.4 Eutrophication

Eutrophication is the excessive development of aquatic plants and algae brought on by sewage's elevated nitrogen and phosphorus levels. Low oxygen levels, or hypoxia, can result from this, endangering fish and other aquatic life (Fig. 3). Detergents and sewage are the primary causes of eutrophication (Ansari et al. 2010). The term eutrophication refers to the natural aging of lake water caused by nutrient enrichment. The water of a young lake is chilly and pure and supports little life. But over time, the streams that empty into the lake add a variety of nutrients to it. Aquatic creatures thrive more readily when additional nutrients like phosphorus and nitrogen are added. The plants and animals that live their flourish as the lake's fertility rises, which leads to an abundance of organic debris accumulating at the lake's bottom. As silt and organic waste build up over time, the lake's depth diminishes and its water temperature rises. Certain aquatic creatures prefer warm water over cold water habitats, and vice versa. The lake basin also starts to fill with marsh plants. In the end, the lake becomes landlocked as it fills with enormous bogs of floating vegetation. A lake's natural ageing process can last thousands of years, depending on a variety of elements like temperature and lake size. But human activities like residential sewerage, industrial detergents, and effluents hasten a lake's aging process, or eutrophication.

Another term for the phenomenon of lakes artificially aging quickly due to a variety of man-made conditions is accelerated eutrophication or cultural aging. The main pollutants that cause eutrophication are detergents and sewage-derived nitrates and phosphates. They encourage the overgrowth of algae, which produces an unattractive slime and offensive odors (O'Sullivan 1995).

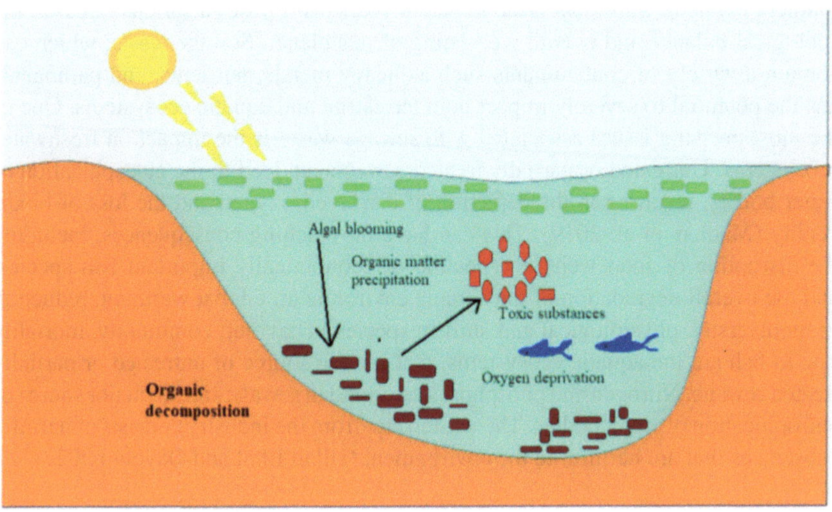

Fig. 3 Process of eutrophication

3.4.2.5 Effects on Soil and Groundwater

In areas with permeable soils or low water tables, sewage water seeping into the soil has the potential to contaminate groundwater. Long-term contamination of agricultural soils and sources of drinking water may result from this pollution (Singh et al. 2012; Pásková et al. 2024).

3.4.2.6 Diminished Water Quality

Water quality can be reduced by increased amounts of suspended solids, turbidity, and organic matter in sewage, which can interfere with natural purification processes in water bodies and render the water unfit for recreational use (Figueroa-Nieves et al. 2014).

3.5 Effects on the Environment

It is necessary to treat the sewage otherwise it may result in bad effects upon the whole environment and human health. The discharge of sewage waste into the environment is one of important issue in recent years, as it poses a serious threat to the ecological balance and overall well-being of our planet. Sewage waste, which can contain a variety of contaminants such as heavy metals, nutrients, and pathogens, has the potential to severely impact both terrestrial and aquatic ecosystems. One of the most pressing issues associated with sewage waste is the impact on freshwater ecosystems. Untreated or partially treated sewage can lead to the eutrophication of water bodies, resulting in the depletion of dissolved oxygen and the loss of biodiversity (Madhav et al. 2019). This can have far-reaching consequences, including the disruption of food webs, the decline of economically important fish species, and the overall degradation of the aquatic environment. Global warming, reduction in biodiversity, physiological and shift in species behaviour, community, mortality rate in fish are the aquatic ecosystems, that are the source of untreated or partially treated sewage. Nitrogen and phosphorus are found in sewage and are major source of eutrophication of water bodies. The wastewater from the industry consists of harmful substances that are harmful to the environment (Ullah Bhat and Qayoom 2022).

3.5.1 Sewage Causes Water Pollution

Sewage holds nitrogen and phosphorus, substances that encourage the growth of algae in water sources. These blooms are dangerous to water life as they use up oxygen in the water, thereby creating dead space that cannot support life. This is generally called Eutrophication. The untreated sewage can pollute the water sources for drinking with some disease-causing organisms, chemicals, and heavy metals. This may result in waterborne disease, including cholera, dysentery, typhoid and so on, Hence the need to improve the sanitation standards in the country. In addition to this, Sewage also contains toxic forms like heavy metals, including lead and mercury; pesticides; and industrial chemicals, that are hazardous to the lives of aquatic beings (Wear et al. 2021b).

3.5.2 Sewage Causes Air Pollution

Organic matter contained in sewage can break down and produce methane, a gas called Methane Emissions, that may lead to climate change. The major component of natural, untreated sewage methane production is due to anaerobic decomposition. In addition, gases such as hydrogen sulfide are released in the process, which cause unpleasant sewer smells and pollute the air and decrease the quality of living for residents in the vicinity (Prasanna Kumar et al. 2024).

3.5.3 Sewage Causes Degradation of Soil

If sewage is applied for irrigation purposes or finds its way into the soil, it brings pathogens, which affect crops, and in the process, infiltrate the food chain to pose health dangers to consumers and animals. The extraction of metals in sewage pollutes and accumulates metals in the soil over time, which leads to poor soil fertility, interferes with microbial life, and adversely affects plant growth (de Guzman et al. 2023). Salts and other substances present in sewage are mixed in the ground, hence the poor yields in agriculture (Jat Baloch et al. 2023).

3.5.4 Sewage Adverse Effects on the Aquatic Ecosystem

Eutrophication reduces dissolved oxygen levels and the toxicity of sewage pollutants that often cause massive death of fish and other aquatic life forms. When water chemistry alters, it is able to affect the balance of an ecosystem in one way or another. Sewage outflows can influence the nature of the aquatic environment and habitat loss by adding sediment and filling up space, changing light conditions, and the temperature regime (Wear et al. 2021b). Some of these changes can result in habitat loss, particularly for those species that are particularly fragile.

3.5.5 Sewage and Health Hazards

Sewage contains pathogens like bacteria, viruses, and parasites, and when humans come into contact with the pathogens through water or food, they easily get infected with diseases. So it causes the spread of various waterborne diseases. Further, Pharmaceutical by-products found in sewage comprise antibiotics. The four substances distort the microbiota and can foster the generation of antibiotic-resistant bacteria in the surrounding environment, which is dangerous for people (Tiwari 2008).

3.5.6 Sewage Shows Adverse Effects on the Climate

Standard sewage treatment processes, if not controlled, can lead to the production of greenhouse gases such as methane and nitrous oxide. The improper treatment of organic wastes in sewage increases the impacts of climate change. It has been reported that even if effluent toxicity was low, the impact seen was lowering of diversity, an increase in primary production as well as herbivory, and a decrease in energy flows related to terrestrial inputs. Combined, the effluent reduced overall through fluxes of organismic energy in stream food chains and demonstrated that treated sewage can cause significant changes on a whole-stream scale for even relatively low concentrations from WWTP discharges. The study indicates that reportedly current procedure for wastewater treatment continues to have impacts on freshwaters, stressing the importance of enhancing the efforts of treating polluted waters to preserve food chains in water bodies (de Guzman et al. 2023).

3.5.7 Sewage Adverse Impact on Biodiversity

The impact of the introduction of sewage into natural ecosystems is often the explosion of particular species that upset the local food chain and decrease the overall number of species so causing a loss of species. Aquatic organisms, especially those categorized as sensitive, may be the first to be affected by changes in water quality and impacts on the habitat. Sewage is full of pollutants that can affect the hormonal balance of water creatures, and in the worst case, it can eliminate some species completely (Wear et al. 2021b). Table 2 explains the effects of sewage water on water and the environment.

Table 2 A summarized table showing the effect of sewage waste on water and the environment

Observations	Method	Outcomes	References
Raw sewage pollutes water sources, polluting them with emerging organic and new pollutants that have a negative impact in water bodies and hence human beings through waterborne diseases and changes in ecosystem balance	(i) Effects of poor facilities in waste disposal, particularly sewage (ii) Analysis of the source of water pollution in Kigali	(i) Sewerage is not centralized and this results in water pollution (ii) The authors provided examples of areawide decentralized treatment systems, which are inadequate in removing new contaminants in water	Bavumiragira et al. (2023)
Emerging contaminants, such as those received from sewage, can be very toxic to aquatic life and pose major risks to health, affecting the overall carrying capacity of the available water resources	(i) Common analytical methods of chemical and biological investigations are described (ii) Some of the techniques include passive biomonitoring, biomarkers such as oxidative stress proteins, and gonadal histopathology	Currently, the knowledge about specific pollutants of emerging concern in African waters is scarce. This came out clearly through passive biomonitoring concluded that this was appropriate in terms of cost when it comes to monitoring of PEC	Siziba and Tapiwa Sero (2023)
Sewage waste triggers coastal eutrophication and algal blooms, which are toxic to aquatic life and other water-dependent organisms and are also a human health hazard	(i) Detailed geographic model (ii) Quantify and locate nitrogen and pathogen sources, human sewerage	(i) Wastewater deposits 6.2 Tg nitrogen to coastal water (ii) Worst degradation scores recorded by 58% of the coral and 88% of the exposed seagrass in wastewater N	Tuholske et al. (2021)

(continued)

Table 2 (continued)

Observations	Method	Outcomes	References
Sewage pollution comes with the effects of chemical elements affecting water sources and causing eutrophication, which in effects water sources to produce hypoxic regions that are detrimental to aquatic species, and loss of biodiversity–water borne diseases and contaminated seafoods are also contributed by sewage pollution	(i) New technologies in the treatment of sewage water (ii) Laws and SOP enforcement and often compliance checks	(i) Discharge of sewage to coastal areas has potential to affect the general population and ecosystems adversely (ii) Global management is best facilitated through expertise in the technologies of advance world, discipline regulations, and cooperation	Rangel-Buitrago et al. (2024)
Thrown sewage waste also shock the microbial population, increase pathogenic bacteria and decline water quality and thereby, cause fatal health impacts on aquatic life and humans dependent on these resources	(i) Sequencing of the variable regions of the bacterial 16S rDNA gene by massively parallel pyrosequencing (ii) Prediction of sample potential by piecrust	(i) The discharge of raw sewage results in poor and low microbial density (ii) The data show the significant correlation of pathogenic bacteria density with frequent sewage discharge	Xie et al. (2022)
Untreated sewage affects aquatic ecosystems by raising nutrient levels and contaminants that are inimical to species diversification and pose risks to health for humans relying on these water sources	(i) Assembled and counted research on effects of untreated sewage (ii) Examined the effect of untreated sewage on Latin America's urban stream	(i) Unprocessed or untreated sewage proves to be a big problem in the Latin American urban streams (ii) Few papers examine the effects of untreated sewage than storm water	Marques and Cunico (2023)

(continued)

Table 2 (continued)

Observations	Method	Outcomes	References
Low concentrations of trace elements and contaminants found in sewage wastes can contaminate groundwater continue to give chronic non-carcinogenic health hazards to human beings and lethal effects on aquatic life forms	(i) A multicomponent strategy for assessing groundwater and surface water quality (ii) MINeral TEQuilibrium aqueous (MINTAQ) is a new geochemical modelling program for predicting trace element distribution using visualization	(i) Wastewater shutdown by trace amounts of metals, simple analysis of groundwater and surface water contamination (ii) Removal through various sorption processes of hazardous trace elements in drinking water is indicated	Kandil et al. (2023)
Sewage without treatment brings pathogens and micropollutants that affect the aquatic ecosystem and humans, especially the most susceptible persons	(i) Automated GIS model incorporates an additional source of fine scale elevation and flowpath tracing (ii) Total risk measures of the land and population health have been computed	(i) Constructed a GIS model for evaluation of the risks associated with sewage spillage (ii) Developed a list of possible measures that may be connected with high risks of land and population health problems	(McDaniel et al. 2023)

(continued)

Table 2 (continued)

Observations	Method	Outcomes	References
Sewage waste enhances the entrance of human enteric viruses and pathogens in water sources and degrades water resources, thus posing threats to human beings and waterborne diseases	(i) in this study, firstly, the sewage samples was concentrated using PEG followed by the extraction of nucleic acids (ii) The nucleic acid detections of norovirus and the virome for the purpose of detecting viral diversity	(i) Enhanced viral detection with dedicated replicate samples in sewage examination (ii) Thus, the system has successfully described the viral diversity and has sequenced several viruses, including complete genomes	(Schaeffer et al. 2023)
Sewage enhances the levels of vibrio vulnificus thereby raising the level of infections, which are harmful to human health in septic individuals and have the potential to alter the balance of bacterial species with the gene transcription and may readjust the entire balance of bacterial flora in aquatic organisations	(i) Effects of the addition of 1% sewage to estuarine water (ii) Organism response to sewage as reflected in gene transcription	(i) Sewage exposure results in a marked congestion of V. Vulnificus in coastal waters (ii) Such changes include the regulation of gene transcription involved in virulence and survival mode of bacteria from the effect of sewage	(Conrad and Harwood 2022)

3.6 Mitigation and Control Measures

Some of the mitigation and control measures that can be put in place to avert more pollution of water resources by sewage or to reduce the effects of Sewage. Such measures are in the form of technological and policy solutions that attempt to improve sewage disposal (Rathinamoorthy and Raja Balasaraswathi 2022). A comprehensive sewage waste management and mitigation strategy is shown in Fig. 4.

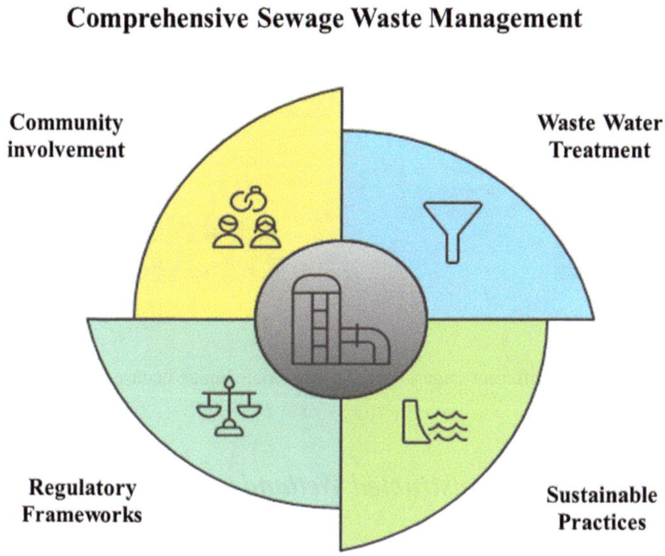

Fig. 4 Mitigation strategies of sewage waste

3.6.1 Treatment of Wastewater

Primary, Secondary, and Tertiary Treatment are the contemporary sewage treatment plant should incorporate all three stages of treatment (Sonune and Ghate 2004). Primary Treatment is a set of procedures that eliminate solids from water through sedimentation. Secondary Treatment encompasses biological treatment necessary to decompose organic substances and eliminate some pathogens. Tertiary Treatment filters out nutrients (nitrogen and phosphorus), pathogens, chemical contaminants, etc., employing ultra-filtration, UV light, or chlorine treatment. Different stage of water treatment is depicted in Fig. 5. Small-scale sewage treatment plants or constructed wetlands are cost-effective treatment systems suitable for decentralized sewage management in rural areas or small communities to reduce pollution of water sources (Phull et al. 1997).

3.6.2 Septic System Maintenance

In areas where the septic system is used, periodic check-ups and maintenance will go a long way in removing contaminants from the groundwater. Inadequate septic systems may discharge sewage effluent onto soil and water sources, polluting local assets. Switching to improved, effective septic systems can greatly decrease the chances of water pollution in an area (Scherba et al. 1991).

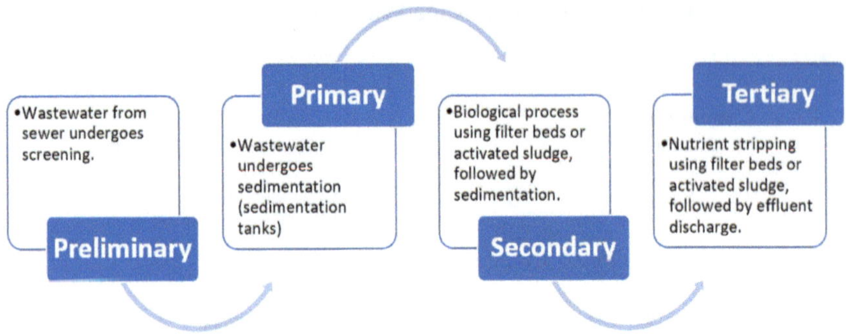

Fig. 5 Water treatment different stages, (open access article image license at https://creativecommons.org/licenses/by/4.0/)

3.6.3 Natural and Constructed Wetlands

They can be used to filter pollutants from sewage water before it spills over to the rivers, lakes, or seas (Wasimuddin et al. 2020). Wetlands reduce nutrient, pathogen, and other pollutant concentrations. These are artificially constructed facilities to perform the role of wetlands in wastewater treatment. They are relatively inexpensive and consume less energy as compared to other wastewater treatment technologies suitable for small communities or decentralized wastewater treatment facilities (Hua 2000).

3.6.4 Recycling and Reusing Water

From laundry baths, showers, and sinks by using grey water from households can be treated and used in irrigation, thus helping to relieve pressure on sewage systems and avoiding polluting water sources. Depending on its quality, treated wastewater can be safely discharged for agricultural purposes, industrial cooling, or, through applying certain sophisticated technologies, for drinking Water purposes, thus minimizing demands on freshwater (Abdel-Fatah 2018).

3.6.5 Legislation and Policy Measures

Integrated water resource management considers the full cycle of water from generation of sewage, treatment the sewage, and even reusing it instead of the conventional methods of water and sewage management (Dang et al. 2019). Like in other areas of development, the governments can avail different types of financial rewards or subsidies for the municipalities and industries that help in up upgradation of sewage treatment plants and implement water-saving technologies. Millions of liters of sewage are released into our waterways every day, and through the use of better policies and incorporating a sustainable attitude towards sewage management, we can reduce the adverse impacts of contained sewage (Zarei 2020).

3.6.6 Sustainable Urban Drainage Systems (SUDS)

Each of these options: rain gardens, permeable pavements, constructed wetlands, can be installed to remove pollutants from the storm water runoff before it flows to the sewage system or to water bodies. They can also manage stormwater and prevent sewage effluent during intense rains without much effort. In many cities, separate sewers for sewage and stormwater cause the mixture to overflow during storms; thus, discharging raw sewage into the water sources (Kesari et al. 2021; Etikala et al. 2022). Separating these systems using the infrastructure reduces the risk of such mishaps.

3.6.7 Regulation and Monitoring

Regulations on sewage discharge should be tightened; authorities should demand that factories and municipalities produce effluent of high quality before it is discharged. This assessment shows that routine surveillance and monitoring of water bodies is critical in order to identify pollution and which compliance has not been met. Water quality monitoring through physical, chemical, and bacteriological investigation can also be done in remote areas, and these will guide the identification of the sources of pollution and the levels of improvement (Foerstner et al. 2006).

3.6.8 Education with Outreach to the Public

To some extent, contamination can be prevented by increasing the level of knowledge and community involvement about the impact of sewage on water and promoting rational use of water resources. People can also help in monitoring water quality and reporting water pollution cases. To rid sewage of harmful substances, the chemical, pharmaceutical, and household waste disposal methods for the proper disposal of Hazardous substances need to be explained to the general public (Montoya-Coronado et al. 2024).

3.7 Conclusion and Future Prospects

There are numerous advantages of using treated wastewater for the purpose of irrigation, but there are drawbacks as well. The efficacy size is significantly impacted by an analysis of the physicochemical and microbiological characteristics of treated wastewater used for irrigation. One sustainable development tactic that aids in addressing the problem of water scarcity is the reutilisation of sewage or wastewater for agricultural irrigation. There are benefits and drawbacks to using treated wastewater for irrigation in terms of the economic, human health, and the environment. One major benefit is that it can boost farmers' profitability. By providing a reliable source of water for their crops, farmers can increase crop yields and produce more profitable crops. The nutrients present in treated wastewater that can be beneficial for the development of plants can reduce the need for costly fertilizers. The treated wastewater can also be used in irrigation because it can preserve water resources, which is important in areas with a high demand for water and a shortage of freshwater resources. Treated wastewater for irrigation leads to the saving of freshwater for other purposes, for domestic and commercial uses. Wastewater can ultimately be used to substitute freshwater in agriculture, which can help in resolving water scarcity. But for a long–term and safe functioning, several issues need to be resolved.

There is a brighter future for meeting water demands through the utilisation of treated wastewater for the purpose of irrigation. Farmers currently utilize wastewater directly for irrigation in underdeveloped nations, which can lead to a number of health problems for both farmers and consumers. Consequently, it is crucial to use both conventional and cutting–edge wastewater treatment techniques. To increase the use of safe water in agriculture and for farmers, it is imperative that certain policies and practices be put in place to promote it.

References

Abascal E, Gómez-Coma L, Ortiz I, Ortiz A (2022) Global diagnosis of nitrate pollution in groundwater and review of removal technologies. Sci Total Environ 810:152233. https://doi.org/10.1016/j.scitotenv.2021.152233

Abdel-Fatah MA (2018) Nanofiltration systems and applications in wastewater treatment: review article. Ain Shams Eng J 9(4):3077–3092. https://doi.org/10.1016/j.asej.2018.08.001

Abedi-Koupai J, Mostafazadeh-Fard B, Afyuni M, Bagheri MR (2006) Effect of treated wastewater on soil chemical and physical properties in an arid region. Plant Soil Environ 52(8):335–344. https://doi.org/10.17221/3450-PSE

Abegunrin TP, Awe GO, Idowu DO, Adejumobi MA (2016) Impact of wastewater irrigation on soil physico-chemical properties, growth and water use pattern of two indigenous vegetables in Southwest Nigeria. Catena 139:167–178. https://doi.org/10.1016/j.catena.2015.12.014

Adjovu GE, Stephen H, James D, Ahmad S (2023) Measurement of total dissolved solids and total suspended solids in water systems: a review of the issues, conventional, and remote sensing techniques. Remote Sens 15(14):3534. https://doi.org/10.3390/rs15143534

Ahamad A, Madhav S, Singh AK, Kumar A, Singh P (2020) Types of water pollutants: conventional and emerging, pp 21–41. https://doi.org/10.1007/978-981-15-0671-0_3

Ahkola H, Lindholm-Lehto P, Perkola N, Välitalo P, Meriläinen P, Mäenpää K, Stelzer JAA, Heiskanen I, Järvistö J, Nuutinen J, Leppänen MT (2021) A preliminary study on the ecotoxic potency of wastewater treatment plant sludge combining passive sampling and bioassays. Sci Total Environ 758:143700. https://doi.org/10.1016/j.scitotenv.2020.143700

Alderton I, Palmer BR, Heinemann JA, Pattis I, Weaver L, Gutiérrez-Ginés MJ, Horswell J, Tremblay LA (2021) The role of emerging organic contaminants in the development of antimicrobial resistance. Emerg Contam 7:160–171. https://doi.org/10.1016/j.emcon.2021.07.001

Anderson A, Nilsson KO (1973) Enrichment of trace elements from sewage sludge fertilizers in soils and plants. Ambio 1:176–179

Aniyikaiye TE, Oluseyi T, Odiyo JO, Edokpayi JN (2019) Physico-chemical analysis of wastewater discharge from selected paint industries in Lagos, Nigeria. Int J Environ Res Public Health 16(7):1235. https://doi.org/10.3390/ijerph16071235

Ansari AA, Gill SS, Khan FA (2010) Eutrophication: threat to aquatic ecosystems. In: Eutrophication: causes, consequences and control. Springer, Netherlands, pp 143–170. https://doi.org/10.1007/978-90-481-9625-8_7

Arora BR, Azad AS, Singh B, Sekhon GS (1984) Pollution potential of municipal waste waters of Ludhiana, Punjab. Indian J Ecol 12:1–7

Balks MR, McLay CDA, Harfoot CG (1997) Determination of the progression in soil microbial response, and changes in soil permeability, following application of meat processing effluent to soil. Appl Soil Ecol 6(2):109–116. https://doi.org/10.1016/S0929-1393(97)00005-X

Bashir I, Lone FA, Bhat RA, Mir SA, Dar ZA, Dar SA (2020) Concerns and threats of contamination on aquatic ecosystems. In: Bioremediation and biotechnology. Springer International Publishing, pp 1–26. https://doi.org/10.1007/978-3-030-35691-0_1

Bavumiragira JP, Yin H, Kipkrui E, Uyisaba P, Ndizeye Y, Bazambanza A (2023) Impact associated with the lack of centralized sewerage treatment systems and inadequate operation of available decentralized treatment systems, a source of emerging contaminants in urban water resources: a review case Study of Kigali City Rwanda, a reflection of other East African Countries' cities. Emerg Issues Environ Geogr Earth Sci 2: 67–93) BP International (a part of SCIENCEDOMAIN International). 10.9734/bpi/eieges/v2/6220E

Bedbabis S, Ben Rouina B, Boukhris M, Ferrara G (2014) Effect of irrigation with treated wastewater on soil chemical properties and infiltration rate. J Environ Manag 133:45–50. https://doi.org/10.1016/j.jenvman.2013.11.007

Begmatov S, Dorofeev AG, Kadnikov VV, Beletsky AV, Pimenov NV, Ravin NV, Mardanov AV (2022) The structure of microbial communities of activated sludgsse of large-scale wastewater treatment plants in the city of Moscow. Sci Rep 12(1):3458. https://doi.org/10.1038/s41598-022-07132-4

Berend JE (1967) An analytical approach to the clogging effect of suspended matter. Int Assoc Sci Hydrol Bull 12(2):42–55. https://doi.org/10.1080/02626666709493523

Boudjabi S, Ababsa N, Chenchouni H (2022) Sewage and sewage treatment. In: The Palgrave handbook of global sustainability. Springer International Publishing, pp 1–27. https://doi.org/10.1007/978-3-030-38948-2_50-1

Bradai M, Han J, El Omri A, Funamizu N, Sayadi S, Isoda H (2016) Effect of linear alkylbenzene sulfonate (LAS) on human intestinal caco-2 cells at non cytotoxic concentrations. Cytotechnology 68(4):1267–1275. https://doi.org/10.1007/s10616-015-9887-4

Cao SPC (2004) Vertical distribution of heavy metal pollutant in soil profiles of Tianjin. Contrib Geol Miner Resour Res 19(4):270–274

Chahal C, van den Akker B, Young F, Franco C, Blackbeard J, Monis P (2016) Pathogen and particle associations in wastewater, pp 63–119. https://doi.org/10.1016/bs.aambs.2016.08.001

Conrad JW, Harwood VJ (2022) Sewage promotes vibrio vulnificus growth and alters gene transcription in vibrio vulnificus CMCP6. Microbiol Spect 10(1). https://doi.org/10.1128/spectrum.01913-21

Dang Q, Tan W, Zhao X, Li D, Li Y, Yang T, Li R, Zu G, Xi B (2019) Linking the response of soil microbial community structure in soils to long-term wastewater irrigation and soil depth. Sci Total Environ 688:26–36. https://doi.org/10.1016/j.scitotenv.2019.06.138

de Guzman I, Elosegi A, von Schiller D, González JM, Paz LE, Gauzens B, Brose U, Antón A, Olarte N, Montoya JM, Larrañaga A (2023) Treated and highly diluted, but wastewater still impacts diversity and energy fluxes of freshwater food webs. J Environ Manag 345:118510. https://doi.org/10.1016/j.jenvman.2023.118510

Dotaniya ML, Rajendiran S, Meena VD, Coumar MV, Saha JK, Kundu S, Patra AK (2018) Impact of long-term application of sewage on soil and crop quality in vertisols of central India. Bull Environ Contam Toxicol 101(6):779–786. https://doi.org/10.1007/s00128-018-2458-6

Etikala B, Madhav S, Somagouni SG (2022) Urban water systems: an overview. Curr Dir Water Scarcity Res 6:1–19

Figueroa-Nieves D, McDowell WH, Potter JD, Martínez G, Ortiz-Zayas JR (2014) Effects of sewage effluents on water quality in tropical streams. J Environ Qual 43(6):2053–2063. https://doi.org/10.2134/jeq2014.03.0139

Foerstner KU, von MC, Bork P (2006) Comparative analysis of environmental sequences: potential and challenges. Philos Trans R Soc B: Biol Sci 361(1467):519–523. https://doi.org/10.1098/rstb.2005.1809

Gleisberg D, Kandler J, Ulrich H, Hartz P (1976) Eutrophication and wastewater purification. Angew Chem Int Ed Engl 15(6):354–365. https://doi.org/10.1002/anie.197603541

Gregory A (2000) Strategic direction of water recycling in Sydney. In: Proceeding of the first symposium water recycling, pp 35–41

Gupta H (2016) Photocatalytic degradation of phenanthrene in the presence of akaganeite nanorods and the identification of degradation products. RSC Adv 6(114):112721–112727

Hashem MS, Qi X (2021) Treated wastewater irrigation—a review. Water 13(11):1527. https://doi.org/10.3390/w13111527

Hua I (2000) Inactivation of *Escherichia coli* by sonication at discrete ultrasonic frequencies. Water Res 34(15):3888–3893. https://doi.org/10.1016/S0043-1354(00)00121-4

Huang M, Li Y, Gu G (2010) Chemical composition of organic matters in domestic wastewater. Desalination 262(1–3):36–42. https://doi.org/10.1016/j.desal.2010.05.037

Hussain A, Kumari R, Sachan SG, Sachan A (2021) Biological wastewater treatment technology: advancement and drawbacks. In: Microbial ecology of wastewater treatment plants. Elsevier, pp 175–192. https://doi.org/10.1016/B978-0-12-822503-5.00002-3

Jat Baloch MY, Zhang W, Sultana T, Akram M, Shoumik BA, Al K, Md Z, Farooq MA (2023) Utilization of sewage sludge to manage saline–alkali soil and increase crop production: is it safe or not? Environ Technol Innov 32:103266. https://doi.org/10.1016/j.eti.2023.103266

Kesari KK, Soni R, Jamal QMS, Tripathi P, Lal JA, Jha NK, Siddiqui MH, Kumar P, Tripathi V, Ruokolainen J (2021) Wastewater treatment and reuse: a review of its applications and health implications. Water Air Soil Pollut 232(5):208. https://doi.org/10.1007/s11270-021-05154-8

Khalid S, Shahid M, Natasha IB, Sarwar T, Shah AH, Niazi NK (2018) A review of environmental contamination and health risk assessment of wastewater use for crop irrigation with a focus on low and high-income countries. Int J Environ Res Public Health 15(5):895. https://doi.org/10.3390/ijerph15050895

Kirkham BM (1974) Disposal of sludge on land-effects on soils, plants and ground water. Compos Sci 15:6–10

Kiziloglu FM, Turan M, Sahin U, Kuslu Y, Dursun A (2008) Effects of untreated and treated wastewater irrigation on some chemical properties of cauliflower (*Brassica olerecea* L. var. botrytis) and red cabbage (*Brassica olerecea* L. var. rubra) grown on calcareous soil in Turkey. Agric Water Manag 95(6):716–724. https://doi.org/10.1016/j.agwat.2008.01.008

Kristiansen R (1981) Sand-filter trenches for purification of septic tank effluent: I. The clogging mechanism and soil physical environment. J Environ Qual 10(3):353–357. https://doi.org/10.2134/jeq1981.00472425001000030020x

Kwiatkowska-Malina J (2018) Functions of organic matter in polluted soils: the effect of organic amendments on phytoavailability of heavy metals. Appl Soil Ecol 123:542–545. https://doi.org/10.1016/j.apsoil.2017.06.021

Kumar N, Kumar B, Gupta H, Gupta A (2023) Development and evaluation of cellulose/graphene-oxide based composite for removing phenol from aqueous solutions. Polymers 15(3):572. https://doi.org/10.3390/polym15030572

Larsen WE, Gilley JR, Linden DR (1975) Consequences of waste disposal on land. J Soil Water Conserv 2:68–71

Lautze J (ed) (2014) Key concepts in water resource management. Routledge. https://doi.org/10.4324/9781315884394

Leverenz HL, Asano T (2011) Wastewater reclamation and reuse system. In: Treatise on water science. Elsevier, pp 63–71. https://doi.org/10.1016/B978-0-444-53199-5.00076-2

Liu C, Cui B, Zeleke KT, Hu C, Wu H, Cui E, Huang P, Gao F (2021) Risk of secondary soil salinization under mixed irrigation using brackish water and reclaimed water. Agronomy 11(10):2039. https://doi.org/10.3390/agronomy11102039

Madhav S, Ahamad A, Singh AK, Kushawaha J, Chauhan JS, Sharma S, Singh P (2019) Water pollutants: sources and impact on the environment and human health. Sens Water Pollut Monit Role Mater:43–62

Marques P, Cunico A (2023) Integrating the influence of untreated sewage into our understanding of the urban stream syndrome. Freshw Sci 42(2):195–203. https://doi.org/10.1086/724823

McAuliffe KW, Scotter DR, MacGregor AN, Earl KD (1982) Casein whey wastewater effects on soil permeability. J Environ Qual 11(1):31–34. https://doi.org/10.2134/jeq1982.00472425001100010009x

McDaniel EL, Atkinson SF, Tiwari C (2023) Quantifying the land and population risk of sewage spills overland using a fine-scale. DEM-Based GIS Model PeerJ 11:e16429. https://doi.org/10.7717/peerj.16429

Metcalf and Eddy TAFBHL (2007) Water reuse: issues, technologies, and applications, 1st edn. New York, NY, USA, McGraw-Hill Education

Mohammad MJ, Mazahreh N (2003) Changes in soil fertility parameters in response to irrigation of forage crops with secondary treated wastewater. Commun Soil Sci Plant Anal 34(9–10):1281–1294. https://doi.org/10.1081/CSS-120020444

Montoya-Coronado VA, Tedoldi D, Lenormand E, Castebrunet H, Molle P, Lipeme Kouyi G (2024) Combined sewer overflow mitigation through SUDS-a review on modelling practices, scenario elaboration, and related performances. J Environ Manag 362:121073. https://doi.org/10.1016/j.jenvman.2024.121073

Namkung E, Rittmann BE (1986) Soluble microbial products (SMP) formation kinetics by biofilms. Water Res 20(6):795–806. https://doi.org/10.1016/0043-1354(86)90106-5

Naseem T, Durrani T (2021) The role of some important metal oxide nanoparticles for wastewater and antibacterial applications: a review. Environ Chem Ecotoxicol 3:59–75. https://doi.org/10.1016/j.enceco.2020.12.001

Nawaz T, Joshi N, Nelson D, Saud S, Abdelsalam NR, Abdelhamid MMA, Jaremko M, Rahman TU, Fahad S (2024) Harnessing the potential of nitrogen-fixing cyanobacteria: a rich bio-resource for sustainable soil fertility and enhanced crop productivity. Environ Technol Innov 36:103886. https://doi.org/10.1016/j.eti.2024.103886

Nevo Z, Mitchell R (1967) Factors affecting biological clogging of sand associated with ground water recharge. Water Res 1(3):231–236. https://doi.org/10.1016/0043-1354(67)90013-9

Obayomi OV, Olawoyin DC, Oguntimehin O, Mustapha LS, Kolade SO, Oladoye PO, Oh S, Obayomi KS (2024) Exploring emerging water treatment technologies for the removal of microbial pathogens. Curr Res Biotechnol 8:100252. https://doi.org/10.1016/j.crbiot.2024.100252

Olds HT, Corsi SR, Dila DK, Halmo KM, Bootsma MJ, McLellan SL (2018) High levels of sewage contamination released from urban areas after storm events: a quantitative survey with sewage specific bacterial indicators. PLoS Med 15(7):e1002614. https://doi.org/10.1371/journal.pmed.1002614

O'Sullivan PE (1995) Eutrophication. Int J Environ Stud 47(3–4):173–195. https://doi.org/10.1080/00207239508710958

Pásková M, Štekerová K, Zanker M, Lasisi TT, Zelenka J (2024) Water pollution generated by tourism: review of system dynamics models. Heliyon 10(1):e23824. https://doi.org/10.1016/j.heliyon.2023.e23824

Paul S, Saha D, Hosen SZ, Emran TB (2012) Overview on water resources and pollution. Res J Sci Technol:108–114

Phull SS, Newman AP, Lorimer JP, Pollet B, Mason TJ (1997) The development and evaluation of ultrasound in the biocidal treatment of water. Ultrason Sonochem 4(2):157–164. https://doi.org/10.1016/S1350-4177(97)00029-1

Popa P, Timofti M, Voiculescu M, Dragan S, Trif C, Georgescu LP (2012) Study of physico-chemical characteristics of wastewater in an urban agglomeration in Romania. Sci World J 2012:1–10. https://doi.org/10.1100/2012/549028

Prasanna Kumar DJ, Mishra RK, Chinnam S, Binnal P, Dwivedi N (2024) A comprehensive study on anaerobic digestion of organic solid waste: a review on configurations, operating parameters, techno-economic analysis and current trends. Biotechnol Notes 5:33–49. https://doi.org/10.1016/j.biotno.2024.02.001

Qiang X, Sun J, Ning H (2022) Impact of subsoiling on cultivated horizon construction and grain yield of winter wheat in the North China plain. Agriculture 12(2):236. https://doi.org/10.3390/agriculture12020236

Ramirez-Fuentes E, Lucho-Constantino C, Escamilla-Silva E, Dendooven L (2002) Characteristics, and carbon and nitrogen dynamics in soil irrigated with wastewater for different lengths of time. Bioresour Technol 85(2):179–187. https://doi.org/10.1016/S0960-8524(02)00035-4

Rangel-Buitrago N, Galgani F, Neal WJ (2024) Addressing the global challenge of coastal sewage pollution. Mar Pollut Bull 201:116232. https://doi.org/10.1016/j.marpolbul.2024.116232

Rathinamoorthy R, Raja Balasaraswathi S (2022) Microfiber pollution prevention—mitigation strategies and challenges, pp 205–243. https://doi.org/10.1007/978-981-19-4185-6_6

Rattan RK, Datta SP, Chhonkar PK, Suribabu K, Singh AK (2005) Long-term impact of irrigation with sewage effluents on heavy metal content in soils, crops and groundwater—a case study. Agric Ecosyst Environ 109(3–4):310–322. https://doi.org/10.1016/j.agee.2005.02.025

Rout PR, Shahid MK, Dash RR, Bhunia P, Liu D, Varjani S, Zhang TC, Surampalli RY (2021) Nutrient removal from domestic wastewater: a comprehensive review on conventional and advanced technologies. J Environ Manag 296:113246. https://doi.org/10.1016/j.jenvman.2021.113246

Samaraweera SAPT, Najim MMM, Alotaibi BA, Traore A (2024) Impacts of a partially connected wastewater treatment plant on the water quality of stormwater drains used as an irrigation source. Front Environ Sci 12. https://doi.org/10.3389/fenvs.2024.1412717

Schaeffer J, Desdouits M, Besnard A, Le Guyader FS (2023) Looking into sewage: how far can metagenomics help to detect human enteric viruses? Front Microbiol 14. https://doi.org/10.3389/fmicb.2023.1161674

Scherba G, Weigel RM, O'Brien WD (1991) Quantitative assessment of the germicidal efficacy of ultrasonic energy. Appl Environ Microbiol 57(7):2079–2084. https://doi.org/10.1128/aem.57.7.2079-2084.1991

Sdiri W, Dabbou S, Chehab H, Selvaggini R, Servili M, Di Bella G, Mansour HB (2020) Quality characteristics and chemical evaluation of Chemlali olive oil produced under dairy wastewater irrigation. Agric Water Manag 236:106124. https://doi.org/10.1016/j.agwat.2020.106124

Sharma S, Bhattacharya A (2017) Drinking water contamination and treatment techniques. Appl Water Sci 7(3):1043–1067. https://doi.org/10.1007/s13201-016-0455-7

Singh PK, Deshbhratar PB, Ramteke DS (2012) Effects of sewage wastewater irrigation on soil properties, crop yield and environment. Agric Water Manag 103:100–104. https://doi.org/10.1016/j.agwat.2011.10.022

Singh S, Gupta H, Dhiman S, Sahu NK (2022) Decontamination of cationic dye brilliant green from the aqueous media. Appl Water Sci 12(4):61. https://doi.org/10.1007/s13201-022-01596-5

Singh S, Parveen N, Gupta H (2018) Adsorptive decontamination of rhodamine-B from water using banana peel powder: a biosorbent. Environ Technol Innov 12:189–195. https://doi.org/10.1016/j.eti.2018.09.001

Siziba N, Tapiwa Sero E (2023) Pollutants of emerging concern in urban wastewater impacted aquatic. Environ Manage Recomm https://doi.org/10.5772/intechopen.106943

Soda N, Ephrath JE, Dag A, Beiersdorf I, Presnov E, Yermiyahu U, Ben-Gal A (2017) Root growth dynamics of olive (*Olea europaea* L.) affected by irrigation induced salinity. Plant Soil 411(1–2):305–318. https://doi.org/10.1007/s11104-016-3032-9

Sonune A, Ghate R (2004) Developments in wastewater treatment methods. Desalination 167:55–63. https://doi.org/10.1016/j.desal.2004.06.113

Sossou K, Bala Prasad S, Agbotsou KE, Saidou Souley H, Mudigandla R (2024) Characteristics of landfill leachate and leachate treatment by biological and advanced coagulation process: feasibility and effectiveness—an overview. Waste Manag Bull 2(2):181–198. https://doi.org/10.1016/j.wmb.2024.04.009

Spietz RL, Williams CM, Rocap G, Horner-Devine MC (2015) A dissolved oxygen threshold for shifts in bacterial community structure in a seasonally hypoxic estuary. PLoS One 10(8):e0135731. https://doi.org/10.1371/journal.pone.0135731

Thapliyal A, Vasudevan P, Dastidar MG, Tandon M, Mishra S (2011) Irrigation with wastewater: responses on the growth and yield of ladyfinger *Assbelmoschus esculentus* and on soil nutrients. J Environ Biol 32:645–651

Thomas RE, Schwartz WA, Bendixen TW (1966) Soil chemical changes and infiltration rate reduction under sewage spreading. Soil Sci Soc Am J 30(5):641–646. https://doi.org/10.2136/sssaj1966.03615995003000050030x

Thomas R, Gough R, Freeman C (2017) Linear alkylbenzene sulfonate (LAS) removal in constructed wetlands: the role of plants in the treatment of a typical pharmaceutical and personal care product. Ecol Eng 106:415–422. https://doi.org/10.1016/j.ecoleng.2017.06.015

Tiwari R (2008) Occupational health hazards in sewage and sanitary workers. Indian J Occup Environ Med 12(3):112. https://doi.org/10.4103/0019-5278.44691

Tuholske C, Halpern BS, Blasco G, Villasenor JC, Frazier M, Caylor K (2021) Mapping global inputs and impacts from of human sewage in coastal ecosystems. PLoS One 16(11):e0258898. https://doi.org/10.1371/journal.pone.0258898

Ullah Bhat S, Qayoom U (2022) Implications of sewage discharge on freshwater ecosystems. In: Sewage-recent advances, new perspectives and applications. IntechOpen. https://doi.org/10.5772/intechopen.100770

Verma N, Singh AK (2013) Development of biological oxygen demand biosensor for monitoring the fermentation industry effluent. ISRN Biotechnol, pp 1–6. https://doi.org/10.5402/2013/236062

Warsinger DM, Chakraborty S, Tow EW, Plumlee MH, Bellona C, Loutatidou S, Karimi L, Mikelonis AM, Achilli A, Ghassemi A, Padhye LP, Snyder SA, Curcio S, Vecitis CD, Arafat HA, Lienhard JH (2018) A review of polymeric membranes and processes for potable water reuse. Prog Polym Sci 81:209–237. https://doi.org/10.1016/j.progpolymsci.2018.01.004

Wasimuddin K, Schlaeppi S, Ronchi F, Leib SL, Erb M, Ramette A (2020) Evaluation of primer pairs for microbiome profiling from soils to humans within the one health framework. Mol Ecol Resour 20(6):1558–1571. https://doi.org/10.1111/1755-0998.13215

Wear SL, Acuña V, McDonald R, Font C (2021) Sewage pollution, declining ecosystem health, and cross-sector collaboration. Biol Conserv 255:109010. https://doi.org/10.1016/j.biocon.2021.109010

Xie Y, Liu X, Wei H, Chen X, Gong N, Ahmad S, Lee T, Ismail S, Ni S-Q (2022) Insight into impact of sewage discharge on microbial dynamics and pathogenicity in river ecosystem. Sci Rep 12(1):6894. https://doi.org/10.1038/s41598-022-09579-x

Zarei M (2020) Wastewater resources management for energy recovery from circular economy perspective. Water-Energy Nexus 3:170–185. https://doi.org/10.1016/j.wen.2020.11.001

Chapter 4
Municipal Sewage Sludge Incineration: Challenges and Strategies for Air Pollution Control

Amanpreet Kaur, Aayasha Negi, Man Vir Singh, Neha Bhatt, and Keshari Nandan

Abstract The formation of municipal sewage sludge, a significant waste product of conventional municipal wastewater treatment processes, has surged dramatically due to increasing urbanization and industrialization. As a result, the safe recycling and resource recovery of municipal sewage sludge has become a pressing environmental challenge. On the other hand, incineration has long been regarded as an effective method for the secure disposal of this sludge, despite its high operational costs and challenges related to the emission of pollutants. Sludge can also be co-incinerated in facilities such as cement kilns, coal-fired power plants, and municipal solid waste (MSW) incinerators. This co-combustion approach is advantageous because it reduces the need for additional flue gas treatment systems and incineration equipment, and in some cases, it can even be more environmentally sustainable. The incineration process itself can be further optimized to reduce nitrogen oxide emissions through methods such as staged air combustion and low-oxygen dilution combustion. Additionally, the removal of sulfur can be enhanced by the use of calcium-based additives, while the volatilization of toxic metals and the formation of harmful components, such as polychlorinated dibenzo-p-dioxins and polychlorinated dibenzofurans, is influenced by chlorine-containing compounds. This chapter will provide an in-depth examination of these issues, with particular emphasis on the ecotoxicological risks associated with pollutants in sewage sludge. It will systematically review and analyze the latest research on pyrolysis of municipal sewage sludge

A. Kaur (✉) · A. Negi
Department of Chemistry, School of Sciences, IFTM University, Moradabad, Uttar Pradesh, India
e-mail: Amanpreet2225@gmail.com

M. V. Singh
School of Applied and Life Sciences, UIT, Uttaranchal University, Dehradun, Uttarakhand, India

N. Bhatt
Department of Chemistry, Pt. L.M.S., Rishikesh Campus, SDS University, Tehri, Uttarakhand, India

K. Nandan
Department of Chemistry, Gurukul Kangri Vishwavidyalaya, Haridwar, Uttarakhand, India

S. Madhav et al. (eds.), *Municipal Sewage and Sludge: Treatment and Disposal Strategies*, Environmental Science and Engineering,
https://doi.org/10.1007/978-3-032-07597-0_4

and the strategies for managing its associated pollution. Furthermore, the chapter will explore potential treatment technologies to facilitate the long-term reuse of sewage sludge within a circular economy framework, as well as the environmental risks related to the utilization of sludge to soils.

Keywords Air pollution · Municipal solid waste sludge · Incineration · Co-combustion

4.1 Introduction

In recent years, both global effluent production and treatment capacities have experienced substantial growth, driven largely by increasing urbanization, industrialization, and the expanding demand for wastewater treatment. Concurrently, the generation of municipal solid waste sludge (MSS), a by-product of wastewater treatment processes, has seen a significant rise. This phenomenon has raised concerns regarding the safe and sustainable management of MSS. For instance, in 2020, China's daily wastewater treatment capacity reached an estimated 2.09×10^8 m^3, with an annual production of MSW surpassing 60 million tons (Liu et al. 2018). Similarly, it was forecasted that by 2020, the European Union's 27 member states (EU27) would collectively produce over 13 million tons of MSS. Despite substantial progress in developing various treatment technologies, the management of MSS remains a formidable environmental challenge. This is because MSS not only contributes to air pollution but also poses significant risks to water and soil quality, making its disposal and resource recovery critical areas of concern. The principal characteristics of MSS include a substantial quantity of volatile carbon (ranging from 60 to 80% of its solid mass), a high organic matter content, and a considerable moisture content, which can reach up to 99%. These features, coupled with large geographical variations in its composition, complicate the effective management of MSS. For instance, organic matter constitutes 30–50% of MSS in China, while in more developed nations, this percentage rises to 60–70% (Xiao et al. 2018). The variability in MSS composition is a major factor influencing the complexity of treatment and disposal strategies, and it further emphasizes the need for region-specific approaches to sludge management.

Hydrothermal carbonization (HTC) has also become a viable substitute for conventional sludge treatment techniques including decomposition, landfilling, and incineration. HTC provides enhanced sustainability advantages and offers a broader range of potential applications. Studies have shown that HTC can produce high-quality carbonized products that can be used as biofuels or soil conditioners, thus contributing to a more circular approach to sludge management. However, Zhou et al. (2020) have shown that the most advantageous environmental, economic, and social results are obtained when treated sludge is applied to land. A multi-criteria decision-making model was developed utilizing LCA data, illustrating that, when managed effectively, land application provides significant advantages in reducing carbon footprints and fostering sustainable agricultural practices. Nonetheless, this approach necessitates thorough evaluation of the possible hazards associated with soil contamination and the buildup of pollutants. Despite these advancements, earlier investigations into the environmental impacts of different wastewater treatment and disposal techniques have faced challenges due to inconsistencies in system boundaries and data assumptions. The inconsistencies observed have posed difficulties in aggregating and contrasting findings from various studies, given that the environmental effects frequently differ based on the particular data inputs utilized. Furthermore, a thorough evaluation of the life cycle of sludge treatment methods utilized in China, including co-incineration for energy generation or the production of construction materials, as well as alkaline thermal hydrolysis for protein recovery, remains to be undertaken. The lack of understanding in this area has constrained the capacity to make well-informed choices about the best sludge management approaches in the region. This study seeks to conduct a comprehensive assessment of the environmental effects associated with different sludge treatment and disposal techniques presently employed in China.

This study also builds upon earlier research by providing an integrated approach to sludge treatment, combining environmental, economic, and social considerations. It aims to provide actionable insights for policy-makers, engineers, and environmental managers seeking to minimize the environmental footprint of municipal sewage sludge while ensuring that treatment technologies remain economically viable and socially acceptable. Through this analysis, it is hoped that more sustainable sludge management practices can be developed, contributing to reduce carbon emissions and pollution, while supporting the long-term goals of a circular economy.

Municipal solid waste incineration (MSWI) within waste-to-energy (WtE) facilities has become a cornerstone of waste management strategies across most developed nations. Over recent decades, technological advancements in energy recovery from MSW have been substantial, leading to the integration of sophisticated air pollution control (APC) systems designed to meet stringent environmental standards. These APC systems are essential in minimizing the environmental impact of WtE processes, ensuring that gaseous emissions comply with rigorous regulatory requirements. This chapter offers a comprehensive examination of MSWI technologies and APC mechanisms, focusing on their function in removing key pollutants such as dioxins and furans, which are pivotal concerns in emission control. Furthermore, the chapter briefly considers the potential health risks associated with these emissions, situating MSWI's role within the broader MSW management framework and underlining its importance in sustainable WtE conversion (Fu et al. 2019). Figure 4.1 showing the impact of sewage sludge on environment sustainability.

4.2 Incineration

A byproduct of treating wastewater in cities or businesses is sludge. The safe and environmentally responsible disposal of sludge has grown in importance as environmental protection has gained more public consciousness. According to China's Ministry

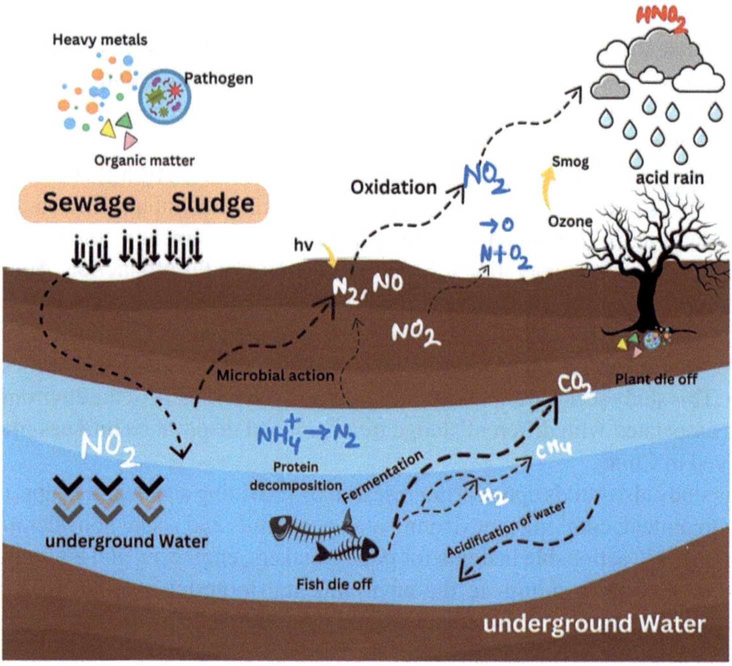

Fig. 4.1 Impact of sewage sludge on environment sustainability

of Environmental Protection's National Standard, the primary ways to dispose of sludge include incineration, land application, sanitary landfilling, and using sludge as construction material (Teoh and Li 2020). Figure 4.2 shows the different fundamental steps for treatment and disposal of Municipal sludge Incineration. Although sanitary landfilling is still the most popular method in China, problems like the sludge's unstable physical characteristics and the scarcity of landfill sites in major cities like Shanghai and Shenzhen cast doubt on its future. The regaining of N_2 from sludge has shown promise as a workable disposal option in terms of land application (Geng et al. 2020).

4.2.1 *Mono-incineration*

Feeding municipal solid waste sludge (MSS) with elevated moisture content directly into an incinerator can lead to combustion challenges, including temperature drops, ignition delays, and variations in furnace temperature. Figure 4.3 illustrates a schematic representation of a standard MSS incineration process. To ensure the stability of MSS incineration, the sludge is typically moved to a dryer following mechanical dewatering before it is subjected to combustion in the incinerator. In

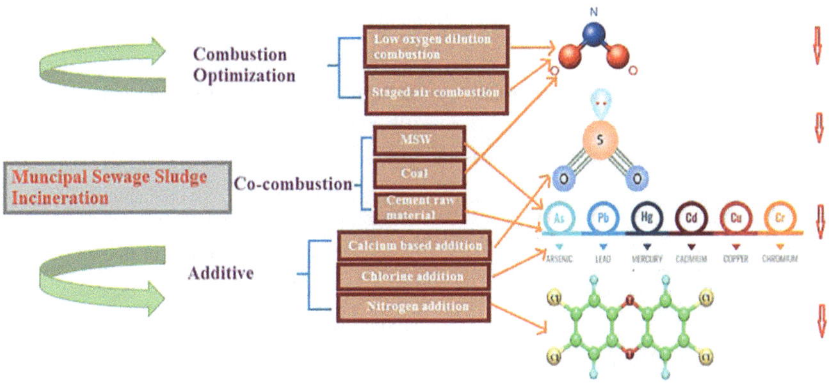

Fig. 4.2 The different fundamental steps for treating and disposing of municipal sludge incineration

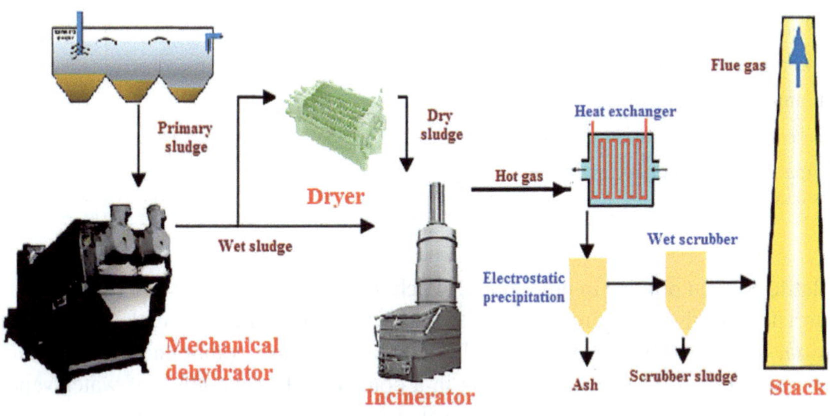

Fig. 4.3 Diagram showing systematic incineration of municipal sewage sludge

certain instances, wet MSW is transported directly to a fluidized bed incinerator, where additional fuel is incorporated. Ensuring that combustion byproducts undergo proper treatment and cleaning prior to their release in the nature is crucial (Akdag et al. 2018). Table 4.1 presents a comprehensive analysis of the benefits and drawbacks related with MSWI (Quina et al. 2011).

4.2.1.1 Combustion of Wet Municipal Sewage Sludge

A study has been carried out on the combustion properties of wet municipal solid waste sludge (MSS) in semi-pilot scale fluidized bed burners. MSS exhibits a greater

Table 4.1 Advantages and disadvantages of municipal solid waste incineration

Advantages	Disadvantages
Manage municipal sewage waste without initially treating it	APC residues are hazardous trash that has to be disposed of safely
Reduce the need for municipal sewage waste to be dumped in landfills	
90% Less waste	Slags of origin (bottom ashes)
70% Less waste weight	Source of a massive amount of flue gasses
Potential for energy recovery (heat or electricity)	High operating and investment expenses
	Expensive upkeep
Low levels of air pollution are emitted when properly regulated	Need qualified personnel
Eliminate hazardous organic pollutants and possible infections	Need the right composition in order to autocombust
May be situated near the MSW generating center of gravity	Unfavorable public opinion
Lower the cost of garbage transportation	
Minimum land requirements	
There are no odors from stack emissions	
Primarily convert organic molecules to CO_2 instead	

volatile content relative to other fuels, which influences its distinct combustion characteristics. With rising temperatures, the moisture within the MSS undergoes evaporation, subsequently leading to the volatilization of organic compounds. The elevated moisture content in wet MSS results in a considerable generation of water vapor, and the endothermic nature of water evaporation indicates that drying and primary devolatilization take place at reduced particle temperatures. Upon reaching a specific temperature threshold following the evaporation of water, these volatile compounds ignite, leading to the final combustion stage on the remaining coke (Akdag et al. 2018). Figure 4.4 showing proposed ways for sewage sludge disposal and utilization.

Huang et al. (2016) performed experiments involving the combustion of MSS for a duration of 3 min at a temperature of 600 °C, followed by rapid cooling in a reducing atmosphere. The observation revealed that dry, devolatilized, and combustion layers existed concurrently without distinct boundaries throughout the four stages of combustion. It is crucial to emphasize that approximately 80% of the carbon in MSS is evaporative, making the gas-phase ignition of these volatiles an essential aspect of the overall ignition procedure. On the other hand, in the case of coal combustion, these percentages represented 9 and 88% of the total duration of the process (Kijo-Kleczkowska et al. 2016).

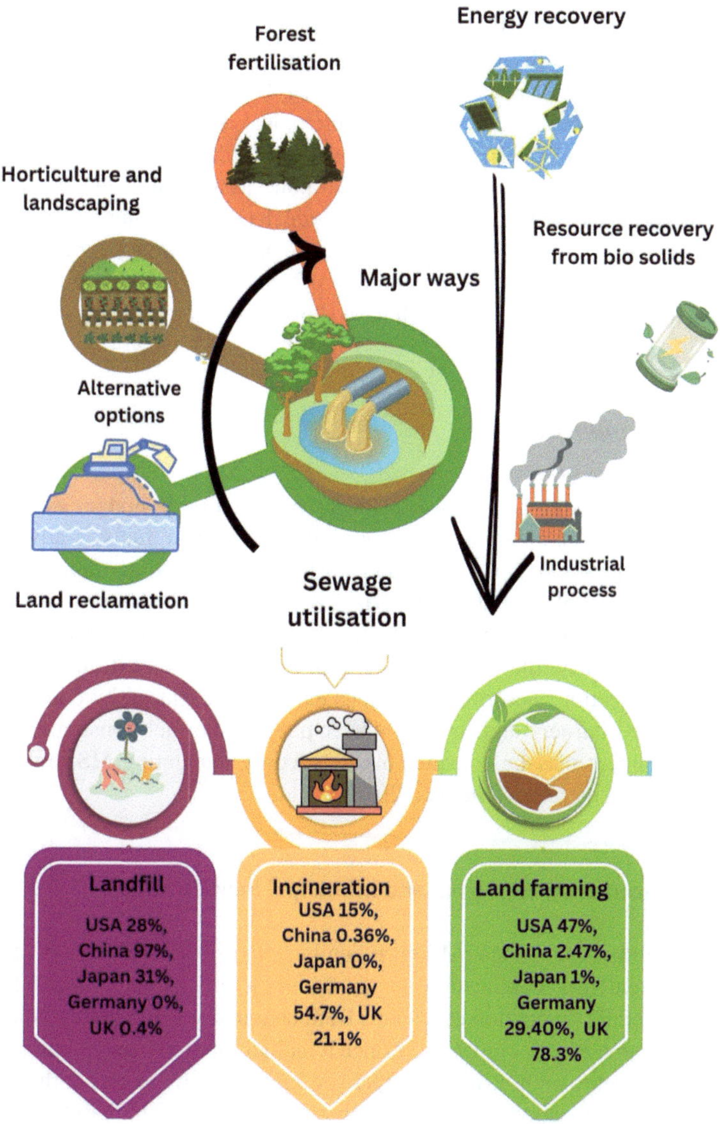

Fig. 4.4 Proposed ways for sewage sludge disposal and utilization

4.2.1.2 Incineration Following the Drying Process

Municipal solid waste sludge (MSS) maintains 80–85% moisture after mechanical dewatering, requiring further drying to reach 10–30% moisture (Ma et al. 2016). High temperatures and high pressures are often used in thermal drying, which improves MSS's dewatering qualities and aids in the discharge of leap water. Nonetheless, traditional thermal drying techniques for water evaporation are energy-intensive. Conversely, biological drying is a more energy-efficient method as it makes use of the metabolic heat produced by aerobic microbes decomposing organic debris in MSS. Cai et al. (2016) state that during biological drying, the first stage's water removal rate makes up 62% of the total removal, while the second stage's rate is 28%. Furthermore, MSS particles develop irregular holes as a result of the breakdown of viscous components during biological drying, which makes it easier for air to transfer for later burning. Biological drying is a promising drying approach for MSS because it successfully lowers the moisture content of MSS while creating a biostabilized and dried material, as several research have shown (Hao et al. 2018).

4.2.2 Co-combustion

Wet or semi-dry sludge has a lower calorific value than dry sludge, which makes steady combustion challenging. Burning sludge with other fuels is advised as a solution to this problem. Municipal solid waste sludge (MSS) is often ignited with coal as a component of cement raw materials, according to which reveals the calorific values of different sludge sorts. By reducing the ignition temperature and increasing combustion efficiency, the use of coal improves MSS's combustion performance. The following sections describe the various emissions produced by co-combustion in fluidized bed burners and thermal power plants. The fuel consumption of the original equipment may be considerably reduced by integrating MSS into cement kilns or MSW incinerators (Hu et al. 2016).

4.2.2.1 Co-combustion of Coal and Municipal Sewage Sludge

Municipal solid waste sludge (MSS) co-combustion affects a number of variables, such as ignition time, combustion length, emissions of pollutants (such SO_2 and NO_x), and overall combustion efficiency. Both char and volatile matter combustion have an impact on MSS's combustion since it includes more volatile chemicals. The volatile components present in municipal sewage sludge (MSS) have the potential to enhance the combustion process by facilitating rapid ignition, thereby potentially reducing combustion time. However, the moisture content in MSS, which can be considerable, plays a crucial role during the evaporation stage of incineration. As moisture evaporates, it absorbs a significant amount of heat, which increases the energy required to reach the ignition point. This results in a mixed fuel that exhibits a prolonged ignition time and a higher ignition temperature compared to dry materials (Kijo-Kleczkowska et al. 2016). Consequently, the presence of moisture in MSS can affect the overall efficiency and energy dynamics during the combustion process. During the incineration of municipal solid sludge (MSS), specific heavy metals, exhibit a higher tendency for volatilization as a result of increased chlorine concentrations in the sludge. Chlorine compounds, especially in substantial amounts, can promote the volatilization of these metals into the gaseous phase, leading to considerable environmental risks, such as air pollution and ecosystem contamination. The significant ash amount of MSS, along with the creation of inert alkali metal aluminosilicates, can obstruct char combustion at elevated temperatures, thereby reducing the potency of the incineration process. The retention of arsenic in combustion systems is affected by specific minerals, including calcite and clay, which notably improve arsenic retention in coal slurry during combustion (Fu et al. 2019). These minerals can bind to arsenic, inhibiting its environmental release; however, their presence complicates the overall process by potentially contributing to the formation of undesirable by-products. The co-combustion of municipal solid waste (MSS) presents a significant drawback due to its elevated levels of copper and chlorine, which can facilitate the production of hazardous substances, including dibenzofurans and polychlorinated dibenzo-p-dioxins (PCDD/Fs). Such compounds exhibit significant environmental persistence and present considerable ecotoxicological risks, including cancer causing and mutagenic impacts. Increasing the concentration of MSS in the co-combustion mix elevates the probability of generating harmful compounds, thereby intensifying the environmental consequences of the incineration process. Furthermore, an increase in the quantity of MSS in the combustion combination correlates with a significant rise in carbon monoxide (CO) emissions, which is a detrimental air pollutant. Akdag et al. (2018) identified a direct correlation between the proportion of MSS in the fuel mix and the rise in CO emissions. When the MSS content expand from 5 to 30%, combustion productiveness reduced from 99.5 to 97.5%. The observed decrease in efficiency indicates incomplete combustion, suggesting that the incorporation of MSS into the incineration process may reduce overall performance, leading to increased emissions and diminished energy recovery efficiency. The significant slagging and scaling potential of MSS complicates its application in the ignition method. The significant content of ash and inorganic material in municipal solid waste (MSS) can result in clinker formation, which accumulates on furnace walls or other heat exchange surfaces, thereby diminishing heat transfer efficiency and posing a risk to equipment integrity. This phenomenon, commonly known as "slagging," presents considerable operational difficulties, necessitating regular maintenance and adjustments in operations.

4.2.2.2 Co-incineration of and Municipal Solid Waste and Municipal Sewage Sludge

The synergistic treatment of MSS and MSW, the removal of the requirement for fossil fuels, and the reduction of Hg emissions are only a few advantages of incineration as an efficient way of treating MSW (Sun et al. 2020). Ignition of MSS with MSW presents a promising solution to generate sufficient thermal energy without increasing fossil fuel use. This process can dry MSS feedstock in situ, facilitate combustion, and support process steam production. However, increasing the proportion of MSS can significantly elevate combustible levels and may lower the combustion temperature, posing challenges for incinerator efficiency. To mitigate these issues, semi-dried

MSS, with its higher heating value and reduced moisture content, proves to be more effective when burned alongside MSW in grate furnace incinerators. With the adoption of a new integrated treatment system, such an incinerator can now process up to 1200 tons of MSW and 300 tons of MSS daily, optimizing the combustion process and energy recovery (Chen et al. 2019). Compared to processing MSS and municipal solid trash separately, this co-combustion technique efficiently uses waste heat to eliminate moisture from MSS, making it an efficient way to dispose.

4.2.2.3 Co-combustion of Raw Cement and Municipal Sewage Sludge

Hazardous materials included in organic waste are efficiently broken down by high-temperature incineration in cement kilns, while inorganic materials, including ash residue, may be fully used in a variety of products. Lime drying is a basic pretreatment method used in wastewater treatment facilities. The calcium found in lime-dried MSS may be used as a raw material to make cement. This suggests that co-processing MSS in cement kilns is feasible. MSS is a feasible substitute fuel for the manufacture of cement because of the organic materials it contains, which also adds extra heating value. With a calorific value of around 8.3 MJ/kg, MSS may replace up to 14% of the raw matter utilized in the formation of cement, meeting the industry's requirement of 6.25 MJ/kg. According to estimates, adding MSS to contemporary cement kilns may cut the use of fossil fuels by around 70% (Xu et al. 2014, 2019). Table 4.2 represents the comparative investigation of different methods applied for sewage sludge utilization.

4.2.2.4 Pyrolysis

For the management of MSS and energy recovery, pyrolysis the thermal breakdown of waste in an oxygen-free environment is a strong substitute for combustion (Dong et al. 2019). In addition to removing pathogens and harmful substances from sewage sludge, this method build chemicals and energy while reducing the amount of carbonaceous waste. Pyrolysis produces high-quality products and has major economic and environmental advantages when it is maximized. Based on heating rates, there are different type of primary forms of pyrolysis: slow pyrolysis and quick pyrolysis. Fast pyrolysis, which is more often employed for biomass processing, completes in seconds and optimizes the output of liquid products, whereas slow pyrolysis takes several hours and mostly creates charcoal. Fast pyrolysis, which is generally accomplished in fluidized bed reactors for sewage sludge, may produce liquid oil at temperatures between 450 and 550 °C that ranges from 30 to 72% (on a dry and ash-free basis) (Wei et al. 2015). Although the primary goal of pyrolysis is to produce pyrolysis oil, other processes that maximize the generation of solids and gases, respectively, include carbonization and gasification. Approximately 90% of the sewage sludge feedstock is converted to char by carbonization, which needs

Table 4.2 Comparative investigation of different methods applied for sewage sludge utilization

Technologies	Target products	Byproduct	Pros	Cons	References
Incineration	Heat and power	Ash	Significant quantity reduction; organics and pathogens are almost entirely removed	1. Dewatering and drying of sewage to a solids matter of 41–65 weight percent is necessary 2. Greenhouse gas emissions (GHG, NO_x, SO_x, and PMs) 3. A low energy efficiency; A waste rate that is not close to zero	Hu et al. (2016)
Gasification	Syngas	Tar char	High carbon balance and thermal efficiency; Syngas is beneficial for chemical synthesis as well as fuel; tars and biochar with possible extra benefits; Steer clear of SO_x and NO_x emissions	1. The need for dewatering and drying to a solids content of more than fifty weight percent 2. The presence of heavy organic pollutant chemicals in tar 3. The presence of extensive syngas cleaning requirements 4. The high investment and operating expenses	Raheem et al. (2018)

(continued)

Table 4.2 (continued)

Technologies	Target products	Byproduct	Pros	Cons	References
Gaseous phase char	·Mostly at benchscale; high	–	A unique reactor and feeding mechanism are needed for the direct utilization of operating pressure; catalyst recovery is challenging, particularly for homogeneous catalysts; and organics are lost in watery byproducts	–	Gollakota et al. (2018)
Hydrothermal	Operating	–	For wet sewage sludge, the temperature range is between 250 and 375 degrees Celsius; the pressure range is between 5 and 20 megapascals; the atmosphere is composed of nitrogen dioxide	Bio-crude oil	Hu et al. (2021)
Pyrolysis	Bio-oil	Gas char	It has the potential to reduce volume; it has lower greenhouse gas emissions than incineration; it can deactivate genes that are resistant to antibiotics; it can remove bioactive substances; it can produce biochar and syngas, all of which have the potential to add value	In comparison to incineration, dewatering and drying are necessary; the process is more complicated; air pollutants and heavy metals are retained in biochar; the expenses of capital and investment are high	Cao et al. (2024)

longer residence periods and lower operating temperatures (approximately 300 °C) (Zheng et al. 2020).

4.2.3 Significance of Controlling Air Pollution from Municipal Sewage Sludge Incineration

Due to land resource limitations and increasing energy demands, incineration is acknowledged as an efficient and broadly applicable pocess for reducing the volume and hazardousness of municipal solid waste (MSW) while recovering energy (Zhang et al. 2020). Currently, MSW incineration surpasses other thermal processing technologies in terms of equipment and techniques. Nonetheless, it is imperative to manage pollution emissions (e.g., SO_2 and NO_x) and mitigate disposal costs arising from the high energy consumption linked to drying processes. Municipal solid waste incineration (MSWI) within waste-to-energy (WtE) systems is a crucial waste management strategy in many developed countries, offering the dual benefits of minimizing landfill volumes and recovering energy from MSW. This method is especially beneficial in areas with high population density and restricted landfill capacity. MSWI operations encounter significant environmental challenges stemming from the emission of various air pollutants and residues, which can adversely affect ecosystems and human health if not properly managed. MSWI produces a range of pollutants, including particulate matter (dust), CO, NO_x, SO_2, hydrochloric acid, hydrofluoric acid, and volatile heavy toxic metals (Hu et al. 2021). These pollutants result from the breakdown and volatilization of materials within MSW during combustion, with heavy metals and dioxins presenting specific challenges due to their toxicity and potential for atmospheric dispersion. Key heavy metals like Pb and Cd are more prone to volatilization at high temperatures, especially in the presence of chlorine, which forms volatile metal chlorides. This necessitates advanced pollution controls to capture these compounds before they enter the environment. To address these challenges, modern MSWI facilities employ a suite of APC technologies that ensure compliance with strict environmental standards. Electrostatic precipitators, fabric filters, and wet scrubbers are widely used to capture particulate matter. To control gaseous emissions, flue gas desulfurization reduces SO_2 levels, and SCR significantly lowers NO_x emissions. Additionally, activated carbon and lime injection systems are utilized to capture heavy metals and acidic gases, converting them into solid residues that can be safely contained rather than emitted as gas (Chen et al. 2019). The behavior of pollutants during incineration is governed by thermodynamic factors such as boiling and sublimation points, which dictate whether a metal or compound will remain in solid ash or volatilize into flue gas. For instance, volatile heavy metals like Pb and Cd readily form chlorides with low sublimation points, enhancing their release into the flue gas unless captured by APC systems. Meanwhile, metals like nickel (Ni) and chromium (Cr), which are less volatile, tend to remain in ash residues, reducing their atmospheric emissions (Milhé et al. 2015).

Economic feasibility remains a critical consideration in adopting these advanced technologies, as they increase the cost per ton of waste processed. However, the health and environmental benefits of reduced emissions often justify these costs, especially as regulations evolve to require lower thresholds. Research indicates that, with adequate APC systems, MSWI poses minimal health risks, demonstrating that current technologies can effectively manage emissions and align with sustainability goals. Where stricter limits are needed, MSWI facilities can enhance gas cleaning systems further, positioning MSWI as an efficient, environmentally sound waste management method in WtE systems.

4.3 Features and Management of Pollutant Emissions

4.3.1 Heavy Metals

Approximately 0.5–2%, and in certain instances up to 4.0%, of the overall dry weight of municipal solid sludge (MSS) consists of heavy metals. MSS exhibits significant content of heavy metals. The primary heavy metals identified in MSS include nickel, zinc, chromium, cadmium, copper, lead, mercury and arsenic, as detailed in Table 4.3. The ignition of MSS results in varying behaviors of heavy metals in the resulting inorganic content and flue gas emissions. Although some toxic metals are volatilized and emitted with the flue gas, the majority are retained in the ash residue. Chimney emissions often contain volatile metals, specifically cadmium, lead, and mercury, owing to their significant volatility. The volatilization of these metals is affected by intricate chemical interactions with various elements and compounds. Additionally, elements such as residence time, temperature of combustion and the chemical composites of the MSS feedstock are also significant contributors. The parameters outlined collectively influence the degree to which heavy metals convert to gaseous or solid phases during the combustion process (Velden et al. 2008; Li et al. 2014).

4.3.1.1 Heavy Metal Properties

The decomposition of toxic metals is strongly affected by the distinct properties of each metal and its compounds at different temperatures. For instance, metals like lead (Pb) and zinc (Zn) are particularly sensitive to high temperatures. As the degree of heat rises from 850 to 1000 °C, their combustion rates increase markedly, with lead nearly doubling its rate. This temperature sensitivity highlights the reactive nature of these metals in combustion environments, where even moderate increases in temperature can significantly enhance their transition to the gaseous phase. (Liu et al. 2015). Similarly, within a comparable temperature range, the volatilization rate of chromium (Cr) increases significantly, from 65.7 to 74.9%. In contrast, transition metals exhibit relatively low sensitivity to temperature fluctuations. Research

Table 4.3 Summarization of metal catalysts benefits and drawbacks

Metal catalysis	Benefits	Drawbacks	References
Alkali metals	The steam reformation of tar is very efficient, and it possesses the ability to enhance the quality of gaseous emission	It is simple to disappear, but difficult to recover	Faba et al. (2015)
Biochar	It is inexpensive, it is produced naturally inside the gasifier, and it has a high tar conversion and is equivalent to dolomite	Consumption as a result of processes involving gasification	Huet al. (2018)
Nickel	Increased activity in the degradation of tar; reversal of the ammonia reaction; reduction of NO_x emission An advantageous condition for the reforming of methane and the water–gas shift process; Cost-effective; simple to generate new	Sintering of metal particles and the deposition of coke both contribute to the rapid deactivation of the material	Noichiet al. (2010)
Noble metal	An very high catalytic activity, together with a strong resistance to sulfur and long-term stability	At a very high cost	Cordero-Lanzacet al. (2018)
transition metals	In situations when there is a large concentration of heavy tar in the tar, it is quickly deactivated by carbon deposition	Moderately active for the breakdown of biomass tar; some exhibited greater catalytic activity compared to the Ni catalyst	Wang et al. (2012)
Zeolites	Excellent thermal stability, in addition to a reasonably high resistance to sulfur and nitrogen chemicals, and a simple regeneration process	Quick loss of activity due to coke buildup	Chenet al. (2018)

indicates that metal cadmium (Cd) demonstrates notably higher volatility than other toxic metal within the 500–1100 °C range, underscoring Cd's propensity for gaseous phase transitions at elevated temperatures.

4.3.1.2 Impact of Chlorine on Heavy Metals

The effect of chlorine on the volatility of heavy toxic metals during incineration processes has been the subject of much research. For instance, it has been shown that addition of $FeCl_3$/CaO to municipal sewage sludge (MSS) increases the volatilization of copper and chromium while concurrently decreasing the levels of lead, zinc and copper in remaining ash (Liu et al. 2016). The significant propensity of chlorine for interacting with toxic metals to generate metal chlorides is the cause of this phenomena. Metal chlorides are more volatile and simpler to discharge into the gas phase during cremation because they typically have lower boiling and sublimation temperatures than their oxide or elemental equivalents. The transition temperatures from gas to solid for some metals, like as arsenic, copper, magnesium, and sodium, are significantly reduced with the addition of 5% calcium chloride to MSS. This decrease in transition temperatures implies that the distribution and volatilization of these metals during incineration are significantly influenced by chlorine. Chlorine compounds change the partitioning behavior of certain metals and increase their volatilization, both of which may have serious environmental effects. The intricacy of chlorine's function in altering heavy metal behavior during incineration is shown by this differential impact. According to Han et al. (2008), Lead volatilization is impacted by chlorine chemicals like PVC and NaCl via several ways. While sodium chloride combines with lead in the gas phase to generate lead chloride ($PbCl_2$), PVC probably reacts with lead to form lead chlorides like $PbCl_2$, $Pb(ClO_4)_2$, and $PbCl_2O_4$.At least 10% of the ash additive mixture must include chlorine in order to transform semi-volatile heavy metals into chlorides for improved volatilization. The incorporation of 10% chlorine does not significantly enhance the volatility of low-volatile heavy metals. Chlorine's ability to increase the volatility of various heavy metals is contingent upon each metal's intrinsic volatility properties. Therefore, in order to better regulate the volatilization of heavy metals, future research should concentrate on improving the process of adding chlorine to MSS with varying compositions. In order to prevent the addition of chlorine from unintentionally increasing the release of hazardous compounds, such study would also seek to reduce the influence on other emissions (Liu et al. 2015).

4.3.2 Polychlorinated Dibenzofurans

Two important harmful air pollutants produced by the burning of municipal sludge (MSS) are dioxins and furans. Dioxins refer to a group of polychlorinated aromatic chemicals that share similar structures and properties, which include 75 varieties of polychlorinated dibenzo-p-dioxins (PCDDs) and 135 types of polychlorinated dibenzofurans (PCDFs). Since PCDD/Fs are polychlorinated chemicals, the emissions of these compounds are greatly impacted by the presence of chlorine in MSS (Chen et al. 2008; Gao et al. 2009). Chlorine and oxygen interact at the carbon surface during burning, resulting in redox reactions that create a variety of organic molecules containing chlorine. The toxicity and damage of the emissions from the incineration process are increased by these substances, which may include chloromethanes, chloroethanes, and chloroaldehydes.

4.3.3 Sulfur Oxides

Researchers have shown significant interest in studying sulfur transformation during the co-combustion of MSW in cement kilns. Key topics include the total sulfur content of the system, the primary chemical processes involved, strategies for mitigating or regulating sulfur incorporation, and the sulfur emission from heated cement basic components in both oxidizing and reducing conditions (Nielsen et al. 2011). Sulfur undergoes transformation into a range of compounds that contain sulfur during MSS incineration and is ultimately emit into the environment as solid residues in MSS ash or as gaseous emissions. Comparative analyses between MSS and coal combustion have shown minor discrepancies, although the comprehensive characterization of sulfur conversion in coal combustion. Additional studies have shown that MSS comprises sulfoxide, sulfone, and sulfate functional groups (Merino et al. 2007).

4.3.3.1 Conversion of Sulfur During the Incineration of Municipal Sewage Sludge

While thiophene and sulfone levels first rise and then decline after coal combustion, mercaptans and sulfides levels fall. During the process, the amount of sulfoxides grows while the concentration of sulfate just slightly changes. The development of functional groups during MSS combustion follows a similar trend. The behavior of thiophenes and sulfoxides varies, but MSS combustion shows consistent patterns in mercaptan, sulfide, sulfone, and sulfate concentration when compared to coal combustion (Li et al. 2009). In MSS combustion, sulfate concentrations are greater and range between 10 and 15%. The production of sulfate from SO_2 generated during combustion is facilitated by the substantial levels of CaO and Mg found in MSS ash (Folgueras et al. 2004). Interestingly, sulfoxide levels increase during the beginning of combustion and decrease at the conclusion. This pattern might be explained by sulfoxides' capacity to change thiols and sulfides at lower temperatures when oxygen is present, then decompose into other forms at higher temperatures as burning proceeds. In contrast to coal burning, MSS combustion results in a reduction in the amount of thiophene. According to researchers, polycyclic structures, which are created during coal formation under circumstances of high temperature, high pressure, and extended time, are often found in thiophenes in coal. Coal and MSS combustions behave differently because MSS primarily comprises monocyclic thiophenes, which are more quickly broken down at the beginning of combustion, whereas coal's polycyclic thiophenes are more stable and likely to stay in coke (Liu et al. 2007).

4.3.3.2 Removal and Capture of Sulfur

Multiple studies have investigated the impact of conditioning agents on the transformation and emission of sulfur (S) during the burning of municipal sewage sludge (MSS), demonstrating efficacy in reducing pollutant emissions in coal combustion. Qin et al. (2017) indicate that calcium-based additives, including CaO, $Ca(OH)_2$, and $CaCO_3$, may eliminate up to 86.9% of SO_2, 38.9% of PAHs, and 75.2% of toxic equivalent quantities (TEQs) at 850 °C when incorporated into an MSS fluidized bed combustor. This suggests that CaO is especially good in lowering emissions of SO_2 and PAHs. Additionally, CaO has been shown to improve the creation of hydrogen during incineration and collect CO_2 (Milhé et al. 2015). Han et al. (2015) created a porous calcium-based adsorbent characterized by a high particular area of surface. They discovered that as porosity and surface area rose, so did the adsorbent's ability to remove SO_2, and that higher reaction temperatures also improved SO_2 reduction. FeO_3 may help SO_2 react with coal's minerals, which increases coal ash's ability to retain sulfur and lowers SO_2 concentrations in the flue gas during coal burning (Elled et al. 2006).

4.3.4 *Nitrogen Oxides*

The burning of municipal sewage sludge (MSS) raises serious concerns about the production of NO_x and other dangerous contaminants. Compared to other fuels like coal (0.5–1.5%) and biomass (0.3–1.0%), MSS has a much greater nitrogen concentration, ranging from 4 to 9% (Zhu et al. 2015). Among these pollutants, N_2O is a powerful greenhouse gas that has a much higher potential to cause global warming than CO_2. The high nitrogen concentration and the presence of Fe_2O_3 and MnO_2, in the feedstock are the main causes of the increased NO_x emissions during MSS combustion. By encouraging the oxidation of substances like HCN and NH_3 into NO_x, these metal oxides have been shown to promote the generation of NO_x during coal combustion (Champion et al. 2013).

4.3.4.1 Features of NO_x Emissions

The combustion and gaseous emission characteristics of municipal sewage sludge (MSS) are profoundly influenced by the water content, which in turn affects both the efficiency of the combustion process and the types and quantities of emissions produced. In dry MSS combustion, akin to coal combustion, the formation of nitrogen oxides (NO_x) generally increases with a higher surplus air ratio. This occurs because the surplus air leads to increased oxygen availability, promoting the oxidation of nitrogen in the fuel to form nitric oxide (NO) and nitrogen dioxide (NO_2). As the combustion temperature escalates, the emission of NO and NO_2 tends to rise due to enhanced thermal formation, while nitrous oxide (N_2O) emissions typically decrease, as N_2O is thermally decomposed into nitrogen and oxygen at high temperatures. This behavior aligns with the established understanding of high-temperature combustion processes, where the formation of NO_x is primarily driven by thermal reactions (thermal NO_x) and the availability of oxygen. In contrast, the combustion characteristics of wet MSS reveal distinct emission patterns due to the substantial moisture content. Despite the higher levels of water-soluble ammonia present in wet MSS compared to dry MSS, the NO_x emissions from wet MSS combustion generally remain below 200 mg/m^3, suggesting a potential mitigating effect of water on NO_x formation. The moisture content in wet MSS acts as a thermal buffer, absorbing heat and lowering the peak combustion temperatures, which can reduce the formation of NO_x. Furthermore, the water-soluble ammonia can undergo reactions with NO_x in the combustion chamber, contributing to the reduction of NO_x emissions via a selective non-catalytic reduction (SNCR) process. This indicates that the presence of moisture in the MSS plays a significant role in moderating the formation of NO_x, providing an advantage in terms of environmental impact, as compared to dry MSS combustion (Mannina et al. 2024). The dynamics of NO_x emissions from wet and dry MSS highlight the importance of water content in the overall environmental footprint of MSS incineration. The water in wet MSS likely facilitates the conversion of NO to nitrogen through reactions with ammonia, a process that reduces the overall NO_x emissions. This phenomenon is further corroborated by the application of advanced combustion techniques, such as staged air combustion and oxygen dilution, which are designed to optimize the oxygen concentration and temperature profiles in the combustion chamber. These methods have been shown to reduce NO_x emissions by limiting the oxygen available at critical stages of combustion, thereby curtailing thermal NO_x formation. However, the combustion of MSS is not without challenges. One significant issue is the accumulation of ash, which can interfere with combustion efficiency and influence NO_x generation. Arrobas et al. (2024) highlighted that the presence of ash buildup, particularly when semi-dry MSS treated with calcium-based inorganic materials is burned in a fluidized bed system, can exacerbate NO_x emissions. Ash particles can act as thermal barriers, reducing the effective temperature in the combustion zone and hindering the complete combustion of organic matter, thus increasing the potential for NO_x formation. Additionally, the calcium-based additives used for ash management may themselves contribute to the generation of NO_x, as they can alter the local chemical environment in the combustion chamber, complicating emission control strategies. Further advancements in understanding the dynamics of nitrogen emissions during MSS combustion have been made through the development of sophisticated mathematical models. Marias et al. created a model that simulates nitrogen emissions based on various operational parameters, enabling the prediction of the concentrations of different nitrogen compounds (NO, N_2O, NO_2, HCN, and NH_3) at different points within the combustion chamber. This model provides valuable insights into the chemical processes that govern the generation and reduction of nitrogen oxides, offering a robust tool for optimizing combustion conditions and reducing nitrogen emissions during MSS incineration. By simulating the behavior of nitrogen species in the reactor, the model helps identify critical parameters that influence the effectiveness of emission control strategies, such as the temperature, oxygen availability, and the presence of ammonia.

4.3.4.2 Technology for Reducing NO_x Emissions

The strategies used to reduce NO_x emissions from MSWI may be divided into two main groups based on how NO_x emissions are managed in coal combustion. The first strategy focuses on flue gas treatment technologies designed to remove NO_x from the exhaust gases. However, they require careful control of operational parameters to maintain their effectiveness and are often costly to implement (Liu et al. 2012). The second strategy aims to reduce NO_x formation during the combustion process itself. This can be accomplished using technologies like air-staged combustion, which divides the air supply into two stages. The initial stage operates under oxygen-limited conditions, which promotes the formation of compounds like HCN and ammonia (NH_3) rather than NO_x. In the second stage, additional air is introduced to further facilitate combustion, allowing HCN and nitrogen radicals to react and form nitrogen (N_2), thereby lowering NO_x emissions. This method can reduce NO_x emissions by 50–80% when the excess air ratio is properly managed. Other methods, such as low-oxygen dilution combustion, have also shown promise in reducing NO_x emissions. Studies have shown that by optimizing burner design and operational conditions, NO_x emissions can be reduced to below 75 ppm at a lower operating cost compared to SCR techniques. Additionally, co-combustion of MSW with coal stabilizes the combustion process and helps modify the production of contaminants. Research indicates that adding coal improves the conversion rate of nitrogen oxides, particularly in fuels with higher fixed carbon content (Tang et al. 2022). Moreover, a two-step incineration and gasification process has demonstrated an average NO_x concentration of around 220 mg/Nm^3, with high combustion efficiency (Zhu et al. 2015). Reducing NO_x emissions from MSW incineration requires optimizing combustion and co-combustion methods.

4.4 Opportunities and Obstacles

Concerns about possible hazardous material emissions are now one of the main obstacles preventing MSWI from being widely used. Furthermore, lowering the expenses related to cleaning the incineration's flue gases might provide a significant benefit for the technology's future use. Notably, co-combustion improves economic efficiency by eliminating the need for separate incineration equipment for flue gas treatment systems and MSW. Moreover, co-combustion is more advantageous for the environment in certain situations. The resultant ash may be used as a raw material for cement manufacture, and MSW can increase the calorific value of cement kilns (Pan et al. 2019). Improving combustion methods, such as low-oxygen dilution combustion, staged air combustion, and flue gas recirculation, may greatly increase incineration efficiency while lowering emissions. By using conditioners, additives, and catalysts, chemical treatments may also improve the dehydration and combustion efficiency of MSW and reduce emissions from its incineration (Yoshida et al. 2018; Roldán et al. 2020).Sludge drying often uses indirect thermal drying, which offers a number of benefits including reduced exhaust gas emission and the capacity to recycle heating media.

It has been shown that these techniques improve pollution control in MSW incineration. However, prior to broad implementation, more operating condition optimization is necessary to improve efficacy and lower costs. A number of crucial areas should be the focus of future advancements in MSW incineration (Yu et al. 2021). Existing cement kilns, municipal solid waste incinerators, and coal power plants shouldn't see significant reductions in emissions or performance as a result of co-incineration. The co-incineration ratios, elevated emissions, and equipment and technology compatibility should be the main topics of future study. The reaction processes in MSW combustion, the impact of equipment, temperature, and other combustion variables on pollutant production, and methods for reducing emissions at the source should all be highlighted in thorough studies into improving combustion technologies. Furthering MSW incineration also requires investigating more efficient and environmentally acceptable additives and creating substitute thermochemical treatment techniques. Each of these research directions seeks to contribute to cleaner manufacturing processes by lowering the expenses and emissions related to MSW incineration. Ash treatment and pollutant gas management are the two primary issues with MSW incineration. Cutting-edge thermochemical treatment techniques like gasification and pyrolysis have drawn interest because of their positive environmental effects. The solution with the greatest calorific value among pyrolysis products may be produced by pyrolysis in a medium-temperature inert environment and include 30–40% organic matter (Zhou et al. 2020). Rapid pyrolysis at medium temperatures is the focus of certain investigations in order to maximize the output of bio-oil (Abraham et al. 2021). Pyrolysis emits less NO_x, PCDD/Fs, and heavy metals than incineration because of its lower temperatures and oxygen shortage (Linkov et al. 2017). However, the intricacy of the process and the system's economic feasibility are issues with pyrolysis. The pyrolysis of MSW may provide gases, liquids, and solids that can be utilized as biofuels, increasing the both quantity and quality of these valuable goods and making the pyrolysis process more economically viable. The effective production of clean combustible gases is the goal of gasification (Liu et al. 2021). Furthermore, the ash generated during gasification may provide a significant recovery of phosphorus (Li and Feng 2018). For sludge management in the future, resource recovery from sludge must be maximized. Before being widely used, emerging methods for recovering sludge resources, such as acidogenic fermentation, polyhydroxybutyrate synthesis, alginate-like exopolymers, and hydrothermal humification, need to be thoroughly assessed from a life cycle viewpoint. Therefore, estimating the environmental advantages of resource goods is a difficult task.

4.5 Conclusion

Municipal solid waste (MSW) burning may successfully prevent the production of nitrogen oxides (NO_x) by using staged air combustion and low-oxygen dilution combustion. Further investigation is necessary regarding the combustion conditions, particularly the sequential conveyance and distribution of air, as well as the establishment of a uniformly distributed temperature field in low-oxygen dilution combustion phenomena. Calcium compounds along with additional calcium-containing regulators or adsorbents, can effectively immobilize sulphur and reduce sulphur dioxide emissions. The incorporation of chlorine in different forms and techniques significantly influences the distribution of highly volatile heavy metals. The incineration of municipal solid waste necessitates attention to PCDD/PCDFs because of their hazardous characteristics as contaminants. The reduction of PCDD and PCDF emissions can be achieved through the application of nitrogen-containing compounds such as CH_4N_2O, ammonia and hydrogen cynide others. Evaluating the coupling effects of specific chemicals on various pollutants is essential in the subsequent work. The advancement in investigation modern technology has led to the development of processes of treatment like combustion, gasification, and biological reusing materials, which demonstrate significant potential for enhancement.

References

Abraham A, Park H, Choi O, Sang B (2021) Anaerobic co-digestion of bioplastics as a sustainable mode of waste management with improved energy production—a review. Biores Technol 322:124537

Akdag AS, Atak O, Atimtay AT, Sanin FD (2018) Co-combustion of sewage sludge from different treatment processes and a lignite coal in a laboratory scale combustor. Energy 158:417–426. https://doi.org/10.1016/j.energy.2018.06.040

Arrobas M, Meneses R, Gusmão AG, da Silva JM, Correia CM, Rodrigues MÂ (2024) Nitrogen-rich sewage sludge mineralized quickly, improving lettuce nutrition and yield, with reduced risk of heavy metal contamination of soil and plant tissues. Agronomy 14(5):924

Cai L, Krafft T, Chen TB, Gao D, Wang L (2016) Structure modification and extracellular polymeric substances conversion during sewage sludge biodrying process. Biores Technol 216:414–421. https://doi.org/10.1016/j.biortech.2016.05.102

Cao J, Jiang Y, Tan X, Li L, Cao S, Dou J, Chen R, Hu X, Qiu Z, Li M, Chen Z (2024) Sludge-based biochar preparation: pyrolysis and co-pyrolysis methods, improvements, and environmental applications. Fuel 373:132265

Champion WM, Cooper CD, Mackie KR, Cairney P (2013) Development of a chemical kinetic model for a biosolids fluidized-bed gasifier and the effects of operating parameters on syngas quality. J Air Waste Manag Assoc 64(2):160–174. https://doi.org/10.1080/10962247.2013.845619

Chen T, Gu Y, Yan J, Li X, Lu S, Dai H, Cen K (2008) Polychlorinated dibenzo-p-dioxins and dibenzofurans in flue gas emissions from municipal solid waste incinerators in China. J Zhejiang Univ Sci A 9(9):1296–1303. https://doi.org/10.1631/jzus.A0720144

Chen G, Yang R, Cheng Z, Yan B, Ma W (2018) Nitric oxide formation during corn straw/sewage sludge co-pyrolysis/gasification. J Clean Prod 197:97–105

Chen G, Wang X, Li J, Yan B, Wang Y, Wu X, Velichkova R, Cheng Z, Ma W (2019) Environmental, energy, and economic analysis of integrated treatment of municipal solid waste and sewage sludge: a case study in China. Sci Total Environ 647:1433–1443. https://doi.org/10.1016/j.scitotenv.2018.08.104

Cordero-Lanzac T, Palos R, Hita I, Arandes JM, Rodríguez-Mirasol J, Cordero T, Bilbao J, Castaño P (2018) Revealing the pathways of catalyst deactivation by coke during the hydrodeoxygenation of raw bio-oil. Appl Catal B 239:513–524

Dong Q, Zhang S, Wu B, Pi M, Xiong Y, Zhang H (2019) Co-pyrolysis of sewage sludge and rice straw: thermal behavior and char characteristic evaluations. Energy Fuels 34(1):607–615

Elled A, Amand L, Leckner B, Andersson B (2006) Influence of phosphorus on sulphur capture during co-firing of sewage sludge with wood or bark in a fluidised bed. Fuel 85(12–13):1671–1678. https://doi.org/10.1016/j.fuel.2006.02.008

Faba L, Díaz E, Ordóñez S (2015) Recent developments on the catalytic technologies for the transformation of biomass into biofuels: a patent survey. Renew Sustain Energy Rev 51:273–287

Folgueras MB, Díaz RM, Xiberta J (2004) Sulphur retention during co-combustion of coal and sewage sludge. Fuel 83(10):1315–1322. https://doi.org/10.1016/j.fuel.2004.01.015

Fu B, Liu G, Mian MM, Zhou C, Sun M, Wu D, Liu Y (2019) Co-combustion of industrial coal slurry and sewage sludge: thermochemical and emission behavior of heavy metals. Chemosphere 233:440–451. https://doi.org/10.1016/j.chemosphere.2019.05.256

Gao H, Ni Y, Zhang H, Zhao L, Zhang N, Zhang X, Chen J (2009) Stack gas emissions of PCDD/Fs from hospital waste incinerators in China. Chemosphere 77(5):634–639. https://doi.org/10.1016/j.chemosphere.2009.08.017

Geng H, Xu Y, Zheng L, Gong H, Dai L, Dai X (2020) An overview of removing heavy metals from sewage sludge: achievements and perspectives. Environ Pollut 266:115375

Gollakota ARK, Kishore N, Gu S (2018) A review on hydrothermal liquefaction of biomass. Renew Sustain Energy Rev 81:1378–1392

Han J, Xu M, Yao H, Furuuchi M, Sakano T, Kim HJ (2008) Influence of calcium chloride on the thermal behavior of heavy and alkali metals in sewage sludge incineration. Waste Manage 28(5):833–839. https://doi.org/10.1016/j.wasman.2007.01.015

Han Y, Hwang G, Kim D, Park S, Kim H (2015) Porous Ca-based bead sorbents for simultaneous removal of SO_2, fine particulate matters, and heavy metals from pilot plant sewage sludge incineration. J Hazard Mater 283:44–52. https://doi.org/10.1016/j.jhazmat.2014.09.009

Hao Z, Yang B, Jahng D (2018) Combustion characteristics of biodried sewage sludge. Waste Manage 72:296–305. https://doi.org/10.1016/j.wasman.2017.11.008

Hu M, Guo D, Ma C, Luo S, Chen X, Cheng Q, Laghari M, Xiao B (2016) A novel pilot-scale production of fuel gas by allothermal biomass gasification using biomass micron fuel (BMF) as external heat source. Clean Technol Environ Policy 18:743–751

Hu M, Laghari M, Cui B, Xiao B, Zhang B, Guo D (2018) Catalytic cracking of biomass tar over char supported nickel catalyst. Energy 145:228–237

Hu M, Ye Z, Zhang H, Chen B, Pan Z, Wang J (2021) Thermochemical conversion of sewage sludge for energy and resource recovery: technical challenges and prospects. Environ Pollut Bioavailab 33(1):145–163

Huang Q, Yu B, Qiu K, Zhou G, Wang S, Chi Y, Yan J (2016) Effect of moisture on sewage sludge combustion temperature profile and heavy metal leaching. Drying Technol 34(15):1810–1819. https://doi.org/10.1080/07373937.2016.1142451

Kijo-Kleczkowska A, Sroda K, Kosowska-Golachowska M, Musial T, Wolski K (2016) Experimental research of sewage sludge with coal and biomass co-combustion, in pellet form. Waste Manage 53:165–181. https://doi.org/10.1016/j.wasman.2016.04.021

Li H, Feng K (2018) Life cycle assessment of the environmental impacts and energy efficiency of an integration of sludge anaerobic digestion and pyrolysis. J Clean Prod 195:476–485

Li PS, Hu Y, Yu W, Yue YN, Xu Q, Hu S, Hu NS, Yang J (2009) Investigation of sulfur forms and transformation during the co-combustion of sewage sludge and coal using X-ray photoelectron spectroscopy. J Hazard Mater 167(1–3):1126–1132. https://doi.org/10.1016/j.jhazmat.2009.01.115

Li S, Li Y, Lu Q, Zhu J, Yao Y, Bao S (2014) Integrated drying and incineration of wet sewage sludge in combined bubbling and circulating fluidized bed units. Waste Manage 34(12):2561–2566. https://doi.org/10.1016/j.wasman.2014.08.018

Linkov I, Trump B, Wender B, Seager T, Kennedy A, Keisler J (2017) Integrate lifecycle assessment and risk analysis results, not methods. Nat Nanotechnol 12:740–743

Liu F, Li W, Chen H, Li B (2007) Uneven distribution of sulfurs and their transformation during coal pyrolysis. Fuel 86(3):360–366. https://doi.org/10.1016/j.fuel.2006.07.021

Liu H, Luo GQ, Hu HY, Zhang Q, Yang JK (2012) Emission characteristics of nitrogen- and sulfur-containing odorous compounds during different sewage sludge chemical conditioning processes. J Hazard Mater 235–236:298–306. https://doi.org/10.1016/j.jhazmat.2012.07.060

Liu J, Fu J, Ning X, Sun S, Wang Y, Xie W, Huang S, Zhong S (2015) An experimental and thermodynamic equilibrium investigation of the Pb, Zn, Cr, Cu, Mn and Ni partitioning during sewage sludge incineration. J Environ Sci 35:43–54. https://doi.org/10.1016/j.jes.2015.01.027

Liu J, Zeng J, Sun S (2016) Combined effects of $FeCl_3$ and CaO conditioning on SO_2, HCl and heavy metals emissions during the DDSS incineration. Chem Eng J 299:449–458. https://doi.org/10.1016/j.cej.2016.04.132

Liu W, Iordan C, Cherubini F, Hu X, Fu D (2021) Environmental impacts assessment of wastewater treatment and sludge disposal systems under two sewage discharge standards: a case study in Kunshan, China. J Clean Prod 287:125046

Liu W, Cao H, Xu J, Liu J, Huang J, Huang X, Li G (2018) Effects of municipal sewage sludge on fixation of Cr, Ni, Cu, and Zn during co-processing of heavy metal-containing waste in cement kilns. J Wuhan Univ Technol Mater Sci 33(4):892–900. https://doi.org/10.1007/s11595-018-1910-0

Ma J, Zhang L, Li A (2016) Energy-efficient co-biodrying of dewatered sludge and food waste: synergistic enhancement and variables investigation. Waste Manage 56:411–422. https://doi.org/10.1016/j.wasman.2016.06.007

Mannina G, Cosenza A, Di Trapani D, Gulhan H, Mineo A, Mofatto PMB (2024) Reduction of sewage sludge and N_2O emissions by an oxic settling anaerobic (OSA) process: the case study of Corleone (Italy) wastewater treatment plant. Sci Total Environ 906:167793

Merino I, Arevalo LF, Romero F (2007) Preparation and characterization of ceramic products by thermal treatment of sewage sludge ashes mixed with different additives. Waste Manage 27(12):1829–1844. https://doi.org/10.1016/j.wasman.2006.10.008

Milhé M, Charlou C, Sauceau M, Arlabosse P (2015) Modeling of sewage sludge flow in a continuous paddle dryer. Drying Technol 33:1061–1067. https://doi.org/10.1080/07373937.2014.982252

Nielsen AR, Larsen MB, Glarborg P, Dam-Johansen K (2011) High-temperature release of SO_2 from calcined cement raw materials. Energy Fuels 25(7):2917–2926. https://doi.org/10.1021/ef2006222

Noichi H, Uddin A, Sasaoka E (2010) Steam reforming of naphthalene as model biomass tar over iron–aluminum and iron–zirconium oxide catalyst catalysts. Fuel Process Technol 91(11):1609–1616

Pan Y, Wang X, Ren Z, Hu C, Liu J, Butler D (2019) Characterization of implementation limits and identification of optimization strategies for sustainable water resource recovery through life cycle impact analysis. Environ Int 133:105266

Qin L, Han J, Chen W (2017) Simultaneous removal of SO_2 and PAHs by adding calcium-based additives during sewage sludge incineration in a fluidized bed incinerator. J Mater Cycles Waste Manage 19(3):1061–1068. https://doi.org/10.1007/s10163-017-0592-6

Quina MJ, Bordado JC, Quinta-Ferreira RM (2011) Air pollution control in municipal solid waste incinerators. In: The impact of air pollution on health, economy, environment and agricultural sources, vol 1, pp 331–358

Raheem A, Sikarwar VS, He J, Dastyar W, Dionysiou DD, Wang W, Zhao M (2018) Opportunities and challenges in sustainable treatment and resource reuse of sewage sludge: a review. Chem Eng J 337:616–641

Roldán M, Bouzas A, Seco A, Mena E, Mayor A, Barat R (2020) An integral approach to sludge handling in a WWTP operated for EBPR aiming phosphorus recovery: simulation of alternatives, LCA and LCC analyses. Water Res 175:115647

Sun Y, Chen G, Yan B, Cheng Z, Ma W (2020) Behaviour of mercury during co-incineration of sewage sludge and municipal solid waste. J Clean Prod 253:119969. https://doi.org/10.1016/j.jclepro.2020.119969

Tang Y, Xie H, Sun J, Li X, Zhang Y, Dai X (2022) Alkaline thermal hydrolysis of sewage sludge to produce high-quality liquid fertilizer rich in nitrogen-containing plant-growth-promoting nutrients and biostimulants. Water Res 211:118036

Teoh SK, Li LY (2020) Feasibility of alternative sewage sludge treatment methods from a lifecycle assessment (LCA) perspective. J Clean Prod 247:119495

Van de Velden M, Dewil R, Baeyens J, Josson L, Lanssens P (2008) The distribution of heavy metals during fluidized bed combustion of sludge (FBSC). J Hazard Mater 151(1):96–102. https://doi.org/10.1016/j.jhazmat.2007.05.056

Wang L, Hisada Y, Koike M, Li D, Watanabe H, Nakagawa Y, Tomishige K (2012) Catalyst property of Co–Fe alloy particles in the steam reforming of biomass tar and toluene. Appl Catal B 121:95–104

Wei L, Wen L, Yang T, Zhang N (2015) Nitrogen transformation during sewage sludge pyrolysis. Energy Fuels 29(8):5088–5094

Xiao L, Lin T, Wang Y, Ye Z, Liao J (2018) Comparative life cycle assessment of sludge management: a case study of Xiamen, China. J Clean Prod 192:354–363. https://doi.org/10.1016/j.jclepro.2018.04.171

Xu W, Xu J, Liu J, Li H, Cao B, Huang X, Li G (2014) The utilization of lime-dried sludge as resource for producing cement. J Clean Prod 83:286–293. https://doi.org/10.1016/j.jclepro.2014.07.070

Xu J, Ping L, Cao H, Liu W, Gu Y, Lin X, Huang J (2019) Application status of co-processing municipal sewage sludge in cement kilns in China. Sustainability 11(12):3315. https://doi.org/10.3390/su11123315

Yoshida H, Hoeve M, Christensen T, Bruun S, Jensen L, Scheutz C (2018) Life cycle assessment of sewage sludge management options including long-term impacts after land application. J Clean Prod 174:538–547

Yu B, Xiao X, Wang J, Hong M, Deng C, Li Y, Liu J (2021) Enhancing phosphorus recovery from sewage sludge using anaerobic-based processes: Current status and perspectives. Biores Technol 341:125899

Zhang Q, Hu J, Lee DJ, Chang Y, Lee YJ (2017) Sludge treatment: current research trends. Biores Technol 243:1159–1172. https://doi.org/10.1016/j.biortech.2017.07.070

Zhang S, Du Q, Cheng K, Antonietti M, Yang F (2020) Efficient phosphorus recycling and heavy metal removal from wastewater sludge by a novel hydrothermal humification technique. Chem Eng J 394:124832

Zheng X, Chen W, Ying Z, Jiang Z, Ye Y, Wang B, Feng Y, Dou B (2020) Structure–reactivity correlations in pyrolysis and gasification of sewage sludge derived hydrochar: effect of hydrothermal carbonization. Energy Fuels 34(2):1965–1976

Zhou G, Gu Y, Yuan H, Gong Y, Wu Y (2020) Selecting sustainable technologies for disposal of municipal sewage sludge using a multi-criterion decision-making method: a case study from China. Resour Conserv Recycl 161:104881

Zhu J, Yao Y, Lu Q, Gao M, Ouyang Z (2015) Experimental investigation of gasification and incineration characteristics of dried sewage sludge in a circulating fluidized bed. Fuel 150:441–447. https://doi.org/10.1016/j.fuel.2015.02.031

Chapter 5
Utilization of Sewage Water in Agriculture for Irrigation of Crop Plants: A Positive Approach for Maximizing the Wastewater Reuse

Saurabh Kumar, Fauzia Naushin, Mudassara Hasan, Umme Kulsoom, Fareed Ahmad Khan, and Arun Lal Srivastav

Abstract Freshwater scarcity and depletion of ground water level in developing countries has made utilizing sewage water in agriculture sector as a common practice. Reusing sewage wastewater is regarded as an old, traditional, cost-efficient as well as economical approach for the irrigation of crop plants. Sewage water is rich in organic nutrients and thus improves the physical, chemical and biological properties of soil if used sustainably. Organic content found in the sewage water positively impact soil salinity, soil fertility, organic carbon and key elements like nitrogen, phosphorus, potassium, calcium and magnesium ions also enhanced to a great content. Sewage water contains high amount of macro and micronutrients and thus resulting in an increase of growth, development and overall productivity of crop plants. However, utilizing or reusing sewage water has few detrimental effect such as accumulation of toxic or heavy metals in crop plants and thus, impacting human health and environment. Therefore, for mitigating the possible hazards of reusing sewage water for irrigation of crop plants in agricultural sector, sewage water should be properly treated before utilizing for irrigation of crop plants.

Keywords Sewage water · Wastewater · Agriculture · Irrigation · Crop plants

---

S. Kumar (✉) · F. Naushin · M. Hasan · U. Kulsoom · F. A. Khan
Environmental Botany Laboratory, Department of Botany, Aligarh Muslim University, Aligarh, Uttar Pradesh, India
e-mail: s.singh100nl@gmail.com

F. Naushin
Botany Section, Women's College, Aligarh Muslim University, Aligarh, Uttar Pradesh, India

A. L. Srivastav
Chitkara University School of Engineering and Technology, Chitkara University, Solan, Himachal Pradesh, India

S. Madhav et al. (eds.), *Municipal Sewage and Sludge: Treatment and Disposal Strategies*, Environmental Science and Engineering,
https://doi.org/10.1007/978-3-032-07597-0_5

5.1 Introduction

Water (H_2O) is crucial for life and essential for all the physiological activities of living organisms which comprises up to 70% of the human body and 50 to 97% of plant biomass (Khurana and Singh 2012; Younas and Younas 2022). Despite of significant value, water is one of the highly exploited and inadequately managed resources by the people as a result of growing population as well as rising rates of illiteracy (Fakayode 2005; Vanham et al. 2021; Younas and Younas 2022). The disparity between the requirement as well as supply of water is widening daily due to the continuous rise in global population (Hussain et al. 2002, 2004). Besides increasing population growth, climate change, urbanisation and excess water usage are major factors for the depletion of freshwater resources (Dixit et al. 2022; Mishra et al. 2023). The World population of human being will increase to 9 billion by 2050 but more than 40% of the global population is expected to resides in areas that are badly stressed in terms of water resource and also highlighting the challenges related to irrigation of crop plants for the productivity of food (Shen et al. 2014) and the two-third world populace will face severe scarcity of water for a minimum one month each year (Mekonnen and Hoekstra 2016). It is also reported that more than 5 million or over 60% individuals of the world population may encounter or experiences challenges related to scarcity of water projected for the years of 2025–2030 (Watts 2018; Qadir et al. 2007; Rijsberman 2006; Wallace 2000). This situation of freshwater scarcity is escalating to concerning levels in certain areas of the globe, presenting a serious threat to human survival (Hussain et al. 2002). FAO (2017) reported that domestic water withdrawal arose from 600 km^3/year during the beginning of the twentieth century to over 330 km^3 to 3800 km^3/year at the onset of the twenty-first century, representing an increase of approximately 6.3 times (Mateo-Sagasta et al. 2015; Egbuikwem et al. 2020). Out of which, more than 80% of water is left untreated and returns to the natural water ecosystem (UN-Water 2014), contaminating a finite quantity of fresh water accessible for human consumption. Thus, rising demand of water consumption and its supply will increase up to a concerning level that will pose a threat to the growth of living organisms including humans in various regions of the world and also to the environment (Islam and Karim 2019; Alobaidy et al. 2010).

The global consumption of water for agricultural purposes constitutes a total of 92% (Clemmens et al. 2008; Hoekstra and Mekonnen 2012; Tanji and Kielen 2002). Since irrigation currently uses more than 69% (approx. 70–80%) of total freshwater withdrawals (Lautze et al. 2014). Because of increasing population and rising demand for food, the irrigated land area continues to grow worldwide in continuation to guarantee food security. The global irrigated land has more than doubled in the last 50 years (UNESCO 2017), creating competition for the finite water resources primarily sourced from rivers and groundwater (Pedrero et al. 2010; Madhav et al. 2018). For this reason, globally, scientists are exploring alternative methods for the conservation of freshwater resources for potable and other important purposes and using alternative sources of water in agriculture for irrigation of crop plants.

In this ongoing present scenario, sewage water can be utilized as a potential resource for agricultural irrigation of crop plants. Utilizing or reusing of treated sewage water represents a top-notch strategy to address prevailing water scarcity issues and capturing worldwide attention (Contreras et al. 2017; Hoekstra and Mekonnen 2012; Mishra et al. 2023). Therefore, reusing sewage wastewater for irrigation of agricultural crops is considered as an ideal resource for replacing the freshwater usage (Contreras et al. 2017). The total agricultural area that is irrigated globally using sewage wastewater is estimated to be approximately 20 million hectares (20 Mha.) (Aydin et al. 2015). This is because sewage wastewater provides both organic and inorganic essential nutrients. Consequently, wastewater use in irrigation of crop plants reduces the need of inorganic fertilizers as well as reduces the expense of fertilizer (Sharma and Singhvi 2017). On a global scale, the crop irrigation using treated sewage wastewater enhances agricultural productivity and supports the numerous small scale farmers (Sato et al. 2013). Between 1.5 and 6.6% treated wastewater is reuse for agriculture applications worldwide (Sato et al. 2013; Ungureanu et al. 2018). Over ten percentages (10%) of the global population depends on agricultural products cultivated by the use of sewage wastewater irrigation (WHO 2006).

However, sewage wastewater represents the major source of toxic or heavy metals and persistent organic pollutants. Inadequate treatment as well the unrestricted discharge of sewage water ultimately contaminate the water bodies, soil and groundwater and thereby negatively impacting the well-being of flora, fauna, human being and aquatic organisms life (Ahmed et al. 2021). The major source of industrial wastewater generation includes tanning of leather, petrochemicals, pharmaceuticals, fertiliser industries, agro-food, colouring and textiling sector etc. (Kumar et al. 2018). Therefore, wastewater must undergo proper treatment prior to being discharged into water channel or utilized for irrigation purposes. Treated wastewater is typically utilized for non-potable applications, such as agricultural activities, irrigation of lands, groundwater replenishment, irrigating golf course, cleaning of vehicle, flushing of toilet, combating fire as well as construction projects. Additionally, it may serve cooling functions in thermal power stations (Katsoyiannis et al. 2017; Mohsen 2004; Smith 1995; Yang et al. 2017).

In addition to toxic contaminants and pollutants, sewage water contains vital nutrients and organic matter that absorbed by crop plants as result of irrigation through wastewater (Zhang and Shen 2019). Among majority of the growing nations, the irrigation of crops such as mangoes, coconuts, lettuce, and tomatoes is being carried out using industrial and sewage wastewater (Anjum et al. 2021; Belhaj et al. 2016; García et al. 2019). El-Nahhal et al. (2013) found that the utilization of treated wastewater significantly enhances plant length, its diameter and both fresh as well as dry weight. Thus, utilizing sewage wastewater in agriculture for irrigation purposes has proved a lot of value for farmers and can lowers crop productivity cost by 10% to 20% (Scott et al. 2010). This book chapter, reviews and highlights the available knowledge about the sewage wastewater, its composition and characteristics, WHO guidelines for wastewater reuse in agriculture, methods of sewage or wastewater irrigation in agriculture, wastewater reuse in agriculture, utilizing sewage water in agricultural irrigation and its positive impact on the yield and growth of crop plants and the sewage water mediated supply of mineral nutrients for growth and development of crop plants and their qualitative as well as quantitative assessment.

5.2 Sewage Wastewater: Composition and Characteristics

Wastewater is defined as generated water that has been utilized in household purposes as well as in industrial processes (Huibers and Van Lier 2005). Wastewater which is released from residential areas, institutions and commercial facilities is commonly referred to as sewage wastewater (Topare et al. 2011). Sewage or wastewater consists approximately 99.9% of water and rest 0.1% are solids that may exist either as dissolved solids or suspended solids. The solid portion of wastewater mainly comprises organic materials and minerals, alongwith colloidal and suspended particles in addition to harmful elements. Furthermore, it contains organic compounds such as micro-molecules like lipids or fats, and macromolecules including carbohydrates and proteins along with pesticides, antibiotics, pharmaceuticals and detergents etc. (Cizmas et al. 2015). The major sources of wastewater generation includes the commercial sector, poultry farming, fisheries, domestic activities, residential blocks and industrial operations etc. (Vymazal 2009). Wastewater can be categorized into three types based on its color: grey, brown and black. Greyish wastewater produced through activities such as dishwashing, laundry and bathing use (except toilet waste) which could be used for irrigation after appropriate wastewater treatment i.e., physical, chemical and biological processes (Boano et al. 2020). A slight variation among brown and black wastewater exist. Brown wastewater consists solely of human faeces, while black wastewater includes both human faeces and urine. This type of wastewater recognized as a breeding ground for pathogens and is restricted to agricultural applications (Younas and Younas 2022).

5.3 World Health Organization (WHO) Guidelines for Agricultural Reuse of Wastewater

In 1971, a committee of experts from the World Health Organization (WHO) conducted an initial evaluation of the health risks associated with utilization of wastewater in agricultural practices. Subsequently, the guidelines for microbial water quality in irrigation were formulated in 1973 (WHO 1973). However, WHO in 1989 updated and established the frameworks for the safe utilization of wastewater in agriculture in their report entitled as "*Guidelines on the safe use of wastewater and excreta in aquaculture and agriculture*". These updated frameworks allow for a maximum of 1000 faecal coliforms/100 ml, based on findings from epidemiological studies related to wastewater irrigation. Furthermore, the quality standard for intestinal nematodes were suggested to be fewer than 1 nematode egg/litre (WHO 1989a, b). WHO framework the agricultural reuse of wastewater which mainly includes irrigation practices for orchards and vineyards as well as for fodder, fibre, seed crops and crops which includes both processed and raw consumption of crops. Additionally, the aquaculture which involves the cultivation of fish or aquatic plants for human consumption as well as watering of green spaces for urban landscaping also fall under the category of wastewater reuse etc. (WHO 1989a, b).

The recent WHO framework, 2006 for the safe use of wastewater in agricultural irrigation practices have undergone significant revisions (WHO 2006). The previous faecal coliform guideline has been substituted with an emphasis in relation to risk that can be linked to specific causes and their impact on health outcomes (Pedrero et al. 2010). In the irrigated regions surrounding Santiago, Chile, it has been observed that 60% of the river water utilized for irrigation contains more than 10,000 faecal coliform bacteria/100 ml, which is 10 times higher than recommended limit of World Health Organization (FAO 1993). Furthermore, United States Environmental Protection Agency (USEPA) sets guideline for watering through wastewater that has been treated to have $\leq$ 2.2 total coliforms/100 ml. In contrast, while using surface water for irrigation, a strict guidelines of $\leq$ 1000 faecal coliforms/100 ml is mandated (USEPA 1973).WHO set microbiological quality standards for the application of treated wastewater in agricultural practices stipulate that for unrestricted irrigation, the limit for faecal coliform should not exceed 1000/100 ml. The nematode egg guideline limit $\leq$ 1 egg/litre (Blumenthal et al. 2000). The WHO set a limit of 4200 faecal coliform/100 ml for irrigation of economically important as well as fodder crops (for which WHO has established a guideline limit regarding the presence of nematode eggs) and no quantifiable faecal coliform are allowed in 100 ml for unrestricted irrigation especially for crops that may consumed as raw or uncooked in contrast to WHO, 1989 standards of 41,000 faecal coliform/100 ml (WHO 1989a, b; Blumenthal et al. 2000).

5.4 Sewage Wastewater Reuse in Agriculture: A Potential Valuable Resource

Agriculture is biggest sector of freshwater user which accounts approximately 70% (more than 90% in some countries) of freshwater extraction across the globe (Gleick 2000; FAO 2002). The accessibility of fresh water sources is diminishing progressively due to rise in population, urban development and ongoing climate changes. By conserving freshwater resource for drinking purpose and other important uses, wastewater offer certain economic advantages such as utilizing or reusing of wastewater in agriculture for irrigating crop plants as well as in aquaculture. Reusing sewage wastewater for irrigating crop plants helps in improving the nutritional status of urban population as well as providing fuel for cooking and heating purposes especially in those countries where reuse is important. About 10% of global people believed to ingest food cultivated using wastewater (Smit and Nasr 1992). Therefore, primary objective of using wastewater is to ensure the rational reuse along with the protection of health at the same time. Through the proper management, yields of crop may be increased by irrigating with raw sewage water or with effluents that have gone primary or secondary treatment. Also, the use of sewage wastewater in irrigating crops reduces application of artificial fertilizer as well as reduces the cost of fertilizers which is important for recycling of nutrients (Sharma and Singhvi 2017).Wastewater serve as a valuable source of both organic as well as inorganic nutrients. Organic material presents in wastewater irrigation acts as a soil ameliorate for increasing water storing capacity of soil, soil texture and soil fertility.

Today, the reuse of wastewater in agriculture rising among both developing and industrialized countries, driven by several key factors (WHO 2006):

i. The scarcity of water and the deterioration of freshwater resources as well as leading to inadequate wastewater management.
ii. The growth in population, which correlates with a heightened demand for food and fibre.
iii. An increasing awareness of intrinsic importance of wastewater as a source, particularly the nutrients it provides.
iv. The Millennium Development Goals (MDGs), particularly those aimed at promoting environmental sustainability and eradicating poverty and hunger.

Therefore, wastewater (both municipal and industrial) is increasingly used as a potential source in agricultural irrigation for particularly in urban and peri-urban areas. Thus, addressing the issue of wastewater management and effluent disposal (Singh et al. 2004; Rattan et al. 2005; Mishra et al. 2009; Gupta et al. 2010).

5.5 Reusing Treated Sewage Wastewater in Agriculture: A Global Scenario

The scenario of reusing sewage water had been evolved in recent period. According to an article published in 2017 by the World Water Development (WWD), wastewater is recognized as an untapped potential valuable resource that has to be fully utilized, leading to a transformative shift in the practices of water reuse (Ryder 2017). Treated wastewater along with nutrients has proved to be an effective way to meet the rising demand of water globally. Thus, contributing towards achieving the objectives of sustainable development i.e., SDG 6 and the principles of circular economy i.e., SDG 12 (Ricart and Rico 2019). Agricultural crop irrigation is among the most common methods for reusing of treated sewage water (Mishra et al. 2023). The practice of utilizing treated and untreated sewage wastewater in agriculture is an oldest traditional method, that likely to be originated during the Indus Valley Civilization. This method may have been adopted in response to water scarcity in various regions, a challenge that has persisted throughout history. For example, the geographic limitations affecting water availability on the island of Crete today bear similarities to those faced by Pakistan (Angelakis et al. 2018). According to United Nations 2017 report, wastewater is utilized for irrigation in nearly 50 countries around the world (Ryder 2017). Approximately, 15 million meter cubic of reclaimed water used each day for agricultural irrigation across the globe (Elifantz et al. 2011). In numerous regions of the globe, particularly in emerging nations like India, China and Mexico, the utilization of wastewater for agricultural irrigation is a common practice. The reusing of treated wastewater has experienced significant growth along annual increases ranging from about 10–20% in Europe, United States, China and reaching as high as 41% in Australia (Aziz and Farissi 2014). Notably, China is recognized as the leading nation in Asia for reusing wastewater, covering an estimated area of 1.3 Mha. including countries like India, Vietnam and Pakistan (Zhang and Shen 2019). At present, it is reported that about 37.6% of urban wastewater in India underwent treatment process (Singh et al. 2019). Israel is stands out as a largest user of treated wastewater for agricultural irrigation, utilizing 90% of its water reuse (Angelakis and Snyder 2015). Many low-income nations like Africa, Asia and Latin America relies on raw wastewater for crop irrigating purposes (Jiménez and Asano 2008). Conversely, the moderately economic nations like Tunisia, Jordan and Saudi Arabia utilize treated wastewater for the irrigation for their agricultural irrigation needs (Al-Nakshabandi et al. 1997; Balkhair and Ashraf 2016; Balkhair 2016; Qadir et al. 2010; Sato et al. 2013).

5.6 Sewage Wastewater Reuse in Agriculture: An Indian Scenario

India generates around 62,000 million litres (ML) of wastewater daily, which is produced from urban areas or cities, out of which 27% is subjected to wastewater treatment plants and 70% of wastewater remains untreated which is discharged into water bodies is dumped into water bodies (CPCB 2009). Scarcity of water in urban edge areas leads to rising use of wastewater in agriculture (Radcliffe 2004). A research conducted indicates that nearly 40,000 acres of agricultural land have been irrigated using effluent diluted with clean water sourced from Musi River flowing from the Hyderabad city (Surinaidu et al. 2023).

According to a latest report prepared for Food and Agricultural Organization (FAO) of the United Nations indicates that India currently has over 200 wastewater irrigation projects in operation which covers an area exceeding 73,000 hectares irrigated with wastewater (Shende 1985; Singh et al. 2022). Majority of these projects are run in a rough and irregular fashion posing significant health risks as a fact that majority of wastewater employed for irrigation remains untreated. Application of wastewater at excessive rates, reaching up to 12 million litres annually, leads to significantly high nutrient loading, with total of 600 kg of N_2/hectare/year. This situation arises as a result of many system prioritize wastewater disposal over the efficient recovery of resources. Irrigation on sewage farms in India is conducted through surface application techniques, which vary from uncontrolled flooding to more regulated methods such as ridge and furrow, border strip as well as check basin irrigation. Subsurface and sprinkler system of irrigation didn't utilize in these settings.

Despite the overall disappointing scenario of present state of sewage wastewater irrigation in India, there are several remarkable achievements to highlights. Insights obtained from the extensive field studies of National Environment Engineering Research Institute (NEERI) suggest that:

i. Wastewater irrigation notably enhances grain yields in comparison to freshwater, even though, when untreated wastewater diluted using two volumes of fresh water. Further, yield can be enhanced by mixing additional NPK fertilizer up to permissible dosage.
ii. The yield of vegetables is also much greater, when wastewater irrigation is place of conventional methods of manuring and freshwater irrigation.
iii. Wastewater irrigation shows higher nutrient efficiency utilization and allowing long term yield stability which should be preserved.
iv. The utilization of untreated wastewater for irrigating trees produces results comparable to those achieved with fresh water, yielding around 55 tons of profitable timber of eucalyptus per hectare after 24 months period, which has a market value of 27,700 rupees i.e., US$ 2170.
v. The effect of wastewater irrigation upon soil characteristics is strongly influenced by the initial characteristics of the soil. However, in numerous instances, even after three decades of wastewater irrigation, soil productivity continues to be remarkably positive (Fig. 5.1).

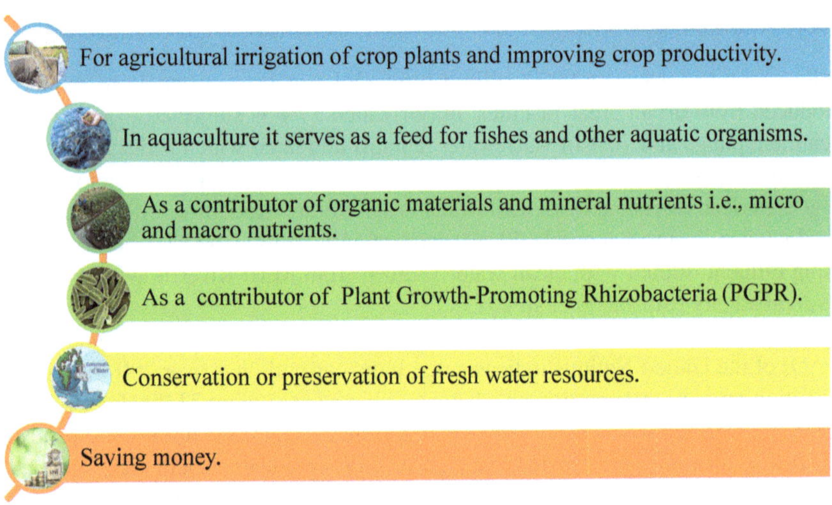

Fig. 5.1 Positive effects of sewage water on agricultural irrigation of crop plants

5.7 Physical, Chemical and Biological Parameters of Sewage Wastewater

Sewage wastewater characteristics typically classify mainly into 3 types i.e., physical, chemical and biological (Younas and Younas 2022). The characteristics of sewage water are influenced by the type of raw materials, the point of origin and the various components present in it (Ilyas et al. 2019).

5.7.1 Physical Parameters

Wastewater parameter that are easily detectable through human sensory organs and measure the physical characteristics of wastewater are known as physical parameters (Metcalf 2003). The key physical parameter of wastewater includes colour, odour, turbidity, temperature and total solid content. The colour of sewage water varies from yellowish brown to greyish black over the time. Wastewater has a characteristic strong unpleasant smell or odour. Temperature is the most crucial physical parameter because of its significant influence on aquatic and biological systems. Temperature of wastewater released into a receiving water body should not surpass 2–3 °C above the ambient temperature to ensure the stability of the aquatic ecosystem population balance in that body of water (Liu 2008). Turbidity is defined as the degree of cloudiness in water, resulting from a diverse array of suspended solids, varying in size from colloidal particles to coarse dispersions (Sawyer et al., 1994). Wastewater will be cloudier, or more turbid, as a result of suspended solids. The total solids (TS) content of wastewater refers to matter which remains left as a residue after evaporating at temperatures ranging from 103 to 105 °C (Gudeta 2010). Further, solids can be categorized into suspended and dissolved types based on their size and settleability, in addition to being divided into volatile i.e., organic as well as fixed i.e., inorganic components.

5.7.2 Chemical Parameters

Chemical properties, on the other hand, describe a substance capacity to combine with or transform into one or more different substances. The chemical parameter of wastewater includes alkalinity, pH, total organic carbon (TOC), total nitrogen and phosphate etc.

pH stands for potential of hydrogen. pH can be defined as the concentration or potency of hydrogen ion (H+) or Hydronium ion (H_3O^+) in the solution. pH is measurement of acidic and basic (alkalinity) of wastewater (Pescod 1992). pH is an indication of growth conditions for microbial communities used during the biological treatment of wastewater (Liu 2008). The pH of water that is used for irrigation should be 6.5–8.4. The values of pH of wastewater is slightly alkaline and depends on the source, environmental conditions, duration of storage of collected wastewater and on the concentration of chemicals, ions and substances present in it (Liu 2008; Younas and Younas 2022). The lower values of pH effects the movement of heavy metals within the soil. Some crops require particular pH levels to achieve their optimum growth.

Total organic carbon (TOC) is a chemical parameter involving the combustion of organic matter present in a wastewater sample, converting them into CO_2 and H_2O. Combustion gases are then dehydrated and passed through an infrared analyzer. This analyzer measures the concentration of CO_2 produced during combustion, which correlates directly with the carbon content present in the wastewater sample.

Wastewater contains various plant nutrients. Numerous results showed the positive impact of wastewater on agricultural yield which can be attributed due to its rich nutritional composition and presence of organic matter (Day et al. 1975; Scott et al. 2000). **Nitrogen** is an essential macronutrient for plants that found in sewage or waste water in various chemical forms and concentrations such as ammonia (NH_3^+), nitrite (NO_2^-), nitrate (NO_3^-) and organic nitrogen (ON). These combined to make the total nitrogen. Mostly plants absorbs nitrate only but the other forms of nitrogen are converted to nitrates in the soil (National Research Council 1996). The organic nitrogen fraction comprises a wide variety of compounds including amino acids, amino sugars and proteins. The presence of nitrogen in wastewater is a sign of decaying organic materials.

Phosphorus is a crucial factor to take into account, particularly regarding the overall phosphorus concentration in wastewater. In wastewater, phosphorus can exist in various forms, including phosphate, ortho-phosphoric acid (H_3PO_4/PO_4-P) or as organically bound phosphate such as nucleic acids and phospholipids. Wastewater generally consists less amount of phosphorus so that it is utilized in agricultural irrigation which is advantageous or does not harm the environment negatively (Girovich 1996).

Potassium is also a macronutrient which is present in higher concentration in the soil but it's not bioavailable since it is attached to others compounds. Wastewater contains low concentration of potassium and using wastewater in agricultural irrigation do not have detrimental environmental effects related with potassium (Mikkelsen and Camberato 1995). Wastewater not only add nutrients (both micro and macro) to the soil but also enhances the humic content by supplying the organic matter that increases soil moisture, retains minerals as well as promoting microbial activity. Organic materials found within wastewater improves soil texture and minimizes the uses of artificial fertilizers. The organic matter varies from $< 1.2\%$ in poor soils to $>$ 5% in rich soils.

Besides all above chemical properties, wastewater also contains inorganic elements such as sodium, calcium, chlorine, sulphate, ammonium salts, carbonates and bicarbonates etc. A number of elements also are found in wastewater in relatively less concentration as low as few mg/l and are called as trace elements such as aluminum (Al), beryllium (Be), cobalt (Co), fluoride (F), iron (Fe), lithium (Li), manganese (Mn), molybdenum (Mo), selenium (Se), tin (Sn), titanium (Ti), Tungsten (W) and vanadium (V). The toxic elements i.e., a unique kind of trace metals present in wastewater are arsenic (As), cadmium (Cd), chromium (Cr), lead (Pb), mercury (Hg), boron (B) and zinc (Zn). Toxic metals such as Cd, Cu, Mo, Ni and Zn are commonly found in wastewater which can be easily mobilized and retained by the plants. The impact of toxic elements on crop plants are complex as there is a negative interaction that impacts their uptake through the plants (Drakatos et al. 2002). According to Kesalkar et al. 2012, changes in chemical properties that above the optimum limit may have various kinds of detrimental impacts on the crop plants, environment, including contamination of water, depletion of oxygen, injury to fish or other aquatic animals and closing of beaches or other recreational areas.

5.7.3 Biological Parameters

Biological parameters of wastewater refers to the existence of various microbes such as bacteria fungi, protozoa and viruses present in the water. The density of these microbial communities can significantly influence water quality, leading to alterations in its overall characteristics. Wastewater professionals utilize specific parameters related to biological wastewater treatment to categorize organic matter. The most commonly used biological parameters are the oxygen demands values. This term denotes amount of O_2 needed for effectively stabilizing organic material found in wastewater. Biological oxygen demand (BOD) and chemical oxygen demand (COD) are the two primary methods used to assess the biological properties as well as and presence of organic materials content in wastewater.

BOD stands for *biological oxygen demands* or *biochemical oxygen demands.* BOD estimates quantity of O_2 required by the aerobic microbes for breaking down the organic material found in the wastewater samples. BOD test is carried out in a dark incubated conditions for a period of about 3 days at 27 °C or for 5 days at 20 °C, so that all biological materials degrade (Quagraine et al. 2017). The change in DO concentration over the given time period represents the oxygen demands for respiration by the aerobic biological microbes in the sample. BOD measurements is indicated as milligrams (mg) of O_2 consumed/litre (lit.) of water sample. There are two methods for determining the BOD: (i) Titrimetric method and (ii) Respirometric method. The main purpose for BOD estimation is to know about the organic load in a water body, for meeting the standard or for framing the guidelines for wastewater use, for designing and to know the efficiency of treatment plants.

COD stands for *chemical oxygen demand.* COD is the estimation of rate of total organic matter present in wastewater. COD is characterized as the quantity of O_2 required for oxidising organic substances in the wastewater by using a strong oxidising agent such as dichromate ions. Dichromate-based COD measurement, now a days is the most commonly utilized technique for the continuous monitoring of biological wastewater treatment system. The quantity of oxidizing agent needed to fully oxidize the organic material present in sample which corresponds to the oxygen demand. The oxidation process involving both organic and inorganic might results in a greater COD value than BOD for a particular wastewater stream. The COD analysis methods are rapid (around two hours) to determine the organic material found in wastewater samples. Most commonly used technique for estimating COD is a closed reflux colorimetric method. The COD values measured in oxygen consumed per litre of solution i.e., mg/l of O_2 or mg/l of COD.

Wastewater is infectious because of the presence of diverse range of pathogenic microbes including bacteria, viruses, fungi, protozoa, and nematodes etc. These pathogens are generally found in wastewater considered as primary health concern when the wastewater is not treated and partially treated which utilized in irrigation. These microbes in sewage or wastewater mainly comes from the human excreta or faecal matter and also from the industrial wastewater. All these types of pathogenic microbes can contaminate crop or soils. Due to presence of these diverse microorganisms, wastewater is infectious, which is responsible for the transmission of numerous diseases (Uyttendaele et al. 2015) (Tables 5.1 and 5.2).

5.8 Sewage Wastewater Application in Agriculture for Irrigation Purposes

For addressing the water scarcity, sewage wastewater is frequently utilized for irrigation especially in developing nations (Drechsel and Evans 2010). Using treated sewage wastewater in irrigation had been utilized mainly for three reasons: (i) as an alternative approach for treatment of wastewater (Bouwer and Chaney 1974); (ii) Utilizing low or poor quality as a viable resource of water for agricultural purposes (Bouwer and Idelovitch 1987; Al-Jaloud et al. 1995; Tanji 1997); (iii) Utilizing sewage wastewater as a source of nutrients (Bouwer and Chaney 1974; Vazquez-Montiel et al. 1996).

Drinking or potable water as well as treated sewage wastewater are rich in different nutrients, including phosphorus, nitrogen, potassium and sulphur. Among these, nitrogen and phosphorus are present in significant quantities in sewage wastewater, making them readily accessible for plant uptake. This characteristic considered a primary factor for worldwide utilization of treated sewage wastewater in irrigation practices (Drechsel and Evans 2010; Mara 2009; Poustie et al. 2020; Sengupta et al. 2015). An abundant amount of mineral nutrients found in recycled water can lead to a reduction in fertilizer usage, favours crop growth, enhances crop yield, and boost soil fertility, while potentially lowering the overall costs linked with crop productivity (Chen et al. 2013; et al. 2016).

Table 5.1 Showing the physical, chemical and biological parameter of sewage wastewater

S. No.	Sewage wastewater parameters	Description	References
1	*Physical parameters*		
i	Colour	Yellowish, Grey, brown and black	Younas and Younas (2022)
ii	Odour	Malodour i.e., unpleasant smell	Onkal-Engin et al. (2005)
iii	Temperature (°C)	6.9–34.4	Hart and Halden (2020)
iv	Total dissolved solids (mg/l)	995–1011	Ghosh et al. (2012)
v	Turbidity (NTU)	4 to 50	Brief (2017)
2	*Chemical parameters*		
i	pH	7.3 to 7.8 or 7.9 (≈8)	Henze et al. (1982), Ghosh et al. (2012) and Singh et al. (2012)
ii	Salinity (%)	1–3.5%	Zhao et al. (2020)
iii	Alkalinity (Eqv/m^3)	1–7	Henze et al. (1982)
iv	Electrical conductivity (dS/m)	0.96–1.09	Ghosh et al. (2012)
v	Total organic carbon (TOC) (mg/l)	10.4–17.6	Shetty and Goyal (2022)
vi	Chemical oxygen demand (COD) in g/m^3	200–570	Henze et al. (1982)
vii	Nitrogen (N) (mg/l)	55.9–68.8	Ghosh et al. (2012)
viii	Phosphorus (P) (mg/l)	6.8–9.7	Ghosh et al. (2012)
ix	Potassium (K) (mg/l)	0.31–4.5	Ghosh et al. (2012) and Singh et al. (2012)
x	Chlorides (Cl) (mg/l)	5.1	Singh et al. (2012)
xi	Sulphates (S) (mg/l)	1.42	Singh et al. (2012)

(continued)

Table 5.1 (continued)

S. No.	Sewage wastewater parameters	Description	References
xii	Sodium (Na) (mg/l)	1.42	Singh et al. (2012)
xiii	Magnesium (Mg) (mg/l)	1.69	Singh et al. (2012)
xiv	Calcium (Ca) (mg/l)	4.76	Singh et al. (2012)
3	*Biological parameters*		
i	BOD at 20 °C for 5 days (mg/l)	140–350	Henze and Comeau (2008)
ii	Viruses	Adenovirus, rotavirus, hepatitis A virus and other enteric virus	Corpuz et al. (2020)
iii	Bacteria	*Escherichia coli*, *Salmonella* spp., *Streptococcus faecalis Staphylococcus aureu*, *Shigella* spp., *Clostridium* spp., *Nitrobacter* sp., *Nitrosomonas* sp.	Mechri et al. (2010) and WHO (2006)
iv	Fungi	*Aspergillus niger*, *Aspergillus flavus*, *Aspergillus fumigatus* and *Actinomycetes*	Mechri et al. (2010) and WHO (2006)
v	Helminths	*Tenia* sp. and *Ancylostoma* sp.	Mechri et al. (2010) and WHO (2006)
vi	Protozoans	*Cryptospridium* spp., *Intestinal giardia*, *Entamoeba* spp.	Mechri et al. (2010) and WHO (2006)
4	*Microbiological parameters*		
i	Coliforms (UFC/100 ml)	0–1000	WHO (2006)
ii	*E. coli* (UFC/100 ml)	0–100	WHO (2006)
iii	Helminths (eggs/litre)	< 1	WHO (2006)

5.9 Methods of Wastewater Irrigation in Agriculture

5.9.1 Surface Irrigation

95% of the global agricultural area is irrigated through surface irrigation system. In this method sewage water is transported from water canal through pipes, ditches or canal pipes in to agricultural fields (Adamala et al. 2014). The inadequate demography characteristics of agricultural area result in inconsistent distribution of wastewater. However, by application of appropriate techniques and utilizing controlled wastewater application methods such as the implementation of gated pipes, flat lay tubes and surge flowing systems such as one can enhance the efficiency of irrigation

Table 5.2 Showing permissible limit (in ppm) of some toxic elements present in sewage wastewater

S. No.	Heavy metals or toxic elements	Optimum or permissible limit (in ppm) of some heavy metals or toxic elements present in sewage wastewater according to			
		WWF (2007)	WHO (2008)	USEPA (2010, 2011)	FAO (2007)
1	Arsenic (As)	–	0.01	0.01	–
2	Cadmium (Cd)	0.01	0.003	0.005	0.01
3	Chromium (Cr)	0.01	0.05	0.1	–
4	Cobalt (Co)	–	0.005	–	–
5	Copper (Cu)	0.2	2	1	–
6	Iron (Fe)	–	–	–	–
7	Lead (Pb)	0.1	0.01	0.015	5
8	Mercury (Hg)	–	0.006	0.002	–
9	Nickel (Ni)	0.2	0.07	–	–
10	Zinc (Zn)	–	–	2	2

methods. This approach facilitates a more even distribution of sewage wastewater, minimizes profound infiltration and concentrates chemicals, nutrients and pathogens effectively (Rodrigues et al. 2013). The utilization of sewage wastewater through surface system of irrigation presents certain limitations. For instance, workers are required to wear protective gloves and footwear. Additionally, this method is not suitable for crops intended for raw consumption. Any fruits or vegetables that come into contact with the agricultural land should not be sold (Sams et al. 2014).

5.9.2 *Sub-surface Irrigation*

In this method, sewage wastewater delivered in a manner which allows it to reach the roots of plants through capillary action. This method of irrigating system reduces exposure of sewage wastewater to field labourer thereby decreasing the potential health risks to humans (Peng et al. 2013). This method is regarded as an appropriate approach for the irrigation of wastewater. However, it presents certain limitations, including the potential for canal or pipe clogging, salt accumulation at the wetting front, emitter blockage caused by root intrusion, and a decrease in pipe diameter resulting from the buildup of microalgae and slime of bacteria.

5.9.3 *Sprinkler Irrigation*

This method refers to the technique of distributing sewage wastewater in the form of fine drops, spray or aerosols. Using this method for applying wastewater may pose potential health risks to humans. In this context, it is essential to design, arrange, and operate a sprinkler system utilizing a low pressurized nozzle positioned at a lower angle for minimizing the effects of airborne mist and fine particles (Li et al. 2012). In a mechanically operated system of sprinkler irrigation it is essential for the sprinkler to be oriented downwardly as well as positioned near to surface of soil for avoiding interaction with droplets of water that were carried by the wind (Megersa and Abdulahi 2015).

5.9.4 Drip Irrigation

Drip irrigation is a technique that involves the gradual application of sewage wastewater in which plants receive wastewater directly at their root system by a system of pipes, valves, emitters and tubes, which may be positioned above the soil or concealed within it. Primary objective of drip system of irrigation is to minimize vaporization and delivering water precisely within the rhizospheric zone (Green et al. 2018). It is regarded as a more efficient technique in comparison to surface and sprinkler method of irrigation, primarily as it minimizes the exposure of workers to wastewater during agricultural activities. This irrigation system offers several benefits, including the partial wetting of soil, which helps inhibit weed growth, limits extensive percolation as well as the leakage of wastewater as a result of gradual release of wastewater and decreases risk associated with drop of water or aerosols (Qi et al. 2016). One of the disadvantages of drip system of irrigation is the possibility of clogging in the emitter caused by accumulation of great amount of solid content. This issue could be mitigated through effective filtration of wastewater aids in the removal of both dissolved and suspended solids (Salgado-Méndez et al. 2019). Therefore, chlorination is a practical and affordable method of preventing these kinds of clogging.

5.10 Utilization of Sewage Wastewater Irrigation on Growth, Development and Productivity of Crop Plants

Utilizing sewage wastewater for irrigating purpose presents numerous functions such as agricultural irrigation of crops, aquaculture practices and irrigation of landscape as well as recharge of artificial groundwater (Asano and Levine 1996). Wastewater consist essential plant nutrients, making its utilization in agriculture serving as a vital nutrient source as well as for irrigation water for crops (Khurana and Singh 2012). In most of the developing nations, industrial and sewage water being utilized for irrigating crops such as mangoes, coconuts, lettuce and tomatoes (Anjum et al. 2021; Belhaj et al. 2016; Garcia et al. 2019). Many studies results showed that the reusing sewage wastewater in irrigation has synergistic impact on the yield of crops. In a research conducted by Mojid et al. (2016) showed in comparison with groundwater irrigation, domestic wastewater irrigation significantly results in notable increase in *Triticum aestivum* (wheat) yield. Similarly, the present study indicate that the wastewater irrigation has showed positive effect on *Oryza sativa* (rice) and *Zea mays* (maize) yield (Tsadilas and Vakalis 2003; Jung et al. 2014; Tran et al. 2019). In spite the fact sewage water has a synergistic effect on the yield of agricultural crop. However, the amount of increase in yield varies greatly, ranging from 2 to 10% (Wang et al. 2022). Sewage usage has positively impacted agricultural yields; if it is applied consistently for a couple of years, top soils may become more enriched (Anderson and Nilssion 1972; Haque and Sharma 1980). Many studies showed that the examination of sewage wastewater is most efficient way for the physical treatment aimed at diminishing heavy metal concentrations (Panicker 1995). The utilization of sewage water has been shown to enhance the productivity of rabi crop in comparison to irrigation with well or ground water, while also contributing a rise in the total NPK and organic carbon content of the soil (Singh et al. 2012). The tested weight of seeds of rabi crop including wheat (*Triticum aestivum*), gram (*Vigna radiate*), palak (*Spinacia oleracia*), fenugreek or methi (*Trigonella foenum-graecum* L.) and berseem (*Trifolium alexan drinum* L.) were notably greater, when irrigated with treated sewage water, exhibiting values between 1.40 and 16.70. In contrast, the tested weight for seeds irrigated with well water ranged from 1.23 to 16.23 (Singh et al. 2012). Application of screened sewage water for irrigation results in a markedly greater yield of grain and straw when compares with the use well or ground water. This is result of high amount of nutrient present in sewage wastewater, which can serve as an effective irrigation source, as demonstrated by the findings of Maiti et al. (1992) and supported by the research conducted (Sreeramulu 1994; Tiwari and Mishra 1996). *Triticum aestivum* (wheat) exhibited the highest grain yield among the crops, achieving 17.37 quintal (q) per hectare (ha) surpassing that of gram, which yielded 13.18 q/ha. In contrast, gram produced the highest straw yield at 25.03 q/ha, outperforming wheat's straw yield of 22.81 q/ha. The lowest grain yield was observed in berseem (*Trifolium alexandrinum*) at 2.68 q/ha and the lowest yields of straw were recorded for Palak and berseem, at 8.57 q/ha and 8.49 q/ha, respectively. Applying screened sewage water resulted in a notable increase in the grain and straw yields of *Trigonella foenum-graecum* (fenugreek or methi) is 9.11 and 10.28 q/ha and *Triticum aestivum* (wheat) is 17.37 and 22.81 q/ha (Singh et al. 2012). The outcomes from the use of whether screened or raw sewage water correlated with the grain and straw yield. Findings revealed that the highest yield of grain (2.68 q/ha) and straw (8.49 q/ha) of berseem (*Trifolium alexandrinum*) resulted by application of untreated sewage water. Additionally the outcomes suggested that berseem (*Trifolium alexandrinum*) exhibits tolerance to sewage water, while wheat (*Triticum aestivum*) and palak (*Spinacia oleracia*) necessitated the use of treated sewage water which indicates or show up the phytotoxicity among these crops to certain heavy metals. In comparison, *Cicer arietinum* (gram) and *Trigonella foenum-graecum* L. (fenugreek or methi) appeared to thrive under irrigation with well water (Singh et al. 2012) (Fig. 5.2).

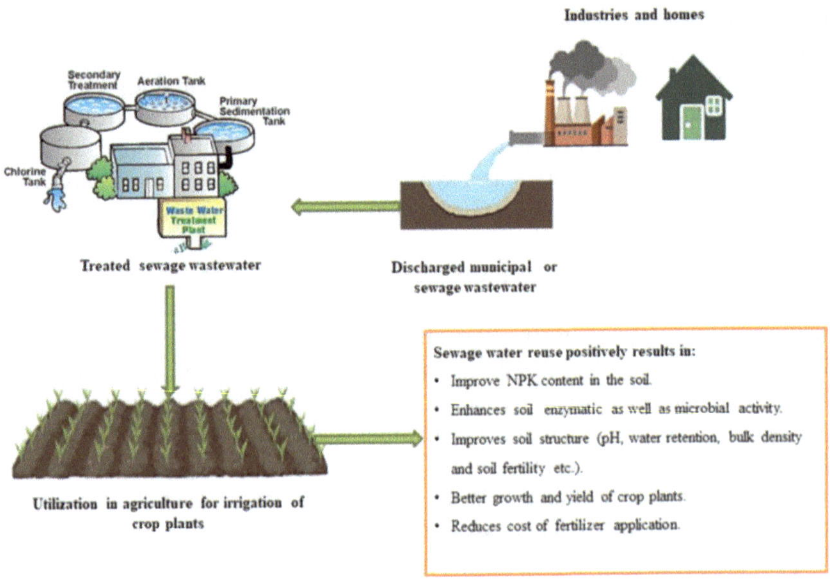

Fig. 5.2 The positive reuse of sewage wastewater in agricultural irrigation of crop plants

Utilizing treated municipal wastewater to grow wheat (*Triticum aestivum*), corn (*Zea mays*), potatoes (*Solanum tuberosum*) and lucerne (*Medicago sativa*) may result in higher yields when compared to control, similar results were obtained as noted (Juwarkar et al. 1990, 1994). When compared to well water, the raw as well filtered sewage water often produced the highest grain and straw production for the whole crop under the study (Singh et al. 2012). Improved crop growth especially for leafy vegetables for example cauliflower, cabbage and spinach has been observed in the fields irrigated with sewage wastewater (Murtaza et al. 2003). A notable rise in yield of onion (*Allium cepa*) and optimal fertilizer utilization efficiency has been observed in plots treated with a combined application of 40 kg of nitrogen, 20 kg of P_2O_5 and 20 kg of K_2O/hectare per hectare along with 25 times diluted distillery effluents, compared to plots irrigated with groundwater (Zalawadia et al. 1996). A field experiment performed for assessing the application of wastewater for irrigation of maize, sunflower, groundnut and soybean demonstrated increases in grain yield of 19.3%, 29.9%, 5.9% and 4.8% respectively when compared by well or ground water irrigation (Udayasoorian et al. 1999). A major increase in dry biomass yield of barseem (*Trifolium* sp.) observed, rising from 3.38 to 8.85 g/pot irrigated through sewage wastewater when compared with those irrigated by groundwater. This indicates the critical nutrient enrichment provided by the sewage wastewater (Singh et al. 1991). The yield has been increased from 6.9 to 13.9% across different varieties of sugarcane cultivated using wastewater irrigation when compared with ground water irrigation. A separate long-term study has demonstrated that irrigation with sewage wastewater resulted in significantly higher grain yields for cotton, rice and wheat with increase of 50%, 46% and 23% respectively compared to groundwater irrigation (Sharma and Kansal 1984). Additionally, a gradual rise in pod yield of peanut had been observed through the application of wastewater at concentrations of up to 50% from the textile industry (Saravanamoorthy and Kumari 2007).

5.11 Wastewater Mediated Availability of Essential Nutrients for Growth of Crop Plants

An amount of essential nutrients available to the crops through wastewater (either municipal and industrial) application in the agriculture through irrigation which must be consider precisely. An amount of primary nutrients i.e., NPK requires for the growth of plants found in municipal as well as industrial wastewater exhibit considerable variation depending on quality of wastewater. Normally, the nitrogen concentrations typically range from 20 to 35 mg per liter, while phosphorus concentrations are observed between 3 and 10 mg per liter (Von Sperling 2007; Metcalf et al. 1991; Vergine et al. 2017a; Lonigro et al. 2016; Gatta et al. 2016) and concentration of potassium varies from 10 to 25 mg/litre (Lonigro et al. 2016; Vergine et al. 2017b). Wastewater discharge contains higher concentrations of mineral nutrients such as boron (B), iron (Fe), copper (Cu), zinc (Zn), manganese (Mn) and molybdenum (Mo) that are necessary for the growth as well as development of crops. An impact of essential nutrients (micro and macro nutrients) provided by wastewater could vary depending on the crop cycle (yearly vegetable crops vs. tree crops) as well as nutrient intake rates (Gatta et al. 2021).

In a research conducted mainly on vegetable crops, the effects of second and fourth stage treated agro-industrial wastewater were evaluated when compared to ground water irrigation. This study specifically examined a series of cultivated crops such as processed tomato (*Lycopersicon esculentum* Mill.) and broccoli (*Brassica oleracia* L. var. *italica*) crops. The results demonstrated the application of treated agro-industrial effluents provided enhanced levels of mineral nutrients specifically ammonium nitrogen ($N\text{-}NH_4$), nitrate nitrogen ($N\text{-}NO_3$) as well as potassium (K^+) to the crops (Vergine et al. 2017b). The authors concluded that utilizing treated industrial effluents for irrigation can lessen the additional requirement of artificial fertilizers by chemical means. Research has indicated that the irrigation of vegetable crops such as lettuce with treated wastewater can effectively enhances the accessibility of irrigation water while also elevating amount of certain soil nutrients including potassium, calcium, hydrogen, aluminum, and sulphur (Urbano et al. 2017).

In a research conducted by Gatta et al. (2016) to see the impact of secondary as well as treated tertiary wastewater on the growth of artichoke found the total inorganic nitrogen (comprising $N\text{-}NH_4$ and $N\text{-}NO_3$) provided for the crops via treated wastewater irrigation was 95 kg per hectare for secondary wastewater and 66 kg per hectare for tertiary wastewater. This accounted for roughly 25% and 17% of nitrogen is needed for the crop of artichoke respectively (Gatta et al. 2021).

Vivaldi et al. (2019) examined impact of utilizing treated wastewater along with a deficit irrigation approach applied on trees of *Prunus dulcis* L. (Almond). Their findings reveals the application wastewater (treated) to soil irrigated provided a nutritional input of 35.8 kg/ha of NO_3^{2-}, 4.41 kg/ha of PO_4^{3-} as well as 149.9 kg/ha of K^+. Research on trees irrigated with wastewater has indicated the elevated levels of mineral nutrients (both macro and meso-nutrients) when introduced in soil from wastewater can results in significant decrease of required fertilizer application (Pedrero et al. 2014; Vivaldi et al. 2017; Montemurro et al. 2017).

An application of wastewater for the provision of mineral nutrients to cultivated species offers considerable agronomic advantages associated with this water resource. Nevertheless, it is essential to evaluate this benefit with careful consideration of potential nutritional imbalances in plants and the risk of phytotoxicity. Each cultivated species possesses distinct nutrient requirements that vary according to its phenological phase (Grattan et al. 2015). Imbalances, whether in the form of excesses or deficiencies of certain mineral elements, can have a considerable impact on crop production, influencing both quantity and quality (Segal et al. 2011).

5.12 Qualitative and Quantitative Response Evaluation of Crop Plants Irrigated with Treated Sewage Wastewater

The grade of water utilized for irrigating crops could significantly limit agricultural productivity as a result of its influence on fertility of soil and crop yield soil fertility. Understanding of how water quality impacts crop yield is essential for establishing effective management practices that ensure long-term productivity. Additionally, this knowledge is vital for formulating an optimal irrigation schedule that maximizes crop production as well as achieves expected financial returns from irrigation. The result of sewage wastewater irrigation on the standard and output of agricultural produce can vary significantly; reports indicate that crops may experience either beneficial or detrimental outcomes, and in some cases, the effects may be neutral (Gatta et al. 2021).

Field experiments were conducted under various pedoclimatic environment in the Apulia area of South Italy, for assessing both qualitative and quantitative responses of many agricultural crops such as vegetables by utilizing treated sewage wastewater and agro-industrial sources for use in irrigation (Gatta et al. 2016; Vergine et al. 2017a; Lonigro et al. 2016; Libutti et al. 2018). The hygienic characteristics of agricultural products such as the detection of coliform bacteria including *E. coli* and *Salmonella typhi* as well as their associated threat to environment and human wellbeing originating from the reuse of wastewater source for irrigation was assessed concurrently. Horticultural crops including processed *Lycopersicon esculentum* Mill. (Tomato), *Cynara cardunculus* L. subsp. scolymus Hayek (Artichoke), *Brassica oleracia* L. var. italica broccoli (Broccoli) as well as fresh vegetable crops such as

Lactuca sativa L. (Lettuce), *Foeniculum vulgare* Mill. (Fennel), *Curcumis melo* L. (Melon) and *Cucumis sativus* L. (Cucumber) was analyzed. These above findings revealed that the qualitative characteristics of crop productivity including dry weight, width, soluble solid material, titrable acidity and pH level was comparable with those observed in vegetables irrigated through traditional water resource (Gatta et al. 2021).

In certain cases, when the level of vital nutrients such as nitrogen (N), phosphorus (P) and Potassium (K) supplied to the crops through sewage wastewater exceeded which is delivered via traditional water sources, these crops exhibited a favorable productive response, reflected in both total and marketable yields (Gatta et al. 2021).

It is generally acknowledged that when municipal wastewater is not subjected to the denitrification processes, it retains significant levels of nitrates, ammonium salts, mineral perphosphates as well as macro-elements compared to traditional water. High availability of nitrogen for crops, particularly at the time of harvest, influences the duration of the crop cycle by prolonging the senescence period and by shortening the juvenile period. Consequently, crop that completes their lifecycle with the reproductive phase by producing seeds when irrigated through sewage wastewater will tend to have an extended lifecycle of crops (Rubino and Lonigro 2015).

A field experiment which is conducted by Qaryouti et al. (2015) demonstrated that reusing of industrial wastewater, which is abundant in organic nutrients and matter especially potassium, positively impacted various parameters of cucumber and tomato crops. Notably, the cucumber plants exhibited significant improvements in plant height, yield of fruit as well as average fruit weight. Additionally, the tomato plants showed marked increases in leaf area and dry weight as a result of substituting K-chemical fertilizers with the wastewater. Al-Lahham et al. (2003) noticed the shape and mass of tomato fruit increased when sprayed using reclaimed domestic wastewater. However, utilizing grey treated wastewater, which has elevated BOD and COD levels, for irrigation resulted in negative effects on the chemical quality of various vegetable crops, including *Abelmoschus esculentus* L. (Okra), *Phaseolus vulgaris* L. (Beans), *Zea mays* L. (Corn), *Helianthus annuus* L. (Sunflowers) (Al-Hamaiedeh and Bino 2010). Sewage wastewater, alongside providing essential plant nutrients, may also harbors a range of possibly toxic heavy metals potentially toxic mineral elements that can negatively that can negatively impact the health of both human and animals (Singh et al. 2011). Notably, elevated levels of toxic elements can leads to diminished yields of marketable crops or a decrease in the amount of agricultural goods available for sale (Khan et al. 2016). Gatta et al. (2018) found the levels of heavy metals including Cd, Cr, Ni, Pb, As and Hg etc. reported that the amount of heavy metals such as cadmium (Cd), chromium (Cr), nickel (Ni), lead (Pb), arsenic (As) and mercury (Hg) etc. among artichoke heads, which was collected following irrigation through secondary and tertiary treated wastewater were below the global threshold limits. Further, they had noticed the bioaccumulation factor in eatable part of the artichoke were relatively low, indicating that there are no further risks of health (hazard index $<$ 1.0) for both adults as well as children resulting from the use of the artichoke heads. Conversely, other researchers (Balkhair and Ashraf 2016) discovered the amount of nickel, lead, cadmium as well as chromium among eatable portion of okra cultivated in treated wastewater irrigated soil considerably exceeded the safety limits, resulting

Table 5.3 Permissible range of some toxic elements in plants (in mg/kg)

S. No	Toxic elements	Permissible range of some metals or toxic elements in plants (in mg/kg)
1	Arsenic	–
2	Cadmium	0.02
3	Chromium	1.3
4	Copper	10
5	Mercury	–
6	Nickel	10
7	Lead	2
8	Zinc	0.6

Source World Health Organization (2006)

in a health risk index (HRI) < 1, which suggests a potential risk of health. Uddin et al. (2015) found the amount of heavy metals such as Zn, Pb and Cd among the edible portion of *Amaranthus cruentus* L. (Amaranth) and tomatoes were watered through industrial wastewater which exceed permissible limits set by the WHO by several times. The findings from various research studies on both untreated and treated wastewaters distinctly demonstrate that these modes of irrigation provides sufficient amount of nutrients. Mostly crops exhibit yields that exceed initial expectations when utilizing wastewater mediated irrigation system and thus reducing the need of artificial fertilizers and lowering the investment costs for farmers. The presence of toxic substances in sewage wastewater is a critical factor to consider, as they can pose serious hazard to both plants and human health when present at elevated concentrations. Consequently, it's essential for assessing the impact of sewage wastewater irrigation on the yield of crop characteristics across various agronomic contexts. This evaluation is vital for the advancement of wastewater irrigation systems, ensuring they are developed in alignment with optimal agronomic and water management practices (Table 5.3).

5.13 Conclusion

Using freshwater in agriculture for irrigating crop plants will results in scarcity of water for potable use and domestic purposes. To conserve the freshwater resources for essential uses, sewage wastewater can be utilized in agriculture as a potential alternative source. The increasing production of municipal, industrial and sewage wastewater in developing nations particularly among Asian countries such as India present an opportunity for utilizing it as a freshwater substitute in agriculture. The idea of utilizing sewage wastewater for agricultural purposes is considered as an acceptable approach by providing a suitable treatment process as well as adopting proper distribution measures. The practice of reusing wastewater is acknowledged as a cost efficient alternative mainly due to its rich content of essential plant nutrients. Response of crop yield varies depending on type of wastewater utilized in irrigation. Irrigation with domestic as well as breeding wastewater can results in substantial increases in crop yields, whereas industrial wastewater generally tends to have a negative, though relatively minor impact. Additionally, higher concentrations of domestic wastewater can notably enhance crop yield. Also, the design of experiments, including factors such as field conditions, the number of replicates, and the application of fertilizers playing a role in variable crop yield outcomes as observed in various studies. The application of wastewater in agriculture increases the chemical fertility of the irrigated soil and availability of NPK nutrient for the irrigated crop plants. However, the major issue regarding the application of wastewater in agriculture is the accumulation or building up of heavy metals or toxic elements such as Cd, Cr, As, Ni and Pb etc. in soil and crops as well as infestation of pathogenic microbes in crops plants. These heavy metals can ultimately transfer to human beings via the food chain, and thus resulting in health complications. Therefore, the precautionary steps including selecting right crop plants as well as proper treatment and dilution of wastewater shall be taken into consideration prior to utilization in agriculture for irrigation purpose. Further, wastewater reuse in agriculture is recommended to adhere to the established techniques and accessible guidelines given by the reputed organizations such as WHO, FAO, EU and USEPA etc. Utilizing wastewater in agriculture for irrigation is regarded both as a challenge and as an opportunity, particularly with respect to its impact on environment as well as implementation. The beneficial aspect of wastewater can be fulfilled by its reuse to achieve certain environmental and social goals. Thus, the sustainable use of sewage water and its application can effectively increases the collection, recycling and reuse of wastewater resources which can be utilized for irrigation of crop plants and positively impact their yields in the present as well in near future.

Acknowledgements Mr. Saurabh Kumar, UGC-JRF (NTA Ref. No. 210510379563) is grateful to his funding agency i.e., UGC for providing him financial support. Authors are thankful to the Chairman, Department of Botany, Aligarh Muslim University, Aligarh-202002 (Uttar Pradesh), India.

References

Adamala S, Raghuwanshi NS, Mishra A (2014) Development of surface irrigation systems design and evaluation software (SIDES). Comput Electron Agric 100:100–109

Ahmed J, Thakur A, Goyal A (2021) Industrial wastewater and its toxic effects

Al-Hamaiedeh H, Bino M (2010) Effect of treated grey water reuse in irrigation on soil and plants. Desalination 256(1–3):115–119

Al-Jaloud AA, Hussain G, Al-Saati AJ, Karimulla S (1995) Effect of wastewater irrigation on mineral com-position of corn and sorghum plants in a pot experiment. J Plant Nutr 18(8):1677–1692

Al-Lahham O, El Assi NM, Fayyad M (2003) Impact of treated wastewater irrigation on quality attributes and contamination of tomato fruit. Agric Water Manag 61(1):51–62

Al-Nakshabandi GA, Saqqar MM, Shatanawi MR, Fayyad M, Al-Horani H (1997) Some environmental problems associated with the use of treated wastewater for irrigation in Jordan. Agric Water Manag 34(1):81–94

Alobaidy AHMJ, Abid HS, Maulood BK (2010) Application of water quality index for assessment of Dokan lake ecosystem, Kurdistan region, Iraq. J Water Resour Prot 2(9):792–798

Anderson A, Nilssion KO (1972) Standard methods for the examination of water and waste water, 21st edn. Washington

Angelakis AN, Snyder SA (2015) Wastewater treatment and reuse: past, present, and future. Water 7(9):4887–4895

Angelakis AN, Asano T, Bahri A, Jimenez BE, Tchobanoglous G (2018) Water reuse: from ancient to modern times and the future. Front Environ Sci 6:26

Anjum MA, Hussain S, Arshad P, Hassan A (2021) Irrigation water of different sources affects fruit quality attributes and heavy metals contents of un-grafted and commercial mango cultivars. J Environ Manage 281:111895

Asano T, Levine AD (1996) Wastewater reclamation, recycling and reuse: past, present, and future. Water Sci Technol 33(10–11):1–14

Aydin ME, Aydin S, Beduk F, Tor A, Tekinay A, Kolb M, Bahadir M (2015) Effects of long-term irrigation with untreated municipal wastewater on soil properties and crop quality. Environ Sci Pollut Res 22:19203–19212

Aziz F, Farissi M (2014) Reuse of treated wastewater in agriculture: solving water deficit problems in arid areas. Ann West Univ Timisoara Ser Biol 17(2)

Balkhair KS (2016) Impact of treated wastewater on soil hydraulic properties and vegetable crop under irrigation with treated wastewater, field study and statistical analysis. J Environ Biol 37(5):1143

Balkhair KS, Ashraf MA (2016) Field accumulation risks of heavy metals in soil and vegetable crop irrigated with sewage water in western region of Saudi Arabia. Saudi J Biol Sci 23(1):S32–S44

Belhaj D, Jerbi B, Medhioub M, Zhou J, Kallel M, Ayadi H (2016) Impact of treated urban wastewater for reuse in agriculture on crop response and soil ecotoxicity. Environ Sci Pollut Res 23:15877–15887

Blumenthal UJ, Mara DD, Peasey A, Ruiz-Palacios G, Stott R (2000) Guidelines for the microbiological quality of treated wastewater used in agriculture: recommendations for revising WHO guidelines. Bull World Health Organ 78:1104–1116

Boano F, Caruso A, Costamagna E, Ridolfi L, Fiore S, Demichelis F, Galvão A, Pisoeiro J, Rizzo A, Masi F (2020) A review of nature-based solutions for greywater treatment: applications, hydraulic design, and environmental benefits. Sci Total Environ 711:134731

Bouwer H, Chaney RL (1974) Land treatment of wastewater. Adv Agron 26:133–176

Bouwer H, Idelovitch E (1987) Quality requirements for irrigation with sewage water. J Irrig Drain Eng 113(4):516–535

Brief A (2017) Water quality and health-review of turbidity: information for regulators and water suppliers. Who/Fwc/Wsh 17:1–10

Chen W, Lu S, Pan N, Jiao W (2013) Impacts of long-term reclaimed water irrigation on soil salinity accumulation in urban green land in Beijing. Water Resour Res 49(11):7401–7410

Cizmas L, Sharma VK, Gray CM, McDonald TJ (2015) Pharmaceuticals and personal care products in waters: occurrence, toxicity, and risk. Environ Chem Lett 13(4):381–394

Clemmens AJ, Allen RG, Burt CM (2008) Technical concepts related to conservation of irrigation and rainwater in agricultural systems. Water Resour Res 44(7)

Contreras JD, Meza R, Siebe C, Rodríguez-Dozal S, López-Vidal YA, Castillo-Rojas G, Amieva RI, Solano-Gálvez SG, Mazari-Hiriart M, Silva-Magaña MA, Vázquez-Salvador N (2017) Health risks from exposure to untreated wastewater used for irrigation in the Mezquital Valley, Mexico: a 25-year update. Water Res 123:834–850

Corpuz MVA, Buonerba A, Vigliotta G, Zarra T, Ballesteros F Jr, Campiglia P, Belgiorno V, Korshin G, Naddeo V (2020) Viruses in wastewater: occurrence, abundance and detection methods. Sci Total Environ 745:140910

CPCB (2009) Status of water supply, wastewater generation and treatment in Class-I cities and Class-II towns of India. Central Pollution Control Board, Government of India New Delhi, India

Day AD, Taher FA, Katterman FRH (1975) Influence of treated municipal waste water on growth, fiber, acid-soluble nucleotides, protein, and amino acid content in wheat grain, vol 4, no 2. American Society of Agronomy, Crop Science Society of America, and Soil Science Society of America, pp 167–169

Dixit A, Madhav S, Mishra R, Srivastav AL, Garg P (2022) Impact of climate change on water resources, challenges and mitigation strategies to achieve sustainable development goals. Arab J Geosci 15(14):1296

Drakatos PA, Kalavrouziotis IK, Hortis TC, Varnanas SP, Drakatos SP, Bladenopoulou S, Fanariotou IN (2002) Antagonistic action of Fe and Mn in Mediterranean-type plants irrigated with wastewater effluents following biological treatment. Int J Environ Stud 59(1):125–132

Drechsel P, Evans AE (2010) Wastewater use in irrigated agriculture. Irrig Drain Syst 24:1–3

Egbuikwem PN, Mierzwa JC, Saroj DP (2020) Assessment of suspended growth biological process for treatment and reuse of mixed wastewater for irrigation of edible crops under hydroponic conditions. Agric Water Manag 231:106034

Elifantz H, Kautsky L, Mor-Yosef M, Tarchitzky J, Bar-Tal A, Chen Y, Minz D (2011) Microbial activity and organic matter dynamics during 4 years of irrigation with treated wastewater. Microb Ecol 62:973–981

El-Nahhal Y, Awad Y, Safi J (2013) Bioremediation of acetochlor in soil and water systems by cyanobacterial mat. Int J Geosci 4(05):880

Fakayode SO (2005) Impact assessment of industrial effluent on water quality of the receiving Alaro River in Ibadan, Nigeria, pp 167–169

FAO (1993) Control de Aguas de Riego Destinadas a la Produccion Hortofruticola: Chile. Technical Report of Project TCP/CHI/2251(A). Food and Agriculture Organization of the United Nations, Rome

FAO (2002) World agriculture: towards 2015/2030. Summary report. food and agriculture organization of the United Nations (FAO), Rome

FAO (2007) The state of food and agriculture 2007: paying farmers for environmental services. Food and Agriculture Organization of the United Nations (FAO), Rome

FAO (2017) The state of food and agriculture 2017: leveraging food systems for inclusive rural transformation

García MG, Fernández-López C, Polesel F, Trapp S (2019) Predicting the uptake of emerging organic contaminants in vegetables irrigated with treated wastewater–implications for food safety assessment. Environ Res 172:175–181

Gatta G, Libutti A, Beneduce L, Gagliardi A, Disciglio G, Lonigro A, Tarantino E (2016) Reuse of treated municipal wastewater for globe artichoke irrigation: assessment of effects on morpho-quantitative parameters and microbial safety of yield. Sci Hortic 213:55–65

Gatta G, Gagliardi A, Disciglio G, Lonigro A, Francavilla M, Tarantino E, Giuliani MM (2018) Irrigation with treated municipal wastewater on artichoke crop: assessment of soil and yield heavy metal content and human risk. Water 10(3):255

Gatta G, Libutti A, Gagliardi A, Disciglio G, Tarantino E, Beneduce L, Giuliani MM (2021) Wastewater reuse in agriculture: effects on soil-plant system properties. In: Interaction and fate of pharmaceuticals in soil-crop systems: the impact of reclaimed wastewater, pp 79–102

Ghosh AK, Bhatt MA, Agrawal HP (2012) Effect of long-term application of treated sewage water on heavy metal accumulation in vegetables grown in Northern India. Environ Monit Assess 184:1025–1036

Girovich MJ (1996) Biosolids treatment and management: processes for beneficial use. CRC Press

Gleick PH (2000) A look at twenty-first century water resources development. Water Int 25(1):127–138

Grattan SR, Díaz FJ, Pedrero F, Vivaldi GA (2015) Assessing the suitability of saline wastewaters for irrigation of *Citrus* spp.: emphasis on boron and specific-ion interactions. Agric Water Manag 157:48–58

Green O, Katz S, Tarchitzky J, Chen Y (2018) Formation and prevention of biofilm and mineral precipitate clogging in drip irrigation systems applying treated wastewater. Irrig Sci 36:257–270

Gudeta DA (2010) School of graduate studies faculty of technology chemical engineering department environmental engineering stream. Doctoral dissertation, Addis Ababa University.

Gupta S, Satpati S, Nayek S, Garai D (2010) Effect of wastewater irrigation on vegetables in relation to bioaccumulation of heavy metals and biochemical changes. Environ Monit Assess 165:169–177

Haque A, Sharma S (1980) Water hyacinth to fight water pollution. Science Reporter 757–762

Hart OE, Halden RU (2020) Modeling wastewater temperature and attenuation of sewage-borne biomarkers globally. Water Res 172:115473

Henze W, Tandberg-Hanssen E, Hagyard MJ, Woodgate BE, Shine RA, Beckers JM, Bruner M, Gurman JB, Hyder CL, West EA (1982) Observations of the longitudinal magnetic field in the transition region and photosphere of a sunspot. Sol Phys 81:231–244

Henze M, Comeau Y (2008) Wastewater characterization. In: Biological wastewater treatment: principles modelling and design, p 27

Hoekstra AY, Mekonnen MM (2012) The water footprint of humanity. Proc Natl Acad Sci 109(9):3232–3237

Huibers FP, Van Lier JB (2005) Use of wastewater in agriculture: the water chain approach. Irrig Drain J Int Comm Irrig Drain 54(S1):S3–S9

Hussain I, Raschid L, Hanjra MA, Marikar F, Van Der Hoek W (2002) Wastewater use in agriculture: review of impacts and methodological issues in valuing impacts

Hussain I, Regassa N, Madar S (2004) Water for food security for the poor, vol 76, pp 90030–90032 (6 Abbreviations 8 Overview)

Ilyas M, Ahmad W, Khan H, Yousaf S, Yasir M, Khan A (2019) Environmental and health impacts of industrial wastewater effluents in Pakistan: a review. Rev Environ Health 34(2):171–186

Islam SMF, Karim Z (2019) World's demand for food and water: the consequences of climate change. In: Desalination-challenges and opportunities

Jeong H, Kim H, Jang T (2016) Irrigation water quality standards for indirect wastewater reuse in agriculture: a contribution toward sustainable wastewater reuse in South Korea.Water 8(4):169

Jiménez B, Asano T (2008) Water reuse: an international survey of current practice, issues and needs. IWA Publishing

Jung K, Jang T, H, Park S (2014) Assessment of growth and yield components of rice irrigated with reclaimed wastewater. Agric Water Manag 138:17–25

Juwarkar AS, Juwarkar A, Deshbratar PB, Bal AS (1990) Exploration of nutrient potential of domestic and sludge through and land application. RAPA Report, pp 178–201

Juwarkar AS, Shende A, Thawale PR, Satyanarayanan PR, Deshbratar PB, Bal AS (1994) In: Tandon HLS (ed) Fertilizers, organic manures, recyclable wastes and biofertilizers, p 72

Katsoyiannis IA, Gkotsis P, Castellana M, Cartechini F, Zouboulis AI (2017) Production of demineralized water for use in thermal power stations by advanced treatment of secondary wastewater effluent. J Environ Manage 190:132–139

Kesalkar VP, Khedikar IP, Sudame AM (2012) Physico-chemical characteristics of wastewater from paper industry. Int J Eng Res Appl 2(4):137–143

Khan I, Ghani A, Rehman AU, Awan SA, Jawed H, Gul R (2016) The analyses of heavy metal concentrations in soil, waste water and *Raphanus sativus* (L.) at three different growth stages. Pyrex J Res Environ Stud 3:42–48

Khurana MPS, Singh P (2012) Waste water use in crop production: a review. Resour Environ 2(4):116–131

Kumar N, Kumar D, Kumar S, Shukla V, Shukla P, Raj B (2018) Spatio-temporal variations in hydro-geochemistry of groundwater at rural, urban and industrial areas of Kanpur, India. Environ Sustain 1:197–208

Lautze J, Stander E, Drechsel P, da Silva AK, Keraita B (2014) Global experiences in water reuse. International Water Management Institute, Colombo, Sri Lanka

Li J, Zhao W, Yin J, Zhang H, Li Y, Wen J (2012) The effects of drip irrigation system uniformity on soil water and nitrogen distributions. Trans ASABE 55(2):415–427

Libutti A, Gatta G, Gagliardi A, Vergine P, Pollice A, Beneduce L, Disciglio G, Tarantino E (2018) Agro-industrial wastewater reuse for irrigation of a vegetable crop succession under Mediterranean conditions. Agric Water Manag 196:1–14

Liu SX (2008) Food and agricultural wastewater utilization and treatment. Wiley

Lonigro A, Rubino P, Lacasella V, Montemurro N (2016) Faecal pollution on vegetables and soil drip irrigated with treated municipal wastewaters. Agric Water Manag 174:66–73

Madhav S, Ahamad A, Kumar A, Kushawaha J, Singh P, Mishra PK (2018) Geochemical assessment of groundwater quality for its suitability for drinking and irrigation purpose in rural areas of Sant Ravidas Nagar (Bhadohi), Uttar Pradesh. Geol Ecol Landsc 2(2):127–136

Maiti PS, Sah KD, Gupta SK, Banerjee SK (1992) Evaluation of sewage sludge as a source of irrigation and manure. J Indian Soc Soil Sci 40(1):168–172

Mara D (2009) Waste stabilization ponds: past, present and future. Desalin Water Treat 4(1–3):85–88

Mateo-Sagasta J, Raschid-Sally L, Thebo A (2015) Global wastewater and sludge production, treatment and use. In: Wastewater: economic asset in an urbanizing world, pp 15–38

Mechri B, Chehab H, Attia F, Mariem FB, Braham M, Hammami M (2010) Olive mill wastewater effects on the microbial communities as studied in the field of olive trees by analysis of fatty acid signatures. Eur J Soil Biol 46(5):312–318

Megersa G, Abdulahi J (2015) Irrigation system in Israel: a review. Int J Water Resour Environ Eng 7(3):29–37

Mekonnen MM, Hoekstra AY (2016) Four billion people facing severe water scarcity. Sci Adv 2(2):e1500323

Metcalf L, Eddy HP, Tchobanoglous G (1991) Wastewater engineering: treatment, disposal, and reuse, vol 4. McGraw-Hill, New York

Metcalf E (2003) Physical unit process. In: Waste water engineering treatment and reuse, 4th edn. Tat-McGraw Hill, New York

Mikkelsen RL, Camberato JJ (1995) Potassium, sulfur, lime, and micronutrient fertilizers

Mishra VK, Upadhyay AR, Tripathi BD (2009) Bioaccumulation of heavy metals and two organochlorine pesticides (DDT and BHC) in crops irrigated with secondary treated waste water. Environ Monit Assess 156:99–107

Mishra S, Kumar R, Kumar M (2023) Use of treated sewage or wastewater as an irrigation water for agricultural purposes-environmental, health, and economic impacts. Total Environ Res Themes 6:100051

Mohsen MS (2004) Treatment and reuse of industrial effluents: case study of a thermal power plant. Desalination 167:75–86

Mojid MA, Wyseure GCL, Biswas SK (2016) Effects of municipal wastewater irrigation on yield and fertilizer requirement of wheat (*Triticum aestivum* L.) in Bangladesh. Agriculturists 14(1):01–14

Montemurro N, Cucci G, Mastro MA, Lacolla G, Lonigro A (2017) The nitrogen role in vegetables irrigated with treated municipal wastewater

Murtaza G, Ghafoor A, Qadir M, Rashid MK (2003) Accumulation and bioavailability of Cd, Co and Mn in soils and vegetables irrigated with city effluent. Pak J Agric Sci 40(1–2):18–24

National Research Council, Global Affairs, Office of International Affairs, Board on Science, & Technology for International Development. (1996). Lost crops of Africa: volume I: grains (Vol. 1). national academies press

Onkal-Engin G, Demir I, Engin SN (2005) Determination of the relationship between sewage odour and BOD by neural networks. Environ Model Softw 20(7):843–850

Panicker PVRC (1995) Recycling of human waste in agriculture. In: Recycling of waste in agriculture. Fort, New Delhi, pp 68–90
Pedrero F, Kalavrouziotis I, Alarcón JJ, Koukoulakis P, Asano T (2010) Use of treated municipal wastewater in irrigated agriculture—review of some practices in Spain and Greece. Agric Water Manag 97(9):1233–1241
Pedrero F, Maestre-Valero JF, Mounzer O, Alarcón JJ, Nicolás E (2014) Physiological and agronomic mandarin trees performance under saline reclaimed water combined with regulated deficit irrigation. Agric Water Manag 146:228–237
Peng G, Bing W, Guangcan Z (2013) Influence of sub-surface irrigation on soil conditions and water irrigation efficiency in a cherry orchard in a hilly semi-arid area of northern China. PLoS One 8(9):e73570
Pescod MB (1992) Wastewater treatment and use in agriculture-FAO irrigation and drainage paper 47. Food and Agriculture Organization of the United Nations, Rome
Poustie A, Yang Y, Verburg P, Pagilla K, Hanigan D (2020) Reclaimed wastewater as a viable water source for agricultural irrigation: a review of food crop growth inhibition and promotion in the context of environmental change. Sci Total Environ 739:139756
Qadir M, Sharma BR, Bruggeman A, Choukr-Allah R, Karajeh F (2007) Non-conventional water resources and opportunities for water augmentation to achieve food security in water scarce countries. Agric Water Manag 87(1):2–22
Qadir M, Wichelns D, Raschid-Sally L, McCornick PG, Drechsel P, Bahri A, Minhas PS (2010) The challenges of wastewater irrigation in developing countries. Agric Water Manag 97(4):561–568
Qaryouti M, Bani-Hani N, Abu-Sharar TM, Shnikat I, Hiari M, Radiadeh M (2015) Effect of using raw waste water from food industry on soil fertility, cucumber and tomato growth, yield and fruit quality. Sci Hortic 193:99–104
Qi Z, Zhang T, Zhou L, Feng H, Zhao Y, Si B (2016) Combined effects of mulch and tillage on soil hydrothermal conditions under drip irrigation in Hetao Irrigation District, China. Water 8(11):504
Quagraine EK, Duncan B (2017) Two decades constructed wetland experience in treating municipal effluent for power plant cooling at the Shand Power Station, SaskPower Part IV: the effect of seasonal changes in temperature, rainfall, plant growth and influent concentration on ammonia and nitrate removal. J Water Sustain 7(2):139
Radcliffe SA (2004) Geography of development: development, civil society and inequality–social capital is (almost) dead? Prog Hum Geogr 28(4):517–527
Rattan RK, Datta SP, Chhonkar PK, Suribabu K, Singh AK (2005) Long-term impact of irrigation with sewage effluents on heavy metal content in soils, crops and groundwater—a case study. Agric Ecosyst Environ 109(3–4):310–322
Ricart S, Rico AM (2019) Assessing technical and social driving factors of water reuse in agriculture: a review on risks, regulation and the yuck factor. Agric Water Manag 217:426–439
Rijsberman FR (2006) Water scarcity: fact or fiction? Agric Water Manag 80(1–3):5–22
Rodrigues GC, Paredes P, Gonçalves JM, Alves I, Pereira LS (2013) Comparing sprinkler and drip irrigation systems for full and deficit irrigated maize using multicriteria analysis and simulation modelling: ranking for water saving vs. farm economic returns. Agric Water Manag 126:85–96
Rubino P, Lonigro A (2015) In: Te.R.R.A. project. Innovation Tecnologiche e di Processo per il Riutilizzo irriguo delle acque Reflue urbane e Agroindustriali ai fini della gestione sostenibile delle risorse idriche. Linee guida per il riuso irriguo delle acque reflue depurate, p 271
Ryder G (2017) The United Nations world water development report, 2017: wastewater: the untapped resource
Salgado-Méndez S, Gilabert-Alarcón C, Daesslé LW, Mendoza-Espinosa L, Avilés-Marín S, Stumpp C (2019) Short-term effects on agricultural soils irrigated with reclaimed water in Baja California, Mexico. Bull Environ Contam Toxicol 102:829–835
Sams JI, Veloski G, Smith BD, Minsley BJ, Engle MA, Lipinski BA, Hammack R, Zupancic JW (2014) Application of near-surface geophysics as part of a hydrologic study of a subsurface drip

irrigation system along the Powder River floodplain near Arvada, Wyoming. Int J Coal Geol 126:128–139

Saravanamoorthy MD, Kumari BR (2007) Effect of textile waste water on morphophysiology and yield on two varieties of peanut (*Arachis hypogaea* L.). J Agric Technol 3(2):335–343

Sato T, Qadir M, Yamamoto S, Endo T, Zahoor A (2013) Global, regional, and country level need for data on wastewater generation, treatment, and use. Agric Water Manag 130:1–13

Sawyer CN, McCarty PL, Parkin GF (1994) Chemistry for environmental engineering and science. McGraw-Hill, New York, Vol. 5, p. 587590

Scott CA, Zarazúa JA, Levine G (2000) Urban-wastewater reuse for crop production in the water-short Guanajuato river basin, Mexico, vol 41. IWMI

Scott CA, Drechsel P, Raschid-Sally L, Bahri A, Mara D, Redwood M, Jiménez B (2010) Wastewater irrigation and health: challenges and outlook for mitigating risks in low-income countries. In: Wastewater irrigation and health: assessing and mitigating risk in low-income countries, pp 381–394

Segal E, Dag A, Ben-Gal A, Zipori I, Erel R, Suryano S, Yermiyahu U (2011) Olive orchard irrigation with reclaimed wastewater: agronomic and environmental considerations. Agr Ecosyst Environ 140(3–4):454–461

Sengupta S, Nawaz T, Beaudry J (2015) Nitrogen and phosphorus recovery from wastewater. Curr Pollut Rep 1:155–166

Sharma VK, Kansal BD (1984) Effect of nitrogen, farm yard manure, town refuse and sewage water on the yield and nitrogen content of maize fodder and spinach

Sharma N, Singhvi R (2017) Effects of chemical fertilizers and pesticides on human health and environment: a review. Int J Agric Environ Biotechnol 10(6):675–680

Shen Y, Oki T, Kanae S, Hanasaki N, Utsumi N, Kiguchi M (2014) Projection of future world water resources under SRES scenarios: an integrated assessment. Hydrol Sci J 59(10):1775–1793

Shende A (1985) Idiopathic thrombocytopenic purpura in children. Pediatr Ann 14(9):609–616

Shetty A, Goyal A (2022) Total organic carbon analysis in water—a review of current methods. Mater Today Proc 65:3881–3886

Singh KP, Mohan D, Sinha S, Dalwani R (2004) Impact assessment of treated/untreated wastewater toxicants discharged by sewage treatment plants on health, agricultural, and environmental quality in the wastewater disposal area. Chemosphere 55(2):227–255

Singh R, Gautam N, Mishra A, Gupta R (2011) Heavy metals and living systems: an overview. Indian J Pharmacol 43(3):246–253

Singh PK, Deshbhratar PB, Ramteke DS (2012) Effects of sewage wastewater irrigation on soil properties, crop yield and environment. Agric Water Manag 103:100–104

Singh A, Sawant M, Kamble SJ, Herlekar M, Starkl M, Aymerich E, Kazmi A (2019) Performance evaluation of a decentralized wastewater treatment system in India. Environ Sci Pollut Res 26:21172–21188

Singh G, Nagora PR, Haksar P, Rani A (2022) Species influenced growth, biomass allocation and productivity in wastewater irrigated plants in sandy soils of Indian desert. Irrig Sci 40(6):829–843

Singh RR, Singh VS, Shukla AK (1991) Yield and heavy metal contents of berseem as influenced by sewage water and refinery effluent

Smit J, Nasr J (1992) Urban agriculture for sustainable cities: using wastes and idle land and water bodies as resources. Environ Urban 4(2):141–152

Smith RG (1995) Water reclamation and reuse. Water Environ Res 67(4):488–495

Sreeramulu US (1994) Utilisation of sewage and sludge for increasing crop production. J Indian Soc Soil Sci 42(4):525–532

Surinaidu L, Gupta PK, Ahmed S, Hussain M, Nandan MJ (2023) Impact of urban wastewater reuse for irrigation on hydro-agro-ecological systems and human health risks: a case study from Musi river basin, South India. Hydro Res 6:122–129

Tanji KK (1997) Irrigation with marginal quality waters: issues. J Irrig Drain Eng 123(3):165–169

Tanji KK, Kielen NC (2002) Agricultural drainage water management in arid and semi-arid areas, no 61, pp xiv+-188

Tiwari RC, Kumar A, Mishra AK (1996) Influence of treated sewage and tubewell water irrigation with different fertilizer levels on rice and soil properties. J Indian Soc Soil Sci 44(3):547–549
Topare NS, Attar SJ, Manfe MM (2011) Sewage/wastewater treatment technologies: a review. Sci Rev Chem Commun 1(1):18–24
Tran LD, Phung LD, Pham DV, Pham DD, Nishiyama M, Sasaki A, Watanabe T (2019) High yield and nutritional quality of rice for animal feed achieved by continuous irrigation with treated municipal wastewater. Paddy Water Environ 17:507–513
Tsadilas CD, Vakalis PS (2003) Economic benefit from irrigation of cotton and corn with treated wastewater. Water Sci Technol Water Supply 3(4):223–229
Udayasoorian C, Devagi P, Ramaswami PP (1999) Case study on the utilization of paper and pulp mill effluent irrigation for field crop. In: Proceedings of workshop on 'bioremediation of polluted habitat's, pp 71–73
Uddin MJ, Khanom S, Al Mamun S, Parveen Z (2015) Effects of irrigation water on some vegetables around industrial areas of Dhaka, Bangladesh. J Sci Res 28(2):151–159
UNESCO (United Nations Educational, Scientific and Cultural Organization) (2017) The United Nations world water development report 2017. UNESCO, Paris, France
Ungureanu N, Vlăduț V, Dincă M, Zăbavă BȘ (2018) Reuse of wastewater for irrigation, a sustainable practice in arid and semi-arid regions. In: Proceedings of the 7th international conference on thermal equipment, renewable energy and rural development (TE-RE-RD), Drobeta-Turnu Severin, Romania, vol 31, pp 379–384
United Nations, World Water Development Report (2014) Water and energy. World water assessment programme, vol 1. UNESCO, pp 1–100
Urbano VR, Mendonça TG, Bastos RG, Souza CF (2017) Effects of treated wastewater irrigation on soil properties and lettuce yield. Agric Water Manag 181:108–115
USEPA (1973) Water quality criteria. National Academy of Sciences Report to the United States Environmental Protection Agency. USEPA, Washington DC, pp 350–366
USEPA (2010) Guidelines for water reuse. United States Environment Protection Agency, EPA/600/R-12/618
USEPA (2011) National recommended water quality criteria. United States Environmental Protection Agency, Office of Water, EPA 822-R-02-047
Uyttendaele M, Jaykus LA, Amoah P, Chiodini A, Cunliffe D, Jacxsens L, Holvoet K, Korsten L, Lau M, McClure P, Medema G (2015) Microbial hazards in irrigation water: standards, norms, and testing to manage use of water in fresh produce primary production. Compreh Rev Food Sci Food Safety 14(4):336–356
Vanham D, Alfieri L, Flörke M, Grimaldi S, Lorini V, de Roo A, Feyen L (2021) The number of people exposed to water stress in relation to how much water is reserved for the environment: a global modelling study. Lancet Planet Health 5(11):e766–e774
Vazquez-Montiel O, Horan NJ, Mara DD (1996) Management of domestic wastewater for reuse in irrigation. Water Sci Technol 33(10–11):355–362
Vergine P, Lonigro A, Salerno C, Rubino P, Berardi G, Pollice A (2017a) Nutrient recovery and crop yield enhancement in irrigation with reclaimed wastewater: a case study. Urban Water J 14(3):325–330
Vergine P, Salerno C, Libutti A, Beneduce L, Gatta G, Berardi G, Pollice A (2017b) Closing the water cycle in the agro-industrial sector by reusing treated wastewater for irrigation. J Clean Prod 164:587–596
Vivaldi GA, Stellacci AM, Vitti C, Rubino P, Pedrero F, Camposeo S (2017) Nutrient uptake and fruit quality in a nectarine orchard irrigated with treated municipal wastewaters. Desalin Water Treat 71:312–320
Vivaldi GA, Camposeo S, Lopriore G, Romero-Trigueros C, Salcedo FP (2019) Using saline reclaimed water on almond grown in Mediterranean conditions: deficit irrigation strategies and salinity effects. Water Supply 19(5):1413–1421
Von Sperling M (2007) Activated sludge and aerobic biofilm reactors. IWA Publishing

Vymazal J (2009) The use constructed wetlands with horizontal sub-surface flow for various types of wastewater. Ecol Eng 35(1):1–17
Wallace JS (2000) Increasing agricultural water use efficiency to meet future food production. Agr Ecosyst Environ 82(1–3):105–119
Wang HJ, Jingjing WANG, Xiaohua YU (2022) Wastewater irrigation and crop yield: a meta-analysis. J Integr Agric 21(4):1215–1224
Watts J (2018) Water shortages could affect 5bn people by 2050, UN report warns. The Guardian, p 19
World Health Organization. Meeting of Experts on the Reuse of Effluents, Methods of Wastewater and Health Safeguards (1973) Reuse of effluents: methods of wastewater treatment and health safeguards, report of a WHO meeting of experts, no 517
World Health Organization ed (2008) World health statistics 2008. World Health Organization
World Health Organization (1989a) Health guidelines for the use of wastewater in agriculture and aquaculture. Technical Report Series 778, Geneva
World Health Organization (1989b) Health guidelines for the use of wastewater in agriculture and aquaculture: Report of a WHO Scientific Group. WHO Technical Report Series 778. World Health Organization, Geneva, Switzerland, 74 pp
World Health Organization (2006) WHO guidelines for the safe use of wasterwater excreta and greywater, vol 4. World Health Organization
World Wildlife Fund (2007) Report on national surface water classification criteria, irrigation water quality guidelines for Pakistan. Waste Water Forum Pakistan
Yang J, Jia RS, Gao YL, Wang WF and Cao, P.Q., 2017, December. The reliability evaluation of reclaimed water reused in power plant project. In: IOP conference series: earth and environmental science, vol 100, no 1. IOP Publishing, p 012189
Younas H, Younas F (2022) Wastewater application in agriculture-a review. Water Air Soil Pollut 233(8):329
Zalawadia NM, Patil RG, Raman S (1996) Effect of distillery waste water with fertilizer on onion and soil properties. J Indian Soc Soil Sci 44(4):802–804
Zhang Y, Shen Y (2019) Wastewater irrigation: past, present, and future. Wiley Interdiscip Rev Water 6(3):e1234
Zhao Y, Zhuang X, Ahmad S, Sung S, Ni SQ (2020) Biotreatment of high-salinity wastewater: current methods and future directions. World J Microbiol Biotechnol 36:1–11

Chapter 6
Contaminant Flow from Sewage and Leaching Potential of Sludge

Sweety Nath Barbhuiya, Dipanwita Chakraborty, and Dharmeswar Barhoi

Abstract The generation and management of sewage sludge are critical to wastewater treatment strategies. This chapter highlights how improper disposal of sludge can release toxic substances over time, that can pose substantial risks to varied ecological landscapes and human health. Through analysis of sludge composition and its influence on environmental and public health, the chapter further highlights the treatment methods and management approaches to minimize contaminant mobility into ecosystems and safeguard ecological integrity as well as public health.

Keywords Contaminant · Heavy metal · Sewage · Sludge · Wastewater

6.1 Introduction

Sewage sludge is produced simultaneously during the treatment of wastewater generated from domestic and industrial sources at wastewater treatment plants (WWTPs) via the sewer system (Fini et al. 2024). Its characteristics can differ widely, depending on various factors including the level of industrial development, population size, per capita water usage, seasonal variations, and other local conditions (Qadir et al. 2020). Wastewater entering WWTPs originates not only from households but also from hospitals and industries such as breweries, dairy processing facilities, and paper manufacturing plants, fertilizer and pesticide industries etc. (Aski et al. 2020). These industrial contributions lead to high levels of polymeric substances such as lipids, proteins, cellulose, and other polysaccharides in both wastewater and the resulting sludge. Since the sludge is generated during wastewater treatment, its composition is directly influenced by the nature of the incoming wastewater. Consequently, it can contain a diverse range of biomolecules such as lipids, pharmaceuticals and

S. N. Barbhuiya
Patharkandi College, Patharkandi, Sribhumi, Assam, India

D. Chakraborty · D. Barhoi (✉)
The Assam Royal Global University, Guwahati, Assam, India
e-mail: barhoidharamesh@gmail.com

S. Madhav et al. (eds.), *Municipal Sewage and Sludge: Treatment and Disposal Strategies*, Environmental Science and Engineering,
https://doi.org/10.1007/978-3-032-07597-0_6

their metabolites, pathogenic elements as well as heavy metals, both in qualitative and quantitative terms (Ćwiertniewicz-Wojciechowska et al. 2023). Being rich in context to nutrient profile, sludge posits as an appropriate raw-material in agricultural practices (Marin and Rusănescu 2023). However, the occurance of heavy metals in sewage sludge poses long-term risks to soil quality, agricultural productivity, and other environmental implications (Tytła 2019; Geng et al. 2020). Sewage sludge also contains diverse microorganisms, including heterotrophs and chemautotrophs, saprophytic organisms and other disease-causing organisms. Common bacterial genera include *Gemmatimonas*, *Nitrosomonas*, and *Thermomonas*, while viral pathogens like Coronavirus HKU1, Adenovirus, and Hepatitis A virus have also been detected (Fijalkowski et al. 2017).

Therefore, sewage sludge, if left untreated, can cause significant health implications because of the occurrence of pathogenic substances, heavy metals, and other pollutants such as parabens and microplastics, among others, that are largely sourced from domestic, industrial, and surface wastewater (Gogina et al. 2024). Therefore, advancement in sludge treatment strategies is crucial to mitigate the contaminant flow into ecosystems, thereby fostering circular economy, safeguarding human health, and improving energy utilization. These initiatives not only enhance resource efficiency and minimize environmental degradation but also strengthen sustainable agricultural practices and contribute to global sustainability targets (Grobelak et al. 2024).

Sewage sludge can be categorised as primary or secondary. Primary sludge is generated during the initial treatment stage, where a settling tank is used to separate and remove sedimented solids from sewage. Despite this initial treatment, the wastewater retains biodegradable matter that needs to be eliminated to prevent pollution in aquatic systems. To address this, the activated sludge process is implemented, thereby decreasing biochemical oxygen demand. However, this process causes an increased accumulation of waste matter, referred to as or secondary sludge (Woo et al. 2022). Primary and secondary sludge are still microbially active and hence referred to as crude (Salva et al. 2025). Globally, sewage sludge is managed by reducing the moisture level using dewatering methods or processes such as thickening and thermal drying (Teoh and Li 2020) or stabilized to reduce the organic content through breakdown of organic matter in absence or presence of oxygen, precisely, anaerobic digestion or composting, respectively or incineration (Fang et al. 2019) followed by its disposal as landfill leachate (Tang et al. 2022). However, majority of these processes require energy input and contribute to rise in temperature across the globe, changes in soil pH, nutrient enrichment in freshwater ecosystem, exposure of carcinogenic substances to humans, and depletion of fossil fuels (Chen et al. 2022; Mayer et al. 2021). Furthermore, sewage sludge treatment accounts for approximately half of the overall operational expenditure of sewage treatment plants, challenging proper handling and disposal (Liu et al 2020).

6.2 Global Scenario of Sewage Sludge Generation

Driven by population explosion, expanding urbanization and industrial growth as well as ongoing technological advancements, the production of sewage sludge on a global scale continues to rise annually (Janaszek and Kowalik 2023). Globally, municipal WWTPs generate 45 million tons (Mt) of sludge per year (Ferrentino et al. 2023). Across the 27 countries in European Union (EU-27), the combined annual production of sewage sludge compounds to 9.25 Mt (Feng et al. 2023). North America is estimated to produce about 85 km^3 of sewage yearly, whereas the Middle East and North Africa produce about 22.3 km^3 of sewage annually from a variety of sources (Zhang et al. 2017). As of 2020, the approximate volume of municipal wastewater produced was estimated to be between 360 and 380 km^3 (Macedo et al. 2022). However, according to recent estimates, sludge production is likely to increase by 24% by 2030 and 51% by 2050 (Qadir et al. 2020). North America, Europe, and, East Asia, also substantially contribute to global sludge production (Shaddel et al. 2019) and estimates suggest that sludge production in Southeast Asian countries could reach 20–24 Mt/year by 2050 (Quan et al. 2022). Bangkok and Thailand accounts for 30,000–350,000 m^3 of municipal sludge per day (Miyaoka et al. 2017) (Fig. 6.1).

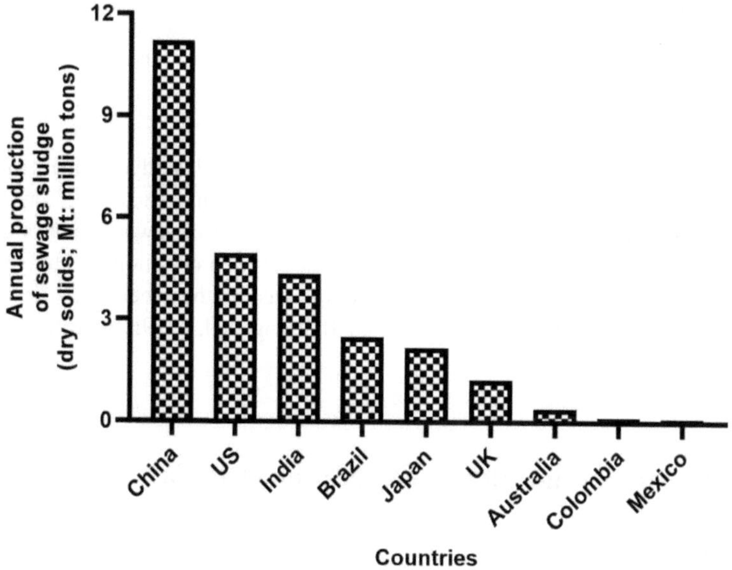

Fig. 6.1 The yearly production of sewage sludge, expressed as dry solids in million tons (Mt) varies worldwide with highest generation compounding to 11.2 Mt in China followed by 4.96 Mt in the United States, 4.36 Mt in India, 2.5 Mt in Brazil, 2.2 Mt in Japan, 1.25 in the United Kingdom and 0.41 Mt in Australia. Further, a very low annual production of sludge has been reported from Colombia (0.15 Mt) and Mexico (0.11 Mt) (Feng et al. 2023)

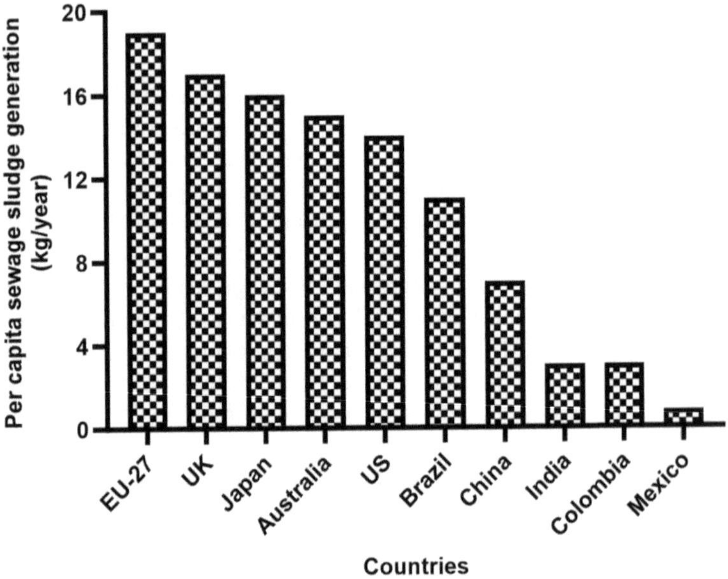

Fig. 6.2 According to reports, the European Union (EU-27) produces the most sewage sludge per capita, with an annual output of 19 kg of dry solid. This is followed by the United Kingdom at 17 kg/year, Japan at 16 kg/year, Australia at 15 kg/year, and the United States at 14 kg/year. Brazil has been reported for a per capita sludge generation of 11 kg/year, followed by China at 7 kg/year. Both India and Colombia generate 3 kg of dry solid yearly, while Mexico has recorded the lowest per capita sludge generation at just 0.8 kg/year (Feng et al. 2023)

Particularly, around 42 and 56 percent of the global sludge come from the BRICS nations, Brazil, Russia, India, China, and South Africa. Further, within Asia, China unaccompanied, produces a greater volume of sewage, compounding to 58.9 km^3 annually (Nguyen et al. 2021). In 2019, production of sludge from municipal waste reached 11.75 Mt in China, followed by the generation of activated sludge exceeding 36 Mt in 2020 (Li et al. 2020; Liang et al. 2019; Zhang et al. 2023) (Fig. 6.2).

6.3 State of Sewage Sludge Generation in India

Approximately 38,354 million litres of wastewater are reported to be released from urban areas across India every day, making it the key contributor of freshwater pollution in the country (CPCB 2009). However, only about 11,786 million litres is treated, while the remaining is discharged without treatment, thereby causing serious pollution to aquatic ecosystems and their catchment zones. Also, only 30% of the 0.04 km^3 of sewage produced by Class I and Class II cities in India is connected to treatment facilities (Thakur et al. 2022).

6.4 Methods of Wastewater Treatment

WWTPs constitute a series of procedures that are used to remove various contaminants from wastewater into a suitably purified state through systematic elimination of the diverse pollutants (Kim et al. 2019; Ahkola et al. 2021). This typically involves three sequential stages in large-scale setups. The first stage involves the use of physical techniques to remove solid materials, followed by chemical and/or biological treatment of wastewater at second stage in bioreactors with activated sludge suspended in it. The final or tertiary stage comprises the disinfection of water to ensure the treated water meets safety standards prior to discharge (Begmatov et al. 2022).

6.5 Physical Method

The primary stage of treatment focuses on eliminating large, suspended, and floating solids from untreated wastewater. This involves mechanical processes such as screening, which captures coarse debris, and sedimentation to remove suspended solids. While primarily mechanical in nature, this step may occasionally involve the use of chemicals to enhance the sedimentation rate. As a result, the biochemical oxygen demand of the wastewater as well as the amount of total suspended solids is typically reduced by nearly 30% and 60%, respectively (Kesari et al. 2021).

6.6 Chemical Method

Chemical treatment, often integrated with other physical and biological processes, utilizes chemicals to increase settleability of pollutants, degradation of organic toxins, pH changes, and rate of decontamination.

6.7 Chemical Precipitation

This technique eliminates dissolved contaminants, particularly heavy metals by converting them into insoluble solid particles (Yadav et al. 2019). It forms solid precipitates by employing counterions to decrease the solubility of dissolved ions (Zueva 2018). Following their separation using techniques like filtration or sedimentation, flocculants may be used to improve the settleability of the precipitate (Ojovan and Lee 2014).

Among these techniques, hydroxide precipitation is the most common, which involves addition of alkaline substances like calcium hydroxide, commonly called lime and sodium hydroxide, preferably, to form metal hydroxides (Dahman 2017). Lime is economically effective and hence preferred over other alternatives in large-scale wastewater treatment processes (Zueva 2018). However, the efficiency of metal removal depends on pH, which can be challenging due to the amphoteric nature of metal hydroxides that may resolubilize at non-optimal pH levels.

$$M^{n+} + OH^- \leftrightharpoons M(OH)_n \downarrow$$

Sulfide precipitation is another effective method, where sulfide ions, derived from compounds like Na_2S, H_2S or CaS, react with dissolved metals to produce metal sulfide precipitates (Estay et al. 2021).

$$M^{n+} + S^{2-} \leftrightharpoons MS \downarrow$$

Further, carbonate precipitation, using agents like sodium carbonate or calcium carbonate, is also employed in removal of metals by forming metal carbonates (Zueva 2018).

$$M^{n+} + CO_3^{2-} \leftrightharpoons MCO_3 \downarrow$$

6.8 Coagulation and Flocculation

Flocculation is frequently misunderstood as being identical to coagulation, though they are distinct and independently occurring processes. Coagulation involves the initial aggregation of particles, where colloidal matter becomes destabilized due to a reduction in electrokinetic potential. Flocculation, in contrast, intensifies sedimentation by promoting the bonding of these destabilized particles into larger flocs (Maćczak et al. 2020). Compounds like aluminum sulfate, polyaluminum chloride, and ferric chloride function in the coagulation process by counteracting the electrical charges on suspended particles, thereby initiating their destabilization. In contrast, flocculants such as polyacrylamide and chitosan assist by linking these destabilized particles, causing the formation of denser flocs that can be easily eliminated. The overall efficiency of this process is influenced by various environmental conditions, including pH levels, dosage of coagulating or flocculating agents and their physical properties such as their particle size and density, mixing speed and duration, and temperature, notably (Badawi et al. 2023). Among commercial flocculants, polyacrylamide and its derivatives are widely used due to the high reactivity of acrylamide, which readily undergoes radical polymerization to form ultra-heavy polymers. In addition to polyacrylamide, non-ionic flocculants may be derived from polyethylene oxide, polyvinyl alcohol, and polyvinylpyrrolidone. Common cationic flocculating agents include polydiallyldimethylammonium chloride, cationic polyacrylamide, and polyethyleneimine. Anionic flocculants typically consist of homopolymers or copolymers of AM with acrylic acid salts. Further, amphoteric flocculants, which possess both cationic and anionic groups, have gained attention as effective agents in flocculation processes (Dao et al. 2015). At present, a there is an evident shift towards the usage of eco-friendly coagulants as promising alternatives for wastewater treatment. Among these, biopolymers, for instance, natural organic polymers derived from plants, animals, or microorganisms, are non-toxic to humans and biodegradable, making them safer and more sustainable than conventional chemical coagulants (Saritha et al. 2014).

6.9 Solvent Extraction

A typical solvent extraction system involves transferring the target solute from untreated wastewater into a solvent, producing two streams, an extract which is rich in solute and a raffinate which is depleted of solute. The extract is then treated with a stripping agent to recover the solute, resulting in its separation from the solvent. In the meantime, the solvent that is used gets separated from the water by a process called solvent recovery. The solvent that has been collected from both streams is recycled for further use in the solute removal phase. If the solute is a metal ion, it can be recovered by electrolysis, precipitated into an insoluble form, or put back into the production process (Hannula et al. 2019). Techniques such as filtration using membranes (Goh and Ismail 2018), exchange of ions (Levchuk et al. 2018), adsorptive methods (Krstić et al. 2018), or oxidation (Lin et al. 2018) might be used to further treat or discard any leftover water. These processes are typically carried out using equipment such as mixer-settlers, structured units such as packed columns, rotating disc systems, and reciprocating plate setups for distillation (Gai et al. 2018) or gas stripping (Pirmoradi and Ashrafizadeh 2017) resulting in solvent recovery.

6.10 Electrochemical Methods

Electrochemical methods for wastewater treatment offer several benefits compared to traditional techniques with respect to high efficiency in removing pollutants, elimination of the need for additional chemicals while processing, insensitivity to toxic substances, and typically little to no generation of secondary waste (Sillanpää and Shestakova 2017a). The widely utilized techniques under this spectrum oxidation, which is applied for the conversion of organic pollutants to inorganic forms, microbial inactivation, and elimination of sulfide or cyanide rich compounds. Additionally, reduction supports both the retrieval of metals as well as the conversion of stable toxins into less toxic derivatives. Other notable methods include electrocoagulation and electroflotation which are employed to remove suspended solids; and electrodialysis, which is primarily used for desalination purposes (Sillanpää and Shestakova 2017b).

6.10.1 Electrochemical Oxidation

Electrochemical oxidation is one of the advanced methods compared to other wastewater treatment strategies and requires electricity to operate (Mirshafiee et al. 2024). Electrochemical oxidation (EO) operates through anodic oxidation, often referred to as direct electro-oxidation or indirect electro-oxidation. In direct EO, pollutant degradation occurs via transfer of electrons directly to the anodic surface from the contaminants. In contrast, in the other mode of EO, i.e., indirect EO, organic pollutants react with oxidizing agents such as chlorine (Cl_2), hydrogen peroxide (H_2O_2), hypochlorous acid (HClO), hypochlorite (ClO^-), sulfate radicals (SO_4^{2-}), ozone (O_3), and others produced during electrolysis, resulting in breakdown of organic toxins in the solution (Najafinejad et al. 2023).

6.10.2 Electrochemical Reduction

Electrodeposition at cathode or electrochemical reduction, is widely used in wastewater treatment processes (Malinović et al. 2022). Further, this technique bears efficacy for eliminating various inorganic and organic toxins with minimal by-product formation—for instance, removal of chloride ions (Lei et al. 2019) as well as reduction of nitrate and nitrite (Kim et al. 2017).

6.10.3 Electrocoagulation and Electroflotation

In electrocoagulation and electroflotation, application of electric current to anode and cathode causes generation of metal cations on the surface of anode while hydroxide ions (OH^-) and hydrogen (H_2) gas on the cathodic surface, the combination of which produces metal hydroxides. which destabilize and aggregate contaminants into flocs, which are then removed by flotation or sedimentation (Kato and Kansha 2024). Metal hydroxides neutralize charged pollutants, which are then assimilated on the surface by sweep coagulation causing the formation of flocs that are removed either by rising with hydrogen bubbles or sedimentation (Emamjomeh and Sivakumar 2009; Das et al. 2022). Similar to the typical coagulation–flocculation process, electrocoagulation can effectively eliminate a range of pollutants from wastewater, for instance, suspended organic and inorganic constituents, oils and heavy metals as well as influence parameters such as biochemical oxygen demand, clarity of water, etc. (Das et al. 2022). This process is efficient at removing pollutants from wastewater generated from various sources, including breweries (Papadopoulos et al. 2020), spaces of automobile repair (Manilal et al. 2017), hospitals (Ebba et al. 2021), paper and pulp manufacturing (Mahesh et al. 2016) as well as mining operations (Touahria et al. 2016).

6.10.4 Electrodialysis

Electrodialysis eliminates pollutants from wastewater using electrically driven ion exchange membranes, where monovalent-selective membranes separate ions based on charge, and bipolar membranes produce H^+ and OH^- ions on the either side of membrane through breakdown of water molecule under the influence of electric current (Costa 2023). Electrodialysis enables the elimination of trace elements by driving the dissolved ions through selectively permeable membranes subjected to electric field (Wang et al. 2021).

6.11 Biological Method

Biological treatment strategies depend majorly on microbial potential of degrading organic pollutants in untreated wastewater (Madondo et al. 2021). These processes are economically effective, compatible towards varied environmental conditions as well as efficient in handling a variety of unwanted substances from wastewater (Martins et al. 2024; Raffa and Chiampo 2021; Sravan et al. 2024). Biological treatment approaches are either aerobic or anaerobic. In aerobic treatment, microbial activity requires oxygen to breakdown organic contaminants into carbon dioxide, water, and biomass which is typically introduced through mechanical agitation. In contrast, anaerobic treatment occurs in absence of oxygen where microbes degrade pollutants through a series of sequential steps such as breakdown of water molecules on application of electric current, followed by formation of fatty acids, acetic acid and ultimately methane, resulting in the formation of biogas (Goswami et al. 2016; Prabhu et al. 2021).

6.12 Aerobic Digestion

Aerobic treatment systems include a variety of processes such as activated sludge, aerated lagoons, trickling filters, and rotating biological contactors, among others (Nguyen et al. 2021).

6.12.1 Activated Sludge

This approach is widely used to remove of organic pollutants from wastewater (Islam and Mahdi 2022). This method involves the use of suspended microorganisms within aeration tanks, where they degrade organic contaminants and assimilate nutrients. After aeration, the treated water is passed into sedimentation tanks, where biomass settles out. A little fraction of this sedimented solid is recycled back into the aeration tank which serves as inoculum to preserve microbial community, while the remaining portion is removed. Following this, the supernatant in the sedimentation tank may either be released directly or subjected to additional treatment before being released (Bajpai 2017; Kato and Kansha 2024).

6.12.2 Aerated Lagoons

Aerated lagoons are biological treatment systems employed for managing wastewater. These systems operate as aerobic suspended growth processes and bear robustness against sudden loads and relatively simple construction (Godini et al. 2021). In aerated lagoons, mechanical aeration is provided by surface aerators that remain floating or submerged diffusers. These systems are typically categorized into two types, viz., complete-mix aerated lagoons, which demand substantial energy input to maintain suspended solids and partial-mix lagoons, which supply oxygen but lack the turbulence needed to keep solids fully suspended (Alvarado et al. 2013).

6.12.3 Trickling Filters

Trickling filters operate by distributing wastewater over solid media such as rocks or plastic, to serve as a surface for microbial growth and biofilm formation. Aerobic conditions are maintained through natural airflow, which may be upward or downward, based on temperature and humidity gradients between the filter and its surroundings. Alternatively, machine-driven ventilation systems with low-pressure blowers can be employed to maintain a uniform direction of air current (Kato and Kansha 2024).

6.12.4 Rotating Biological Contactors

Rotating biological contactors are simple and low energy consuming systems that are utilized in the treatment of wastewater in diverse settings, varying from decentralized units to municipal treatment facilities because of their high efficiency in eliminating organic pollutants and elements like nitrogen or phosphorus. These systems are valued for their durability and capacity to withstand significant variations in influent quality. As an attached-growth biological treatment method, this system facilitates colonization of microbes by use of rotating discs and enhances elimination of organic toxins and other nutrients (Waqas et al. 2023).

6.13 Anaerobic Digestion

This method is widely regarded as an economically effective and ecologically sustainable method to treat sewage sludge (Nasir et al. 2012). Under anaerobic conditions, organic compounds are biologically degraded into energy-rich biogas, predominantly composed of methane and other by-products (Funmi et al. 2021). Globally, more than 1,300 anaerobic digestion facilities are either operational or being developed (Hanum et al. 2019). Germany leads in annual biogas output with approximately 87 TWh, followed closely by China at around 81 TWh. Further, recent trends in regions like Europe, Brazil, and North America show a growing emphasis on upgrading biogas to biomethane for injection into gas grids or as a transportation fuel (Gustafsson et al. 2024).

6.13.1 Upflow Anaerobic Sludge Blanket (UASB)

In this type of reactor, wastewater flows from the base in upward direction through granulated sludge blanket. Biogas, produced during anaerobic digestion supports the development and stability of these granular particles (Daud et al. 2018). Sludge retention in these systems depends on the formation of sedimented sludge aggregates as well as the use of a gas–liquid-solid separation device with a reverse funnel-shaped design (van Lier et al. 2015).

6.14 Composition of Sewage Sludge

Sewage sludge may include a substantial concentration of both organic and inorganic constituents, a rich nutrient profile as well as potentially hazardous compounds (Ekka et al. 2024; Uddin et al. 2025). The organic macromolecular compounds that vary in concentration and composition, may accumulate in sludge because of their lipophilic nature and poor ability of being degraded during the wastewater treatment processes (Wluka et al. 2021). Pollutants, including substances like polychlorinated biphenyls (PCBs), polycyclic aromatic hydrocarbons (PAHs), dioxins, furans, organochlorine pesticides, chlorinated chemicals, phenolic compounds, phthalates, among others, are some notable organic toxins, encountered in sludge (Uddin et al. 2025). Among these, the most commonly detected organic substances are absorbable organic halogens, linear alkylbenzenesulfonates, nonylphenols and their ethoxylated forms, di-ethylhexylphthalate, PAHs, PCBs, as well as dioxins and furans (Kapanen et al. 2013). Further, sewage sludge also includes inorganic pollutants, predominantly heavy metals (Qasem et al. 2021). The nutrient profile of sewage sludge incorporates both macronutrients and micronutrients (Feng et al. 2023). Also, elevated levels of nutrients and heavy metals have also been documented in sludge generated from

Table 6.1 Composition of sewage sludge in India

	Elements/compounds	Concentration	Source
Heavy metals	Arsenic	8–23 mg/kg	Dubey et al. (2006)
	Cadmium	2–9 mg/kg	
	Chromium	66–1098 mg/kg	
	Mercury	7–32 mg/kg	
	Nickel	12–596 mg/kg	
	Lead	26–154 mg/kg	
Nutrients	Nitrogen	4600–6300 mg/kg	
	Phosphorous	44–60 mg/kg	
	Potassium	290–410 mg/kg	
Organic compounds	Polychlorinated biphenyls	497–800 μg/kg	Patel et al. (2015)
	Polycyclic aromatic hydrocarbons	2110–2330 μg/kg	Gupta (2018)

municipal treatment plants (Kumar and Chopra 2013; Kumar et al. 2016). In present times, endocrine-disrupting compounds, persistent organic pollutants, microplastics, as well as pharmaceutical residues and personal care products, that are potentially toxic, have been detected in various environmental matrices, including water, soil, and air due to landfill leachate (He et al. 2024) (Table 6.1).

6.15 Environmental and Health Implications of Contaminants from Sewage

Wastewater commonly contains various chemical pollutants, including heavy metals, nitrogen-based compounds, pesticides, pharmaceutical residues, detergents, and other organic as well as inorganic compounds (Agoro et al. 2020). But, unlike other pollutants, heavy metals pose a significant risk as these do not degrade over time, leading to soil contamination that affects agricultural yield and promotes biomagnification through trophic transfer, ultimately disrupting environmental balance (Angulo-Bejarano et al. 2021).

The continuous input of both essential and non-essential elements into the food chain is largely driven by the overuse of agro-based chemicals, municipal and industrial discharges, and untreated sewage for land application. This further challenges the well-being of living systems, as these contaminants can migrate from polluted soil and water sources into the food chain, causing potential toxicity (Mitra et al. 2022). The use of sewage sludge in agricultural setups helps recycle nutrients like nitrogen and phosphorus while enhancing soil quality and crop productivity. However, despite its benefits, this practice may lead to medium and long-term implications on the trophic levels and soil contamination that threatens essential ecosystem functions and human well-being (Milinovic et al. 2014; Santana et al. 2021).

Impact of sewage on environmental matrices

Wastewater irrigation in agriculture poses significant environmental risks, particularly soil health. It alters soil sustainability by affecting fertility, productivity, and microbial composition. Heavy metals from wastewater, accumulate in soil and disrupt organic matter decomposition, enzyme activity, and microbial balance. Globally, around 20 million hectares across 50 countries or 10% of irrigated farmland receive untreated or inadequately treated wastewater. In India, urban farms often rely on untreated sewage due to limited treatment infrastructure or distant farmlands. While trace heavy metals are essential for ecosystems, excessive concentrations pose severe ecological threats, endangering biota and ecosystem stability (Ghanem et al. 2023).

WWTPs serve as major entry points for toxic chemicals into freshwater ecosystems, subjecting aquatic systems to continuous micropollutant exposure. Beyond chemical contamination, WWTPs can significantly modify stream flow and temperature regimes, especially when wastewater constitutes a substantial portion of river flow (Rani et al. 2022). These changes create environmental boundaries, often leading to species turnover and biodiversity loss. Further, the decline of local biotic communities, partly driven by chronic chemical exposure, disrupts ecological interactions and endangers vital services such as pollination, nutrient cycling, and pest control (Enns et al. 2023).

Coral reefs, oyster reefs, and salt marshes provide essential coastal protection and biodiversity but are increasingly threatened by sewage pollution. Coral reefs suffer from inhibited growth, disease, bleaching, and even death due to excess nutrients, which lead to algal overgrowth and oxygen-depleted dead zones. Additionally, sewage introduces heavy metals and endocrine disruptors that further harm reef ecosystems. Oyster reefs, crucial for water filtration, are also impacted, as contaminated waters expose oysters to toxins, heavy metals, and pharmaceuticals, affecting their reproduction and the broader marine food web. Salt marshes, once considered resilient, experience structural degradation from nutrient pollution, leading to plant die-off, erosion, and biodiversity loss. Addressing sewage pollution is vital to safeguarding these ecosystems and their ecological functions (Wear et al. 2021).

Pharmaceuticals can affect living organisms across various ecosystems, however, the level of exposure through these pathways across many taxonomic groups and ecosystem is still largely unknown (Carter et al. 2020). Drugs like diclofenac and ibuprofen that may pass into the soil either remain unchanged and move through the soil or undergo transformations before reaching groundwater (Xu et al. 2010). Carbamazepine is another pharmaceutical known for its high persistence, as it resists breakdown in conventional biological wastewater treatment systems (Martín et al. 2012). Hormones and endocrine-disrupting chemicals have also drawn concern, particularly due to their role in causing feminization in fish (Fent et al. 2006). These pollutants are now focal areas of discussion because of their continuous release into the environment, potential long-term damage to the range of environmental matrices, as well as the increasing rates at which they are being produced and used.

Human health risks

Of all toxic forms, inorganic pollutants, being non-biodegradable, possess the tendency to persist in the ecosystem and, by making their way through the food chain, lead to its bioconcentration. Among these, heavy metals, characterised by a density exceeding 4.5 g/cm^3, are particularly significant and greatly concerning in the context of human health (Kabata-Pendias 2010; Fijalkowski et al. 2017). Human exposure to heavy metals, via breathing of toxic fumes, accidental consumption of tainted food and water, or contact through skin, can result in the bioaccumulation of such toxins in fatty tissues, adversely affecting the nervous and endocrine systems, as well as immunological and hematopoietic function (Tytła and Widziewicz-Rzońca 2023; Yakamercan and Aygün 2023). Further, heavy metals may activate signalling cascades that are critical in the regulation of different cellular processes, such as immune response, malignant transformation, and apoptosis, among others. Disruptions in these pathways can promote cellular dysfunction that are linked to the development of chronic conditions such as diabetes, cardiovascular problems, neurodegenerative conditions, carcinogenesis, disruption in endocrine and reproductive functions as well as conditions involving inflammation and allergic responses (Haidar et al. 2023). The interaction of heavy metals with biological systems, commonly occur through generation of metal cations, which have a high affinity for nucleophilic sites on vital macromolecular compounds. The production of reactive oxygen species (ROS), reduction of enzyme function, and suppression of the body's antioxidant defense mechanisms are some of the other ways in which heavy metals affect living systems (Balali-Mood et al. 2021).

Although many organic compounds are easily degradable and pose little risk to the environment, some are more persistent and can build up in trophic chains over time, eventually leading to negative effects on humans, wildlife, and the various ecological matrices (Steele et al. 2022). In addition to being detected as contaminants in effluents (Rivoira et al. 2019), polychlorinated biphenyls (PCBs) and polycyclic aromatic hydrocarbons (PAHs) have also been identified in sludge (Mailler et al. 2014, 2017). Precise monitoring of these substances is crucial to prevent the seepage of pollutants from sludge into soil and water, as well as their potential uptake by plants (Renai et al. 2021; Song et al. 2016).

6.16 Leaching of Heavy Metals

The leaching potential of metals from sewage sludge depends on multiple factors. Studies have shown a substantial effect of pH with maximum release of metals at acidic conditions (Zhang et al. 2016). Cationic elements typically show reduced leachability at higher pH levels, while for amphoteric elements, leaching pattern display parabolic relation with pH, where solubility increases at both low and high pH due to their transformation into metal hydroxide forms. Furthermore, leaching potential is significantly influenced by the mineral composition of the waste, which can potentially immobilize hazardous elements through their assimilation into mineral structure or superficial adsorption (Kasina et al. 2021). Further, redox potential has also been shown as a key factor determining the leaching behaviour of metals (Zhang and Lin 2020). A study identified Cu as the most mobile and bioavailable metal, while Hg and Zn were also predominantly associated with increased mobility, suggesting a high potential for environmental release. In contrast, Ni and Cd were largely bound to the residual fraction, indicating lower mobility (Janas et al. 2018).

Further, bioleaching has been employed as an alternative strategy which is eco-friendly as well as economical in context to elimination these trace elements in sludge (Pathak et al. 2009). The primary microorganisms responsible for leaching are *Acidithiobacillus thiooxidans* and *A. ferrooxidans* (Liu et al. 2021). A study investigating the role of iron-oxidizing bacteria in the bioleaching of heavy metals from sewage sludge demonstrated the potential of microbial processes in enhancing sludge treatment. Specifically, *Acidithiobacillus ferrooxidans* was employed to facilitate the removal of Al, Cu, Pb and Zn over a 15-day period. The results showed removal efficiencies of 62.7% for Al, 80.7% for Cu, 43% for Pb, and 75.5% for Zn. This microbial process outperformed control samples, in which negligible metal removal was observed, highlighting the importance of both the bacterial presence and energy source in bioleaching (Shokoohi et al. 2025).

Bioleaching is more advantageous compared to chemical methods of leaching because of increased bioavailability of metals and sludge detoxification during microbial activity. Its efficiency to reduce metal levels as well as sludge toxicity has been confirmed through various tests, making the treated sludge safer for use on land (Jung and Lovitt 2011). In one comparative study, the bacterial action facilitated the dissolution of Cd, Zn and Mn by 71.90%, 89.14%, and 92.50%, respectively. In contrast, chemical leaching achieved only 22.03% for Cd, 25.06% for Mn, and 14.23% for Zn by simply reducing the sludge pH, revealing the critical role of bacterial activity in maximizing metal solubilization, beyond the effect of acidic conditions alone. While acidification contributes to metal mobilization from the solid phase, it is markedly less efficient than microbially mediated processes. Bioleaching benefits from the synergistic action of both bacterial metabolism and low pH, thereby improving overall dissolution efficiency. Moreover, the rate and extent of metal leaching were found to vary on the basis of sludge composition, indicating that matrix characteristics influence the effectiveness of both biological and chemical leaching strategies. These observations highlight the need for future investigations to assess the performance of various bacterial species and target metals for further optimization of the bioleaching process (Ghavidel et al. 2018).

Moreover, studies have reported that in comparison to iron-oxidizing bacteria, sulfur-oxidizing bacteria prove to be more efficient and cost-effective at biosolubilization of heavy metals (Narayanan and Sreekrishnan 2008). Some key sulfur-oxidizing strains commonly involved in bioleaching include *Acidithiobacillus ferrooxidans*, *A. caldus*, *A. thiooxidans*, *A. albertensis*, and *Acidiphilium acidophilum* (Qiu et al. 2005; Prabhu and Baskar 2015). In sulfur-based bioleaching, elemental sulfur serves as an energy source for microbial oxidation, resulting in sulfuric acid production, which causes metal solubilization. The concentration of sulfur is an important parameter that affects the kinetics of the leaching process (Ko et al. 2013). At suboptimal levels, sulfur may lead to insufficient acid production, whereas an excessive amount can result in unoxidized residues, increasing treatment costs and posing environmental risks including secondary pollution. Therefore, optimizing sulfur dosage is vital for effective and sustainable bioleaching (Prabhu and Baskar 2015).

Also, filamentous fungi, notably *Aspergillus niger* and *Penicillium simplicissimum*, have demonstrated the ability of leaching heavy metals from various substrates through the secretion of weak organic acids, that combine with metal ions to form soluble complexes (Dusengemungu et al. 2021). Furthermore, fungal strains isolated from contaminated sites have been reported to show adaptive traits that enhance their survival and leaching efficiency under metal stress, making them promising candidates for bioremediation applications (Qayyum et al. 2019).

6.17 Conclusion

Sewage sludge poses significant threats to ecological matrices and public health due to the occurrence of toxic contaminants including organic and inorganic matter, heavy metals, pharmaceutical residues, and pathogenic substances. Unlike organic contaminants, heavy metals are non-biodegradable and tend to enter the food chain which ultimately poses long-term ecological threats as well as public health implications. The use of wastewater on land in untreated state can disturb the nutrient availability from soil, thereby impacting agricultural productivity, as well as disrupt microbial communities. Ecosystems such as freshwater bodies, coral reefs, salt marshes, and oyster beds suffer from nutrient overload, hypoxia, heavy metal contamination, and structural degradation due to sewage influx. WWTPs, while reducing the contaminant load, also act as vectors for introducing micropollutants into aquatic systems, altering biodiversity and ecological functions.

Human exposure to sludge toxins may occur through multiple ways, for instance, inhalation of toxic vapours, consumption of tainted food and water, or through skin contact, leading to bioaccumulation in tissues. This can impair neurological, immune, endocrine, and reproductive systems, in addition to heightening the risk of conditions such as metabolic issues, cardiovascular diseases, and carcinogenesis. The toxicological mechanisms largely involve oxidative stress, enzyme inhibition, and disruption of critical cellular signaling pathways. In particular, the presence of PAHs, PCBs, and heavy metals further enhance the toxicity and environmental persistence of sewage sludge. Sludge exposed to varying environmental conditions can release trapped heavy metals, especially under acidic or oxidizing environments, thus increasing their mobility and bioavailability. Therefore, techniques and technologies to ensure proper monitoring, treatment, and management are imperative to reduce the release of sewage sludge into ecological systems, thereby minimizing the toxic burden of sludge contaminants on the ecosystem as a whole.

References

Agoro MA, Adeniji AO, Adefisoye MA, Okoh OO (2020) Heavy metals in wastewater and sewage sludge from selected municipal treatment plants in Eastern Cape Province, South Africa. Water 12(10):2746. https://doi.org/10.3390/w12102746

Ahkola H, Lindholm-Lehto P, Perkola N, Välitalo P, Meriläinen P, Mäenpää K, Stelzer JAA, Heiskanen I, Järvistö J, Nuutinen J, Leppänen MT (2021) A preliminary study on the ecotoxic potency of wastewater treatment plant sludge combining passive sampling and bioassays. Sci Total Environ 758:143700. https://doi.org/10.1016/j.scitotenv.2020.143700

Alvarado A, Vesvikar M, Cisneros JF, Maere T, Goethals P, Nopens I (2013) CFD study to determine the optimal configuration of aerators in a full-scale waste stabilization pond. Water Res 47(13):4528–4537. https://doi.org/10.1016/j.watres.2013.05.016

Angulo-Bejarano PI, Puente-Rivera J, Cruz-Ortega R (2021) Metal and metalloid toxicity in plants: an overview on molecular aspects. Plants 10(4):635. https://doi.org/10.3390/plants10040635

Aski AL, Borghei A, Zenouzi A, Ashrafi N, Taherzadeh MJ (2020) Steam explosion pretreatment of sludge for pharmaceutical removal and heavy metal release to improve biodegradability and biogas production. Fermentation 6(1):34. https://doi.org/10.3390/fermentation6010034

Badawi AK, Salama RS, Mostafa MMM (2023) Natural-based coagulants/flocculants as sustainable market-valued products for industrial wastewater treatment: a review of recent developments. RSC Adv 13(28):19335–19355. https://doi.org/10.1039/d3ra01999c

Bajpai P (2017) External treatment technologies used for pulp and paper mill effluents. In: Pulp and paper industry. Elsevier eBooks, pp 71–92. https://doi.org/10.1016/b978-0-12-811099-7.00006-x

Balali-Mood M, Naseri K, Tahergorabi Z, Khazdair MR, Sadeghi M (2021). Toxic mechanisms of five heavy metals: mercury, lead, chromium, cadmium, and arsenic. Front Pharmacol 12. https://doi.org/10.3389/fphar.2021.643972

Begmatov S, Dorofeev AG, Kadnikov VV, Beletsky AV, Pimenov NV, Ravin NV, Mardanov AV (2022) The structure of microbial communities of activated sludge of large-scale wastewater treatment plants in the city of Moscow. Sci Rep 12(1). https://doi.org/10.1038/s41598-022-07132-4

Carter LJ, Agatz A, Kumar A, Williams M (2020) Translocation of pharmaceuticals from wastewater into beehives. Environ Int 134:105248. https://doi.org/10.1016/j.envint.2019.105248

Chen R, Yuan S, Chen S, Ci H, Dai X, Wang X, Li C, Wang D, Dong B (2022) Life-cycle assessment of two sewage sludge-to-energy systems based on different sewage sludge characteristics: energy balance and greenhouse gas-emission footprint analysis. J Environ Sci 111:380–391. https://doi.org/10.1016/j.jes.2021.04.012

Costa JM (2023) Considerations on electrochemical technologies for water purification and wastewater treatment. Int J Environ Res Public Health 20(12):6140. https://doi.org/10.3390/ijerph20126140

CPCB (2009) Status of water supply, wastewater generation and treatment in Class I cities and Class II towns of India. Series: CUPS/70/2009-10. Central Pollution Control Board, India

Ćwiertniewicz-Wojciechowska M, Cema G, Ziembińska-Buczyńska A (2023) Sewage sludge pretreatment: current status and future prospects. Environ Sci Pollut Res 30(38):88313–88330. https://doi.org/10.1007/s11356-023-28613-7

Dahman Y (2017) Nanopolymers. In: Dahman Y, Deonanan K, Dontsos T, Iammatteo A (eds) Elsevier eBooks, pp 121–144. https://doi.org/10.1016/b978-0-323-51256-5.00006-x

Dao VH, Cameron NR, Saito K (2015) Synthesis, properties and performance of organic polymers employed in flocculation applications. Polym Chem 7(1):11–25. https://doi.org/10.1039/c5py01572c

Das PP, Sharma M, Purkait MK (2022) Recent progress on electrocoagulation process for wastewater treatment: a review. Sep Purif Technol 292:121058. https://doi.org/10.1016/j.seppur.2022.121058

Daud MK, Rizvi H, Akram MF, Ali S, Rizwan M, Nafees M, Jin ZS (2018) Review of upflow anaerobic sludge blanket reactor technology: effect of different parameters and developments for domestic wastewater treatment. J Chem 2018:1–13. https://doi.org/10.1155/2018/1596319

Dubey SK, Yadav RK, Chatuvedi PK, Goyel B, Yadav R, Minhas PS (2006) Agricultural uses of sewage sludge and water and their impact on soil water and environmental health in Haryana, India. In: Abstract of 18th World Congress of Soil Science, Philadelphia, 9–15 July 2006

Dusengemungu L, Kasali G, Gwanama C, Mubemba B (2021) Overview of fungal bioleaching of metals. Environ Adv 5:100083. https://doi.org/10.1016/j.envadv.2021.100083

Ebba M, Asaithambi P, Alemayehu E (2021) Investigation on operating parameters and cost using an electrocoagulation process for wastewater treatment. Appl Water Sci 11(11). https://doi.org/10.1007/s13201-021-01517-y

Ekka B, Mierina I, Zarina R, Mezule L (2024) Efficient Removal of lipophilic compounds from sewage sludge: comparative evaluation of solvent extraction techniques. Heliyon 10(23):e40749. https://doi.org/10.1016/j.heliyon.2024.e40749

Emamjomeh MM, Sivakumar M (2009) Review of pollutants removed by electrocoagulation and electrocoagulation/flotation processes. J Environ Manage 90(5):1663–1679. https://doi.org/10.1016/j.jenvman.2008.12.011

Enns D, Cunze S, Baker NJ, Oehlmann J, Jourdan J (2023) Flushing away the future: the effects of wastewater treatment plants on aquatic invertebrates. Water Res 243:120388. https://doi.org/10.1016/j.watres.2023.120388

Estay H, Barros L, Troncoso E (2021) Metal sulfide precipitation: recent breakthroughs and future outlooks. Minerals 11(12):1385. https://doi.org/10.3390/min11121385

Fang C, Huang R, Dykstra CM, Jiang R, Pavlostathis SG, Tang Y (2019) Energy and nutrient recovery from sewage sludge and manure via anaerobic digestion with hydrothermal pretreatment. Environ Sci Technol 54(2):1147–1156. https://doi.org/10.1021/acs.est.9b03269

Feng J, Burke IT, Chen X, Stewart DI (2023) Assessing metal contamination and speciation in sewage sludge: implications for soil application and environmental risk. Rev Environ Sci Bio/technol 22(4):1037–1058. https://doi.org/10.1007/s11157-023-09675-y

Fent K, Weston A, Caminada D (2006) Ecotoxicology of human pharmaceuticals. Aquat Toxicol 76(2):122–159. https://doi.org/10.1016/j.aquatox.2005.09.009

Ferrentino R, Langone M, Fiori L, Andreottola G (2023) Full-scale sewage sludge reduction technologies: a review with a focus on energy consumption. Water 15(4):615. https://doi.org/10.3390/w15040615

Fijalkowski K, Rorat A, Grobelak A, Kacprzak MJ (2017) The presence of contaminations in sewage sludge—the current situation. J Environ Manage 203:1126–1136. https://doi.org/10.1016/j.jenvman.2017.05.068

Fini EH, Kazemi M, Poulikakos L, Lazorenko G, Akbarzade V, Lamanna A, Lammers P (2024) Perspectives on innovative non-fertilizer applications of sewage sludge for mitigating environmental and health hazards. Commun Eng 3(1). https://doi.org/10.1038/s44172-024-00298-x

Funmi AE, Suleiman MA, Deborah OI, Dorcas AT (2021) Biogas production as energy source and strategy for managing waste and climate change. SN Appl Sci 3(1). https://doi.org/10.1007/s42452-020-03973-8

Gai H, Zhong C, Qiao L, Chen S, Xiao M, Song H (2018) Extraction of 1-amino-2-Naphthol-4-Sulfonic acid from wastewater using trioctylamine (N, N-dioctyloctan-1-amine) in methyl isobutyl ketone. J Clean Prod 201:774–782. https://doi.org/10.1016/j.jclepro.2018.08.082

Geng H, Xu Y, Zheng L, Gong H, Dai L, Dai X (2020) An overview of removing heavy metals from sewage sludge: achievements and perspectives. Environ Pollut 266:115375. https://doi.org/10.1016/j.envpol.2020.115375

Ghanem MFSA, Afzal S, Nesar H, Imran Z, Ahmad W (2023) Impact of metal polluted sewage water on soil nematode assemblages in agricultural settings of Aligarh, India. Soil Ecol Lett 6(1). https://doi.org/10.1007/s42832-023-0193-3

Ghavidel A, Rad SN, Alikhani HA, Sharari M, Ghanbari A (2018) Bioleaching of heavy metals from sewage sludge, direct action of *Acidithiobacillus ferrooxidans* or only the impact of pH? J Mater Cycles Waste Manage 20(2):1179–1187. https://doi.org/10.1007/s10163-017-0680-7

Godini K, Azarian G, Kimiaei A, Dragoi EN, Curteanu S (2021) Modeling of a real industrial wastewater treatment plant based on aerated lagoon using a neuro-evolutive technique. Process Saf Environ Prot 148:114–124. https://doi.org/10.1016/j.psep.2020.09.057

Gogina E, Makisha N, Gulshin I (2024) Comparative analysis of sewage sludge characteristics after natural deposition, accelerated aging, and composting. Appl Sci 14(22):10446. https://doi.org/10.3390/app142210446

Goh P, Ismail A (2018) A review on inorganic membranes for desalination and wastewater treatment. Desalination 434:60–80. https://doi.org/10.1016/j.desal.2017.07.023

Goswami R, Chattopadhyay P, Shome A, Banerjee SN, Chakraborty AK, Mathew AK, Chaudhury S (2016) An overview of physico-chemical mechanisms of biogas production by microbial communities: a step towards sustainable waste management. 3 Biotech 6(1). https://doi.org/10.1007/s13205-016-0395-9

Grobelak A, Całus-Makowska K, Jasińska A, Klimasz M, Wypart-Pawul A, Augustajtys D, Baor E, Sławczyk D, Kowalska A (2024) Environmental impacts and contaminants management in sewage sludge-to-energy and fertilizer technologies: current trends and future directions. Energies 17(19):4983. https://doi.org/10.3390/en17194983

Gupta H (2018) PAH determination in effluent and sludge samples of paper industry. Environ Technol Innov 9:115–121. https://doi.org/10.1016/j.eti.2017.11.009

Gustafsson M, Meneghetti R, Souza Marques F, Trim H, Dong R, Al Saedi T, Rasi S, Thual J, Kornatz P, Wall D, Berntsen C, Saxegaard S, Lyng KA, Nägele HJ, Heaven S, Bywater A (2024) A perspective on the state of the biogas industry from selected member countries. In: Gustafsson M, Liebetrau J (eds) IEA Bioenergy Task 37, 2024:2

Haidar Z, Fatema K, Shoily SS, Sajib AA (2023) Disease-associated metabolic pathways affected by heavy metals and metalloid. Toxicol Rep 10:554–570. https://doi.org/10.1016/j.toxrep.2023.04.010

Hannula P, Khalid MK, Janas D, Yliniemi K, Lundström M (2019) Energy efficient copper electrowinning and direct deposition on carbon nanotube film from industrial wastewaters. J Clean Prod 207:1033–1039. https://doi.org/10.1016/j.jclepro.2018.10.097

Hanum F, Yuan LC, Kamahara H, Aziz HA, Atsuta Y, Yamada T, Daimon H (2019) Treatment of sewage sludge using anaerobic digestion in Malaysia: current state and challenges. Front Energy Res 7. https://doi.org/10.3389/fenrg.2019.00019

He D, Zhu T, Sun J, Pan X, Li J, Luo H (2024) Emerging organic contaminants in sewage sludge: current status, technological challenges and regulatory perspectives. Sci Environ 955:177234. https://doi.org/10.1016/j.scitotenv.2024.177234

Islam MD, Mahdi MM (2022) Evaluation of micro-pollutants removal from industrial wastewater using conventional and advanced biological treatment processes. In: Biodegradation and detoxification of micropollutants in industrial wastewater. Elsevier eBooks, pp 1–26. https://doi.org/10.1016/b978-0-323-88507-2.00011-7

Janas M, Zawadzka A, Cichowicz R (2018) The influence of selected factors on leaching of metals from sewage sludge. Environ Sci Pollut Res 25(33):33240–33248. https://doi.org/10.1007/s11356-018-3094-8

Janaszek A, Kowalik R (2023) Analysis of heavy metal contaminants and mobility in sewage sludge-soil mixtures for sustainable agricultural practices. Water 15(22):3992. https://doi.org/10.3390/w15223992

Jung I, Lovitt R (2011) Leaching techniques to remove metals and potentially hazardous nutrients from trout farm sludge. Water Res 45(18):5977–5986. https://doi.org/10.1016/j.watres.2011.08.062

Kabata-Pendias A (2010) Trace elements in soils and plants. In: CRC Press eBooks. https://doi.org/10.1201/b10158

Kapanen A, Vikman M, Rajasärkkä J, Virta M, Itävaara M (2013) Biotests for environmental quality assessment of composted sewage sludge. Waste Manage 33(6):1451–1460. https://doi.org/10.1016/j.wasman.2013.02.022

Kasina M, Kajdas B, Michalik M (2021) The leaching potential of sewage sludge and municipal waste incineration ashes in terms of landfill safety and potential reuse. Sci Total Environ 791:148313. https://doi.org/10.1016/j.scitotenv.2021.148313

Kato S, Kansha Y (2024) Comprehensive review of industrial wastewater treatment techniques. Environ Sci Pollut Res 31(39):51064–51097. https://doi.org/10.1007/s11356-024-34584-0

Kesari KK, Soni R, Jamal QMS, Tripathi P, Lal JA, Jha NK, Siddiqui MH, Kumar P, Tripathi V, Ruokolainen J (2021) Wastewater Treatment and reuse: a review of its applications and health implications. Water Air Soil Polluti 232(5). https://doi.org/10.1007/s11270-021-05154-8

Kim W, Son D, Yun C, Kim D, Chang D, Sunwoo Y, Hong K (2017) Performance assessment of electrolysis using copper and catalyzed electrodes for enhanced nutrient removal from wastewater. J Electrochem Sci Technol 8(2):124–132. https://doi.org/10.33961/jecst.2017.8.2.124

Kim YK, Yoo K, Kim MS, Han I, Lee M, Kang BR, Lee TK, Park J (2019) The capacity of wastewater treatment plants drives bacterial community structure and its assembly. Sci Rep 9(1). https://doi.org/10.1038/s41598-019-50952-0

Ko M, Park H, Kim K, Lee J (2013) The role of *Acidithiobacillus ferrooxidans* and *Acidithiobacillus thiooxidans* in arsenic bioleaching from soil. Environ Geochem Health 35(6):727–733. https://doi.org/10.1007/s10653-013-9530-2

Krstić V, Urošević T, Pešovski B (2018) A review on adsorbents for treatment of water and wastewaters containing copper ions. Chem Eng Sci 192:273–287. https://doi.org/10.1016/j.ces.2018.07.022

Kumar V, Chopra AK (2013) Accumulation and translocation of metals in soil and different parts of French bean (*Phaseolus vulgaris* L.) amended with sewage sludge. Bull Environ Contamin Toxicol 92(1):103–108. https://doi.org/10.1007/s00128-013-1142-0

Kumar V, Chopra AK, Srivastava S (2016) Assessment of heavy metals in spinach (*Spinacia oleracea* L.) grown in sewage sludge amended soil. Commun Soil Sci Plant Anal 47(2):221–236. https://doi.org/10.1080/00103624.2015.1122799

Lei C, Liang F, Li J, Chen W, Huang B (2019) Electrochemical reductive dechlorination of chlorinated volatile organic compounds (Cl-VOCs): effects of molecular structure on the dehalogenation reactivity and mechanisms. Chem Eng J 358:1054–1064. https://doi.org/10.1016/j.cej.2018.10.105

Levchuk I, Márquez JJR, Sillanpää M (2018) Removal of natural organic matter (NOM) from water by ion exchange—a review. Chemosphere 192:90–104. https://doi.org/10.1016/j.chemosphere.2017.10.101

Li Y, Zhu Y, Wang D, Yang G, Pan L, Wang Q, Ni B, Li H, Yuan X, Jiang L, Tang W (2020) Fe(II) catalyzing sodium percarbonate facilitates the dewaterability of waste activated sludge: performance, mechanism, and implication. Water Res 174:115626. https://doi.org/10.1016/j.watres.2020.115626

Liang J, Huang J, Zhang S, Yang X, Huang S, Zheng L, Ye M, Sun S (2019) A highly efficient conditioning process to improve sludge dewaterability by combining calcium hypochlorite oxidation, ferric coagulant re-flocculation, and walnut shell skeleton construction. Chem Eng J 361:1462–1478. https://doi.org/10.1016/j.cej.2018.10.143

Lin S, Mackey HR, Hao T, Guo G, Van Loosdrecht MC, Chen G (2018) Biological sulfur oxidation in wastewater treatment: a review of emerging opportunities. Water Res 143:399–415. https://doi.org/10.1016/j.watres.2018.06.051

Liu H, Li Y, Fu B, Guo H, Zhang J, Liu H (2020) Recovery of volatile fatty acids from sewage sludge through anaerobic fermentation. Elsevier eBooks, pp 151–175. https://doi.org/10.1016/b978-0-444-64321-6.00008-2

Liu H, Yang K, Luo L, Lu Q, Wu Y, Lan M, Luo Y, Liang W (2021) Study on bioleaching of heavy metals and resource potential from tannery yard sludge. Environ Sci Pollut Res 28(29):38867–38879. https://doi.org/10.1007/s11356-021-13425-4

Maćczak P, Kaczmarek H, Ziegler-Borowska M (2020) Recent achievements in polymer bio-based flocculants for water treatment. Materials 13(18):3951. https://doi.org/10.3390/ma13183951

Macedo HE, Lehner B, Nicell J, Grill G, Li J, Limtong A, Shakya R (2022) Distribution and characteristics of wastewater treatment plants within the global river network. Earth Syst Sci Data 14(2):559–577. https://doi.org/10.5194/essd-14-559-2022

Madondo NI, Tetteh EK, Rathilal S, Bakare BF (2021) Synergistic effect of magnetite and bioelectrochemical systems on anaerobic digestion. Bioengineering 8(12):198. https://doi.org/10.3390/bioengineering8120198

Mahesh S, Garg KK, Srivastava VC, Mishra IM, Prasad B, Mall ID (2016) Continuous electrocoagulation treatment of pulp and paper mill wastewater: operating cost and sludge study. RSC Adv 6(20):16223–16233. https://doi.org/10.1039/c5ra27486a

Mailler R, Gasperi J, Chebbo G, Rocher V (2014) Priority and emerging pollutants in sewage sludge and fate during sludge treatment. Waste Manage 34(7):1217–1226. https://doi.org/10.1016/j.wasman.2014.03.028

Mailler R, Gasperi J, Patureau D, Vulliet E, Delgenes N, Danel A, Deshayes S, Eudes V, Guerin S, Moilleron R, Chebbo G, Rocher V (2017) Fate of emerging and priority micropollutants during the sewage sludge treatment: case study of Paris conurbation. Part 1: contamination of the different types of sewage sludge. Waste Manage 59:379–393. https://doi.org/10.1016/j.wasman.2016.11.010

Malinović BN, Markelj J, Gotvajn AŽ, Cigić IK, Prosen H (2022) Electrochemical treatment of wastewater to remove contaminants from the production and disposal of plastics: a review. Environ Chem Lett 20(6):3765–3787. https://doi.org/10.1007/s10311-022-01497-8

Manilal AM, Nampoothiri MGH, Soloman PA (2017) Removal of oil and grease from automobile garage wastewater using electrocoagulation. IOP Conf Ser Mater Sci Eng 206:012082. https://doi.org/10.1088/1757-899x/206/1/012082

Marin E, Rusănescu CO (2023) Agricultural use of urban sewage sludge from the wastewater station in the Municipality of Alexandria in Romania. Water 15(3):458. https://doi.org/10.3390/w15030458

Martín J, Camacho-Muñoz MD, Santos JL, Aparicio I, Alonso E (2012) Distribution and temporal evolution of pharmaceutically active compounds alongside sewage sludge treatment. Risk assessment of sludge application onto soils. J Environ Manage 102:18–25. https://doi.org/10.1016/j.jenvman.2012.02.020

Martins RA, Salgado EM, Gonçalves AL, Esteves AF, Pires JCM (2024) Microalgae-based remediation of real textile wastewater: assessing pollutant removal and biomass valorisation. Bioengineering 11(1):44. https://doi.org/10.3390/bioengineering11010044

Mayer F, Bhandari R, Gäth SA (2021) Life cycle assessment of prospective sewage sludge treatment paths in Germany. J Environ Manage 290:112557. https://doi.org/10.1016/j.jenvman.2021.112557

Milinovic J, Vidal M, Lacorte S, Rigol A (2014) Leaching of heavy metals and alkylphenolic compounds from fresh and dried sewage sludge. Environ Sci Pollut Res 21(3):2009–2017. https://doi.org/10.1007/s11356-013-2100-4

Mirshafiee A, Nourollahi M, Shahriary A (2024) Application of electro oxidation process for treating wastewater from petrochemical with mixed metal oxide electrode. Sci Rep 14(1). https://doi.org/10.1038/s41598-024-52201-5

Mitra S, Chakraborty AJ, Tareq AM, Emran TB, Nainu F, Khusro A, Idris AM, Khandaker MU, Osman H, Alhumaydhi FA, Simal-Gandara J (2022) Impact of heavy metals on the environment and human health: novel therapeutic insights to counter the toxicity. J King Saud Univ Sci 34(3):101865. https://doi.org/10.1016/j.jksus.2022.101865

Miyaoka Y, Yoochatchaval W, Sumino H, Banjongproo P, Yamaguchi T, Onodera T, Okadera T, Syutsubo K (2017) Evaluation of the process performance of a down-flow hanging sponge reactor for direct treatment of domestic wastewater in Bangkok, Thailand. J Environ Sci Health Part A 52(10):956–970. https://doi.org/10.1080/10934529.2017.1324708

Najafinejad MS, Chianese S, Fenti A, Iovino P, Musmarra D (2023) Application of electrochemical oxidation for water and wastewater treatment: an overview. Molecules 28(10):4208. https://doi.org/10.3390/molecules28104208

Narayanan R, Sreekrishnan T (2008) A two-stage process for simultaneous thermophilic sludge digestion, pathogen control and metal leaching. Environ Technol 30(1):21–26. https://doi.org/10.1080/09593330802422514

Nasir IM, Ghazi TIM, Omar R (2012) Production of biogas from solid organic wastes through anaerobic digestion: a review. Appl Microbiol Biotechnol 95(2):321–329. https://doi.org/10.1007/s00253-012-4152-7

Nguyen VK, Chaudhary DK, Dahal RH, Trinh NH, Kim J, Chang SW, Hong Y, La DD, Nguyen XC, Ngo HH, Chung WJ, Nguyen DD (2021) Review on pretreatment techniques to improve anaerobic digestion of sewage sludge. Fuel 285:119105. https://doi.org/10.1016/j.fuel.2020.119105

Ojovan M, Lee W (2014) Treatment of radioactive wastes. Elsevier eBooks, pp 171–203. https://doi.org/10.1016/b978-0-08-099392-8.00014-0

Papadopoulos KP, Economou CN, Tekerlekopoulou AG, Vayenas DV (2020) Two-step treatment of brewery wastewater using electrocoagulation and cyanobacteria-based cultivation. J Environ Manage 265:110543. https://doi.org/10.1016/j.jenvman.2020.110543

Patel KS, Ramteke S, Sahu BL, Nayak Y, Sharma S, Hung C (2015) Polychlorinated biphenyls contamination of sludge in India. Am J Anal Chem 06(11):867–877. https://doi.org/10.4236/ajac.2015.611082

Pathak A, Dastidar M, Sreekrishnan T (2009) Bioleaching of heavy metals from sewage sludge: a review. J Environ Manage 90(8):2343–2353. https://doi.org/10.1016/j.jenvman.2008.11.005

Pirmoradi M, Ashrafizadeh SN (2017) Removal of nitrate from water by bulk liquid membrane. Desalin Water Treat 58:228–238. https://doi.org/10.5004/dwt.2017.0194

Prabhu AV, Sivaram A, Prabhu N, Sundaramahalingam A (2021) A study of enhancing the biogas production in anaerobic digestion. Mater Today Proc 45:7994–7999. https://doi.org/10.1016/j.matpr.2020.12.1009

Prabhu SV, Baskar R (2015) Kinetics of heavy metal biosolubilization from electroplating sludge: effects of sulfur concentration. J Kor Soc Appl Biol Chem 58(2):185–194. https://doi.org/10.1007/s13765-015-0027-9

Qadir M, Drechsel P, Cisneros BJ, Kim Y, Pramanik A, Mehta P, Olaniyan O (2020) Global and regional potential of wastewater as a water, nutrient and energy source. Nat Res Forum 44(1):40–51. https://doi.org/10.1111/1477-8947.12187

Qasem NAA, Mohammed RH, Lawal DU (2021) Removal of heavy metal ions from wastewater: a comprehensive and critical review. Npj Clean Water 4(1). https://doi.org/10.1038/s41545-021-00127-0

Qayyum S, Meng K, Pervez S, Nawaz F, Peng C (2019) Optimization of pH, temperature and carbon source for bioleaching of heavy metals by *Aspergillus flavus* isolated from contaminated soil. Main Group Met Chem 42(1):1–7. https://doi.org/10.1515/mgmc-2018-0038

Qiu M, Xiong S, Zhang W, Wang G (2005) A comparison of bioleaching of chalcopyrite using pure culture or a mixed culture. Miner Eng 18(9):987–990. https://doi.org/10.1016/j.mineng.2005.01.004

Quan LM, Kamyab H, Yuzir A, Ashokkumar V, Hosseini SE, Balasubramanian B, Kirpichnikova I (2022) Review of the application of gasification and combustion technology and waste-to-energy technologies in sewage sludge treatment. Fuel 316:123199. https://doi.org/10.1016/j.fuel.2022.123199

Raffa CM, Chiampo F (2021) Bioremediation of Agricultural soils polluted with pesticides: a review. Bioengineering 8(7):92. https://doi.org/10.3390/bioengineering8070092

Rani L, Srivastav AL, Kaushal J, Grewal AS, Madhav S (2022) Heavy metal contamination in the river ecosystem. In: Ecological significance of river ecosystems. Elsevier, Amsterdam

Renai L, Tozzi F, Scordo CV, Giordani E, Bruzzoniti MC, Fibbi D, Mandi L, Ouazzani N, Del Bubba M (2021) Productivity and nutritional and nutraceutical value of strawberry fruits (*Fragaria*

× *ananassa* Duch.) cultivated under irrigation with treated wastewaters. J Sci Food Agric 101(3):1239–1246. https://doi.org/10.1002/jsfa.10737
Rivoira L, Castiglioni M, Kettab A, Ouazzani N, Al-Karablieh E, Boujelben N, Fibbi D, Coppini E, Giordani E, Bubba MD, Bruzzoniti MC (2019) Impact of effluents from wastewater treatments reused for irrigation: strawberry as case study. Environ Eng Manag J 18:2133–2143
Salva J, Sečkár M, Schwarz M, Samešová D, Mordáčová M, Poništ J, Veverková D (2025) Analysis of the current state of sewage sludge treatment from the perspective of current European directives. Environ Sci Europe 37(1). https://doi.org/10.1186/s12302-025-01097-7
Santana J, Fraga S, Zanatta M, Martins, Pires M (2021) Characterization of organic compounds and drugs in sewage sludge aiming for agricultural recycling. Heliyon 7(4), e06771. https://doi.org/10.1016/j.heliyon.2021.e06771
Saritha V, Srinivas N, Vuppala NVS (2014) Analysis and optimization of coagulation and flocculation process. Appl Water Sci 7(1):451–460. https://doi.org/10.1007/s13201-014-0262-y
Shaddel S, Bakhtiary-Davijany H, Kabbe C, Dadgar F, Østerhus S (2019) Sustainable sewage sludge management: from current practices to emerging nutrient recovery technologies. Sustainability 11(12):3435. https://doi.org/10.3390/su11123435
Shokoohi R, Najafi-Vosough R, Shabanloo A, Torkshavand Z, Mahmoudi MM, Abasi MA (2025) Removing heavy metals and improving the dewaterability of sewage sludge with the bioleaching process by *Thiobacillus ferrooxidans* bacteria. Appl Water Sci 15(2). https://doi.org/10.1007/s13201-025-02365-w
Sillanpää M, Shestakova M (2017a) Introduction. In: Electrochemical water treatment methods. Elsevier eBooks, pp 1–46. https://doi.org/10.1016/b978-0-12-811462-9.00001-3
Sillanpää M, Shestakova M (2017b) Electrochemical water treatment methods. In: Electrochemical water treatment methods. Elsevier eBooks, pp 47–130. https://doi.org/10.1016/b978-0-12-811462-9.00002-5
Song C, Qiang F, Tianxiao L, Dong L, Yifan L, Min W (2016) Transfer and migration of polycyclic aromatic hydrocarbons in soil irrigated with long-term wastewater. Int J Agric Biol Eng 9(5):83–92
Sravan JS, Matsakas L, Sarkar O (2024) Advances in biological wastewater treatment processes: focus on low-carbon energy and resource recovery in biorefinery context. Bioengineering 11(3):281. https://doi.org/10.3390/bioengineering11030281
Steele JC, Meng X, Venkatesan AK, Halden RU (2022) Comparative meta-analysis of organic contaminants in sewage sludge from the United States and China. Sci Total Environ 821:153423. https://doi.org/10.1016/j.scitotenv.2022.153423
Tang Y, Sun J, Dong B, Dai X (2022) Thermal hydrolysis pretreatment-anaerobic digestion promotes plant-growth biostimulants production from sewage sludge by upregulating aromatic amino acids transformation and quinones supply. Environ Sci Technol 56(3):1938–1950. https://doi.org/10.1021/acs.est.1c06506
Teoh SK, Li LY (2020) Feasibility of alternative sewage sludge treatment methods from a lifecycle assessment (LCA) perspective. J Clean Prod 247:119495. https://doi.org/10.1016/j.jclepro.2019.119495
Thakur H, Dhar A, Powar S (2022) Biogas production from anaerobic co-digestion of sewage sludge and food waste in continuously stirred tank reactor. Results Eng 16:100617. https://doi.org/10.1016/j.rineng.2022.100617
Touahria S, Hazourli S, Touahria K, Eulmi A, Aitbara A (2016) (2016) Clarification of industrial mining wastewater using electrocoagulation. Int J Electrochem Sci 11(7):5710–5723. https://doi.org/10.20964/2016.07.51
Tytła M (2019) Assessment of heavy metal pollution and potential ecological risk in sewage sludge from municipal wastewater treatment plant located in the most industrialized region in Poland—case study. Int J Environ Res Public Health 16(13):2430. https://doi.org/10.3390/ijerph16132430
Tytła M, Widziewicz-Rzońca K (2023) Ecological and human health risk assessment of heavy metals in sewage sludge produced in Silesian Voivodeship, Poland: a case study. Environ Monitor Assess 195(11). https://doi.org/10.1007/s10661-023-11987-z

Uddin S, Zaman M, Martínez-Guijarro K, Al-Murad M, Behbehani M, Habibi N, Al-Mutairi A (2025) Sewage sludge as soil amendment in arid soils—A trace Metal, nutrient and trace organics perspective. Emerg. Contamin 100420. https://doi.org/10.1016/j.emcon.2024.100420

van Lier JB, van der Zee FP, Frijters CTMJ, Ersahin ME (2015) Celebrating 40 years anaerobic sludge bed reactors for industrial wastewater treatment. Rev Environ Sci Biotechnol 14(4):681–702. https://doi.org/10.1007/s11157-015-9375-5

Wang C, Li T, Yu G, Deng S (2021) Removal of low concentrations of nickel ions in electroplating wastewater by combination of electrodialysis and electrodeposition. Chemosphere 263:128208. https://doi.org/10.1016/j.chemosphere.2020.128208

Waqas S, Harun NY, Sambudi NS, Bilad MR, Abioye KJ, Ali A, Abdulrahman A (2023) A review of rotating biological contactors for wastewater treatment. Water 15(10):1913. https://doi.org/10.3390/w15101913

Wear SL, Acuña V, McDonald R, Font C (2021) Sewage pollution, declining ecosystem health, and cross-sector collaboration. Biol Cons 255:109010. https://doi.org/10.1016/j.biocon.2021.109010

Wluka A, Huang Y, Coenen L, Dsikowitzky L, Schwarzbauer J (2021) Structural diversity of organic contaminants in sewage sludge: a comparison of sewage fingerprints from Germany and China. Discover Water 1(1). https://doi.org/10.1007/s43832-021-00004-4

Woo DCY, Goh QH, Poh PE, Chew IML (2022) A technoeconomic analysis of sewage sludge valorization for carbon emission reduction. Biomass Convers Biorefinery 13(15):13591–13604. https://doi.org/10.1007/s13399-022-02922-2

Xu J, Wu L, Chen W, Chang AC (2010) Leaching potential of nonsteroidal anti-inflammatory drugs in soils. Environ Toxicol Chem 29(4):800–807. https://doi.org/10.1002/etc.107

Yadav M, Gupta R, Sharma RK (2019) Green and sustainable pathways for wastewater purification. In: Ahuja S (ed) Advances in water purification techniques, pp 355–383. https://doi.org/10.1016/B978-0-12-814790-0.00014-4

Yakamercan E, Aygün A (2023) Health risk assessment of metal(loid)s for land application of domestic sewage sludge in city of Bursa, Türkiye. Environ Monitor Assess 195(6). https://doi.org/10.1007/s10661-023-11302-w

Zhang B, Zhou X, Ren X, Hu X, Ji B (2023) Recent research on municipal sludge as soil fertilizer in China: a Review. Water Air Soil Pollut 234(2). https://doi.org/10.1007/s11270-023-06142-w

Zhang Q, Hu J, Lee D, Chang Y, Lee Y (2017) Sludge treatment: Current research trends. Biores Technol 243:1159–1172. https://doi.org/10.1016/j.biortech.2017.07.070

Zhang W, Lin M (2020) Influence of redox potential on leaching behavior of a solidified chromium contaminated soil. Sci Total Environ 733:139410. https://doi.org/10.1016/j.scitotenv.2020.139410

Zhang Y, Cetin B, Likos W, Edil T (2016) Impacts of pH on leaching potential of elements from MSW incineration fly ash. Fuel 184:815–825. https://doi.org/10.1016/j.fuel.2016.07.089

Zueva SB (2018) Current legislation and methods of treatment of wastewater coming from waste electrical and electronic equipment processing. Elsevier eBooks, pp 213–240. https://doi.org/10.1016/b978-0-08-102057-9.00009-3

Chapter 7
Recovery of Nutrients and Metals from Sewage

Prakash B. Rathod, Mahendra Pratap Singh, and Akshya C. Dhayagude

Abstract Sewage treatment and the removal of nutrients and metals from sewage are increasingly in demand as part of sustainable resource management. Innovative technologies are being developed to extract nutrients such as potassium, nitrogen, phosphorus, and metals like zinc, iron, and copper from sewage. These processes help mitigate environmental pollution and create opportunities for using the recovered materials in agriculture and industry. This chapter aims to compile information on the processes involved in recovering nutrients and metals from sewage, as well as to explore the superiority and drawback of these procedure. Additionally, the chapter will discuss the future needs for further development in sewage treatment processes.

Keywords Nanomaterial · Sewage treatment · Recovery of nutrients · Remediation · Biochar · Membrane filtration

7.1 Introduction

Wastewater generated by human activity is referred to as sewage. It is the liquid waste that is released from homes, businesses, and farms. Although its main component is water, it also includes a variety of contaminants, including as chemicals, oils, food scraps, pathogens, nutrients (such as phosphate and nitrogen), and human

P. B. Rathod (✉)
School of Chemical Sciences, Kavayitri Bahinabai Chaudhari North Maharashtra University, Jalgaon, Maharashtra, India
e-mail: prakash.hariom2011@gmail.com; drprakashrathod2011@gmail.com; pbrathod@nmu.ac.in

P. B. Rathod · M. P. Singh
Department of Engineering Chemistry, Dr. Vishwanath Karad MIT–World Peace University, Pune, Maharashtra, India

A. C. Dhayagude
Department of Chemistry, MVP's S. S. S. M. ASC College, Saikheda, Niphad, Nashik, Maharashtra, India

S. Madhav et al. (eds.), *Municipal Sewage and Sludge: Treatment and Disposal Strategies*, Environmental Science and Engineering,
https://doi.org/10.1007/978-3-032-07597-0_7

waste (Schweitzer and Noblet 2018). Usually, sewage is moved via a system of sewers (Heberer et al. 2002), where it is either processed at sewage treatment facilities or, in certain situations, released straight into bodies of water. Water sources become contaminated when sewage is dumped into rivers, lakes, or the ocean without being properly treated first. Fish habitats may be harmed by this (Mahmoud et al. 2013). Overnutrients in sewage, such as nitrogen and phosphorus, encourage the extention of algae (algal blooms) (Lau et al. 1995). These algae create dead zones where aquatic life cannot thrive because they take up oxygen in the water as they decay. Water supplies can get contaminated by bacteria, viruses, and parasites found in sewage, which puts both human and animal health at risk. Hazardous materials including pesticides, solvents, and heavy metals are frequently found in industrial sewage. These materials can damage aquatic life and upset food networks (Sharma et al. 2024). The sources of sewage can be roughly divided into three categories: industrial sewage, which contains heavy metals, oils, toxic chemicals, and other dangerous material; domestic sewage, which contains human waste, food particles, soap, and detergents and agricultural sewage, which contains chemicals and pesticides (Zhang et al. 2022b). Treatment of these sewage is important and the proper disposal strategies are required to adapt. There are numerous traditional and conventional technologies are adapted for the treatment of sewage. These are more traditional, simpler, and frequently inexpensive systems that depend more on organic processes (Rathod et al. 2018b). Their low energy requirements and cost make them still popular in underdeveloped nations and rural areas. Traditional technologies includes preparation of septic and imhoff tanks, waetlands and pit latrines, but the traditional technologies has effective for the small towns and limited to industrial sewage management. To treating high amount of municipal and industrial sewage conventional methods were used and it is work well. Conventional methods are contemporary, widely-used systems with engineered solutions and scientific foundations. Conventional process includes the use of physical, biological, and chemical process. Physical action cover the use of sedimentation process, biological process includes the use of microorganism to treat the waste water and chemical processes includes the sorption, catalysis, photocatalysis, degradation and conversion.

Recovering valuable resources from sewage—such as metals and nutrients like phosphorus and nitrogen—becomes more crucial, supporting the circular economy and sustainability. This procedure produces resources that can be used again in business and agriculture, in addition to aiding in environmental protection. Given its potential to address important environmental and resource sustainability challenges, the recovery of nutrients and metals from sewage is of utmost relevance on a global scale. Sewage systems are becoming overburdened due to increased urbanization and industrialization, which results in the loss of important nutrients like phosphate and nitrogen as well as the discharge of dangerous metal contaminants into water bodies. In addition to lowering pollution, recovering these resources allows us to reuse valuable materials and support the circular economy. Additionally, this approach can help promote sustainable agriculture, lessen reliance on virgin resources, and alleviate the global shortage of vital metals and fertilizers. Additionally, it contributes significantly to reducing the environmental impact, promoting ecological health and human welfare.

The chemical process of treatment of sewage has useful in the recovery of nutrients, removal of toxic metals (Rathod et al. 2018b, 2018a) and degradation of toxic chemicals (Rathod et al. 2024). Currently there are various procedure were used to recover nutrients from sewage such as Precipitation, Ion exchange, Microbial use (Margonwar et al. 2024; Singh et al. 2024), Biosorption (Rathod et al. 2018b, 2019), Pyrolysis and Incineration, Membrain filteration, (Yakkala et al. 2021) biochar, hydrothermal carbonization. Nanomaterial (Rathod et al. 2016), Metal/Metal oxide (Singh et al. 2022), Rhizoremediation (Maske et al. 2025) and metal–organic frameworks (Rathod et al. 2023) are used for sorption of metals and degradation of toxic dyes are the promising technology for water remediation and sewage treatment. Microbial communities are also utilized for this motive (Singh et al. 2024; Margonwar et al. 2024). These technologies promote the circular economy and sustainable wastewater management by converting waste into useful resources. Maintaining ecological balance, cutting pollution, and protecting natural resources will all be made possible by the ongoing development of recovery strategies (Rathod et al. 2025; Maske et al. 2025).

This chapter has aim to provide a brief information about the process or nutrient and metal recovery from the sewage and their applicability in the prevention of water pollution. This chapter will help to understand the advantages of recovering nutrients and the process that are used in the sewage treatment. Additionally it include the future prospective of the development needed within this area.

7.2 Source of Sewage

Sewage is categorized broadly in Domestic sewage, Agricultural sewage, Industrial sewage and Commercial sewage (Fig. 7.1).

Domestic sewage: Wastewater produces at homes, mostly from cooking, cleaning, bathing, home garage, and using the restroom, is known as domestic sewage. Domestic sewage contains disease causing pathogenic microorganisms such as bacteria, viruses, and parasites, along with biodegradable substances like food scraps and human excreta. Nutrients like nitrogen and phosphorus are mostly derived from the cleaning products and human waste can cause eutrophication and water pollution if discharged into water bodies without proper treatment. Hair and other tiny debris, like food waste, do not dissolve in water. Household cleaning, disinfection and personal hygiene products, often include hazardous ingredients namely phosphates, ammonia, and chlorine. The management of domestic sewage is essential to preserving environmental quality and public health since it constitutes a sizeable portion of urban wastewater (Aiyuk et al. 2006).

Agricultural sewage: Field runoff includes sediments from agricultural operations, animal waste, chemicals, and fertilizers. Livestock manure from animal farms, which

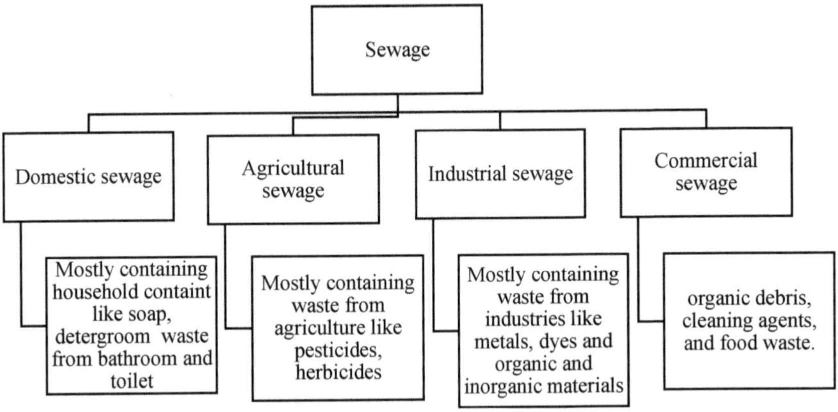

Fig. 7.1 Sewage types and their contains

contaminate water sources with infections, nutrients (like phosphorus and nitrogen), and organic debris.

Industrial Waste: Production facilities and manufacturing sites wastewater from industrial operations includes oils, hazardous effluents, toxic compounds, and heavy metals.

Process and cooling water: Occasionally tainted with heat or chemicals, this water is used in manufacturing and other industries to cool machinery (Zhang et al. 2022b).

Commercial Sewage: Schools, Hospitals, and Restaurants Large-scale commercial activities wastewater, which contains organic debris, cleaning agents, and food waste.

These sources add to the total amount of sewage, which has to be appropriately handled and cleaned in order to avoid polluting the environment and safeguard public health. In order to clean up wastewater and release it into rivers, lakes, or the ocean, sewage treatment plants are necessary.

7.3 Need of Recovery of Nutrients and Metals from Sewage

A number of environmental, financial, and sustainability issues have been drawn a lot of attention to the recovery of metals and minerals from sewage. Recovery is crucial because of the jump up worldwide call for these commodities and the negative environmental effects of their loss from the discharge of untreated wastewater. Nutrient and metal recovery from sewage is crucial for the following main reasons Table 7.1 represents the important contents of sewage.

Table 7.1 Represents the Sewage and containing nutrients and their uses

Sewage	Nutrients in the sewage	Uses of recovered nutrients
Domestic sewage	• Nitrogen, phosphorus • Carbon, potassium • Calcium, magnesium, sulfur	Nutrients were used to increase soil fertility
Agricultural sewage	• Nitrogen, phosphorus, potassium, calcium, manganese, boron, copper • Iron, nickel, zinc	Used to increase the plant growth
Industrial sewage	• Nitrogen, phosphorus, • Potassium, calcium, magnesium	Used as fertilizers for crops
Commercial sewage	• Nitrogen, phosphorus, potassium	Used as fertilizers for crops

7.3.1 *Environmental Protection*

7.3.1.1 Prevention of Eutrophication

Water contamination is mostly caused by nutrients found in domestic sewage, such as phosphorus and nitrogen. Untreated sewage containing these nutrients can cause eutrophication, an algae bloom that lowers oxygen levels in water and damages the aquatic life of flora and fauna, when it is dumped into bodies of water. This biological imbalance can be avoided by recovering nutrients before they enter into the environment, protecting aquatic habitats and preserving water quality (Khan and Mohammad 2014).

7.3.1.2 Reduction of Toxic Metal Contamination

Metals are frequently detected in sewage from industrial and urban runoff, particularly heavy metals like cadmium, lead, and mercury. These metals can build in soils and aquatic bodies if improper management is not practiced, causing serious health dangers to both people and wildlife. By ensuring that these potentially harmful materials are identified and eliminated, metal recovery from sewage helps to lower the environmental influence of these pollutants (Camargo et al. 2016).

7.3.2 *Resource Conservation*

7.3.2.1 Phosphorus Scarcity

As an important ingredient of fertilizers, phosphorus is a narrow, non-renewable assets which is necessary for agricultural sector. The main supply of phosphorus, phosphate rock, is steadily declining globally, which raising concerns about its future availability.

By removing phosphorus from sewage, one may ensure long-term agricultural productivity and reduce reliance on mining by finding a sustainable alternative source of this essential nutrient (Taware et al. 2024).

7.3.2.2 Sustainable Nitrogen Use

Another nutrient which is essential for plant growth is nitrogen, but conventional techniques for producing nitrogen, including the Haber–Bosch method, is energy-concentrated and based on fossil fuels. We can lessen the need for energy-intensive industrial nitrogen production by recovering nitrogen from sewage in forms like ammonia, which will help save energy and cut greenhouse gas emissions (Sugurbekova et al. 2023).

7.3.2.3 Conservation of Metal Resources

Metals like copper, zinc, and even traces of precious metals like gold and silver present in sewage, are important to many industries.

Recovering these metals promotes the growth of a circular economy—one in which waste products are reused—reduces the demand for mining, and helps preserve natural resources.

7.3.3 Financial Gains

7.3.3.1 Lowering the Production Cost of Fertilizer

It is possible to reduce expenses for farmers and lessen their dependency on chemical fertilizers made from mined minerals by recovering nutrients like nitrogen and phosphorus from sewage. This is because sewage is a cheap and sustainable supply of fertilizers.

7.3.3.2 Profits from Metals Recovered

The repurposing or selling of recovered metals opens up new markets for wastewater treatment facilities to make money. This can increase the financial viability of sewage treatment, particularly for facilities that handle industrial effluent (Otoo et al. 2015).

7.3.3.3 Economical Wastewater Management

Some of the operational expenses related to wastewater treatment can be compensated for by resource recovery. Treatment facilities can provide marketable byproducts and lower the cost of managing trash by recovering valuable elements.

7.3.4 Ecological Disposal and Recycling

7.3.4.1 Reducing Waste

Recovering nutrients and metals decreases the quantity of sewage sludge that require to be processed or discard, which helps minimize waste. As a result, there is less garbage going to landfills, disposal expenses go down, and the environmental impact goes down.

7.3.4.2 Encouragement of a Circular Economy

The recuperation of assets from garbage is consistent with the circular economy's tenets, which advocate for converting waste into useful resources rather than discarding it. This strategy encourages sustainable consumption and production, conserves natural resources, and lowers waste overall (Otoo et al. 2015).

7.3.5 Reducing the Risk to Public Health:

The regaining of nutrients and metals from garbage lowers the hazards to public health associated with contaminated water sources by ensnaring and eliminating hazardous microorganisms, toxic metals, and excess nutrients. Sewage that is improperly managed or left untreated can transmit waterborne illnesses like typhoid, dysentery, and cholera. Communities will have safer water and cleaner wastewater when nutrients and metals are recovered properly.

7.3.6 Energy Efficiency and the Depletion of Greenhouse Gas Emissions

Mining and chemical manufacture are two examples of traditional ways of nutrient and metal extraction that are frequently energy-intensive and have a substantial impact on greenhouse gas emissions. It is possible to recover these resources from sewage in a far more energy-efficient manner. Further reducing emissions and saving energy are procedures like anaerobic digestion, which is used in sewage treatment, that generate biogas, which can be utilized as an energy origin.

7.3.7 Demand for Sustainable Agriculture Practices Worldwide

The need for sustainable farming practices is rising as the world's population rises. By recovering nutrients like phosphorus and nitrogen from sewage, fertilizers that are renewable and environmentally benign can be made, promoting sustainable food supply and lessening the negative result of agriculture on the environment.

Sustainable waste management, resource conservation, economic gains, and environmental protection all depend on the recovery of metals and nutrients from sewage. By considering sewage as an asset rather than a waste product, we can lower pollution levels, protect vital resources like metals and phosphorus, and advance a circular economy that is more sustainable.

7.4 Processes Involved in the Regaining of Nutrients and Metals from Sewage

7.4.1 Regaining of Nutrients

Numerous chemical, biological, and physical processes are involved in the regaining of nutrients—mainly nitrogen and phosphorus from sewage. The goal of these procedures is to remove beneficial nutrients from wastewater streams so that they can be used again as fertilizer or for other purposes. The following are the primary procedures in nutrient recovery from sewage:

7.4.1.1 Struvite Precipitation (Regaining of Phosphorus and Nitrogen)

Struvite, or Magnesium ammonium phosphate, is a crystalline substance that is rich in nitrogen and phosphorus that may be extracted from sewage. Because struvite is a slow-release fertilizer, the technique is frequently employed.

$$\mathrm{NH_4} + \mathrm{PO_4^{3-}} + \mathrm{Mg^{2+}} \rightarrow \mathrm{MgNH_4PO_4} \cdot 6\mathrm{H_2O}(\mathrm{Struvite})$$

In wastewater, phosphate $\left(\mathrm{PO_4^{3-}}\right)$, ammonium $\left(\mathrm{NH_4^+}\right)$, and magnesium ($\mathrm{Mg^{2+}}$) ions can combine to make struvite in specific ratios. Precipitation is started by adjusting the pH to about 8–9 and adding magnesium salts. The wastewater contains struvite crystals that precipitate out and can be collected as a solid that is subsequently turned into fertilizer (Wei et al. 2018).

7.4.1.2 Enhanced Biological Removal of Phosphorus (EBPR)

EBPR is a biological technique intended to eliminate phosphorus from effluent. In this method particular kinds of bacteria (phosphorus-accumulating organisms (PAOs)) that store phosphorus inside their cells were used. Extra phosphorus from the wastewater is absorbed by PAOs under aerobic circumstances, where it is stored as polyphosphate within their cells. In anaerobic conditions these phosphorus-accumulating organisms (PAOs) release the phosphorus that has been stored in their cells.

7.4.1.3 Stripping of Ammonia (Recovery of Nitrogen)

Nitrogen is extracted from wastewater as ammonia gas using a technique called ammonia stripping. In industries where nitrogen removal or recovery is required, this technique is commonly utilized. During this procedure, alkaline materials such as sodium hydroxide or lime are utilized to elevate the pH of the effluent (above 10). After that, the wastewater is put into an aeration tank or stripping tower and forced through with steam or air. Ammonium ions $\left(NH_4^+\right)$ are transformed into ammonia gas (NH_3) at high pH levels, and the air or steam removes this gas. The ammonia gas has been collected and can be treated with H_2SO_4, to produce ammonium sulfate $(NH_4)_2SO_4$ of fertilizer-grade.

7.4.1.4 Ion Exchange

A physico-chemical method called ion exchange is utilized to eliminate ammonium ions from effluent in a targeted manner. This technique works especially well for recovering concentrated types of nitrogen. Ion-exchange columns with resins specific for ammonium ions $\left(NH_4^+\right)$ are used to filter wastewater. These resins bind ammonium from wastewater. After the resins are saturated with ammonium ion's they are regenerated by passing through a solution that releases ammonium ions, such as a salt solution.

After that, the ammonium-rich regeneration solution can be treated to create fertilizer-grade ammonium salts, such as ammonium sulfate.

7.4.1.5 Anammox Process

Anammox bacteria perform the process under anaerobic conditions, converting ammonium and nitrite directly into nitrogen gas. The nitrogen gas produced is harmless and is released into the atmosphere.

$$NH_4^+ + NO_2^- \rightarrow N_2 + 2H_2O$$

7.4.1.6 Membrane Filtration (Phosphorus and Nitrogen Recovery)

Reverse osmosis and nanofiltration are two membrane filtration techniques that are utilized to regain nutrients from wastewater by removing impurities and concentrating nutrients. Wastewater is directed via membranes that hold onto bigger molecules, such as nitrogen and phosphorus compounds, while allowing only water and tiny ions to flow through. It is possible to recover the nutrients from the concentrated, nutrient-rich stream that has been kept. After that, concentrated solutions containing nitrogen and phosphorus can be recovered and processed to create fertilizers that are useful.

7.4.1.7 Chemical Precipitation (Recovery of Phosphorus)

Chemical precipitation is the process of introducing chemicals into wastewater in order to precipitate recoverable and distinct phosphorus components. The effluent is treated using chemicals such as lime, ferric chloride, or aluminum sulfate (alum) to cause phosphate ions to react. These substances combine to generate insoluble phosphate compounds that precipitate out of the solution, such as calcium phosphate or aluminum phosphate. The sludge that results from the precipitation of phosphorus compounds can be treated to extract phosphorus for use as fertilizer.

7.4.1.8 Hydrothermal Carbonization Process (HTC)

Sewage sludge is converted into hydrochar, a carbon-rich solid that may also contain phosphorus and nitrogen, using a process known as hydrothermal carbonization (HTC). After processing, the nutrients in the hydrochar can be extracted. Hydrochar is obtained by keeping sewage sludge to a high pressure and moderate temperature in water. From the hydrochar, phosphorus and nitrogen can be separated and utilized to make fertilizer or other goods.

7.4.2 Recovery of Metals

Sewage sludge frequently contains a range of metals, including both valuable and hazardous ones. Sludge can contain a variety of metals from home sources, urban runoff, and industrial discharges, including nickel (Ni), copper (Cu), zinc (Zn), cadmium (Cd), chromium (Cr), and lead (Pb). Precious metals like gold (Au) and

Table 7.2 Represents the metals to be recovered by using different processes

Process	Recovered metals
Acid leaching	Iron, zinc, nickel and copper
Biological leaching	Zinc, nikel and potassium
Electrokinetic separation	Potassium
Pyrometallurgical processing	Zinc, lead and manganese
Supercritical water oxidation	Potassium and sodium
Wet oxidation	Potassium
Phytomining	Copper
Electrowinning	Gold, silver, copper, cadmium and zinc

silver (Ag) are also likely to be present. By reusing the metals, recovering metals from sewage sludge may both protect the environment and generate revenue (Table 7.2).

7.4.2.1 Acid Leaching

Acid leaching is a chemical method that allows metals to be recovered from sludge by dissolving the metals in acidic solutions (Jung and Lovitt 2011). This method works very well for recovering metals like nickel, zinc, and copper. The sludge is treated with acids, usually hydrochloric acid (HCl) or sulfuric acid (H_2SO_4).

$$\text{Metal(M)} + \text{HCl} \rightarrow \text{Metal chloride complex } (\text{MCl}_2)$$

The acid dissolves the metals into a liquid solution, separating them from the sludge's solid matrix. The metal-rich liquid is then separated from the residual sludge. The metals can be recovered from the leachate (the acidic solution containing dissolved metals) by additional processing, such as solvent extraction, precipitation, or electrowinning.

7.4.2.2 Biological Leaching

The environmentally benign process of bioleaching, or biological Leaching, is a popular mining technique that uses microorganisms, specifically acidophilic bacteria, to mobilize and extract metals from sewage sludge. It is a kin to natural weathering (Pathak et al. 2009).

The sludge is mixed with acidophilic bacteria, such as *Acidithiobacillus thiooxidans* or *Thiobacillus ferrooxidans*. Sulfuric acid, which is created when these bacteria oxidize sulfide compounds, dissolves metals like copper, zinc, and nickel. Similar to acid leaching, sulfuric acid, which is produced by bacterial activity, dissolves metals into the liquid phase.

$$4FeS_2 + 15O_2 + 8H_2O \rightarrow 2Fe_2(SO_4)_3 + 8H_2SO_4$$

Using methods like electrowinning or precipitation, metals are extracted from the leachate.

7.4.2.3 Electrokinetic Separation

Metals are mobilized and extracted from sewage sludge via electrokinetic separation, which makes use of electric fields. This procedure is especially helpful for recovering and removing heavy metals. Metal ions move toward the electrodes when an electric current is run through the sludge, producing an electric field. The negative electrode, or cathode, is approached by positively charged metal ions, or cations, whereas the positive electrode, or anode, is approached by negatively charged metal ions. After migrating, metals may precipitate from the solution or gather close to the electrodes (Wang et al. 2022).

7.4.2.4 Pyrometallurgical Processing (Thermal Treatment)

In pyrometallurgical processing, metals are recovered from sewage sludge by treating it at a high temperature. The metallurgical and mining sectors make extensive use of this technique for metal extraction. High temperatures (600–1200 °C) are used to pyrolyze or burn sewage sludge, which concentrates the metals in the ash and lowers its organic content. The resulting ash has a high metal content. There are several ways to remove the metals from the ash, such as: Smelting (for lead, zinc, and other non-ferrous metals). Leaching (using acidic solutions to extract metals from ash). Hydrometallurgical treatment: the extraction of metals from ash by aqueous chemistry (Jiabao et al. 2023).

7.4.2.5 Supercritical Water Oxidation (SCWO)

The sophisticated oxidation method known as supercritical water oxidation (SCWO) oxidizes organic content in sludge while retaining inorganic components like metals. It works by operating at high temperatures and pressures. Supercritical Water Oxidation: When sludge is subjected to temperatures and pressures higher than 374 °C and 22 MPa, water becomes supercritical, meaning it is neither liquid nor gas. In this condition, organic molecules are soluble in water and quickly oxidized by oxygen. Inorganic metals remain as residue after the sludge's organic components have entirely oxidized. Leaching and heat processing are two techniques used to gather and process the residue that is rich in metals in order to recover the metal (Griffith and Raymond 2002).

7.4.2.6 Wet Oxidation

In the wet oxidation process, sewage sludge is treated with an oxidizing agent typically air or oxygen at a moderate temperature and pressure. Organic components are oxidized, and metal-containing residue is left behind. Wet oxidation is a process that allows organic materials to be oxidized by treating sludge in a liquid phase under pressure and at temperatures between 150 and 320 °C. A residue rich in metals is left over after the process, which can be treated to extract and recover valuable metals (Munir et al. 2019).

7.4.2.7 Phytomining

Phytomining is a new technique that pulls metals out of contaminated soil or sludge by using plants known as hyperaccumulators. This approach is more environmentally friendly and sustainable. Some plants are grown in places treated with metal-rich sludge, such as *Pteris vittata* (brake fern) and *Brassica juncea* (Indian mustard). Metals from the sludge seep into the tissues of these plants. After the plants have gathered metals, they are harvested and burned, and the concentrated metals in the ash that remains can be retrieved. Plant ash can have its metals removed through chemical leaching or other methods of recovery (Tognacchini et al. 2020).

7.4.2.8 Electrowinning

Metals are recovered from sludge via the electrochemical process of electrowinning, which occurs after the metals have been dissolved in a solution (typically by leaching or bioleaching). Following the dissolution of the metals in a solution, metal ions are drawn to the cathode by an electric current, where they are reduced and solidified. After that, the recovered metals are gathered as a solid on the cathode plates, where they can be cleaned and utilized again.

7.5 Current Progress in the Regaining of Metals and Nutrients from Sludge

Sludge, especially wastewater sludge, can include nutrients and metals that can be recovered. This is becoming more and more of a priority for resource recovery and environmental sustainability. This approach can help the circular economy by turning garbage into useful items. An outline of recent developments in this field is provided below.

7.5.1 Nutrients Removal

Removal of nutrients like phosphorous, Nitrogen and Potassium is the important to get environmental sustainability.

7.5.1.1 Phophorous Removal

The process of eutrophication, which causes an excessive development of algae and other photosynthesizing water plants, can be prevented in large part by removing phosphorus from household and commercial wastewater (Cloete and Bosch 1994). For such a reason, the amount of phosphate in effluents from dams and rivers is limited in many nations. Detection of phosphorus from garbage is only achievable via fusion of phosphate into a solid form, that may finally be isolate from the water form. Biological phosphate elimination and chemical precipitation are two examples of such processes (Cloete 2001).

Using sewage mud as a direct fertilizer on agricultural land is one method of recycling phosphorus. Because of this high phosphorus amount, sewage mud might be a good substitute supply of phosphate fertilizer. However, because sewage sludge contains organic contaminants as well as potentially harmful substances like heavy metals, its direct application to farms and agricultural operations is debatable. It is imperative to remove any trace amounts of phosphorus from the sewage sludge and purify it (Pasalari et al. 2024).

The report by Hasan Pasalari and colleagues, evaluates both chemical and thermal technologies for regaining phosphorus from sewage mud. The research addressing the growing need for phosphorus recovery due to its finite nature and essential role in agriculture. The conclusion of the study highlights that chemical methods such as precipitation using magnesium and calcium compounds and thermal processes like pyrolysis and incineration are effective at recovering phosphorus from sewage sludge. However, these methods also present challenges, particularly in terms of cost, efficiency, and scalability. Although thermal methods are effective, they demand high energy consumption and often generate harmful by-products that require further management. The research emphasizes that while both methods have significant potential, there is still room for optimization to make phosphorus recovery more cost-efficient and environmentally sustainable (Pasalari et al. 2024).

The report by Cieslik and Konieczka (2017) provides a comprehensive review of phosphorus regaining processes used during different stages of wastewater action towards and sewage mud administration. The article highlights the pressing need for phosphorus recovery due to the environmental concerns associated with its loss (eutrophication) and the economic need to recover this finite resource used in agriculture. The review covers both conventional methods like chemical precipitation and innovative approaches such as biological phosphorus removal and struvite crystallization (Cieślik and Konieczka 2017).

The report by Zhang et al. (2022a, b), focuses on an innovative method to regain phosphorus (P) from sewage slurry and wastewater by promoting the making of vivianite ($Fe_3(PO_4)_2{\cdot}8H_2O$). The importance of this work be found in the ability of vivianite as a valuable P-recovery product, which can be utilized as a slow-liberate fertilizer and help address the global phosphorus scarcity problem. The study highlights the efficiency of vivianite crystallization, which occurs in the presence of iron under anaerobic conditions, as a method for phosphorus recovery. The conclusion emphasizes that vivianite recovery not only offers a viable pathway for sustainable phosphorus recycling but also contributes to reducing the environmental impact of wastewater treatment processes (Zhang et al. 2022a).

The article by Meng et al. (2019) focuses on phosphorus regain from various thermal actioned products of sewage mud, emphasizing the environmental and economic importance of phosphorus as a non-renewable resource. The study examines various methods like incineration, pyrolysis, and hydrothermal treatments, which transform sewage sludge into ash or biochar, offering potential pathways for phosphorus recovery. The review concludes that various thermal treatments can be effective in recovering phosphorus, the option of process turn on several causes, such as energy consumption, cost-efficiency, and the quality of the recovered phosphorus. In particular, Pyrolysis method is known for producing biochar with high phosphorus content, which is suitable for the use in agriculture field. However, challenges such as heavy metal pollutant and the complexity of phosphorus extraction from incinerated ash remain key issues to address before extensive execution (Meng et al. 2019).

The study by Havukainen et al. (2016) explores the capacity of phosphorus regaining from sewage mud and compost using thermochemical action, specifically focusing on the Ash Deactivation and Enrichment Cycle (ASH DEC) method. The ASH DEC method, in particular, shows efficiency in removing heavy metals from ash and producing a phosphorus-rich product (Havukainen et al. 2016).

The study by J. D. Marin-Batista et al. focuses on the capability of hydrothermal carbonization (HTC) to convert digestate from anaerobic digestion of sewage mud into valuable substances for phosphorus (P) recovery and energy production. The digestate was subjected to hydrothermal action at 180–240 °C ranging temperatures. The study observed that while higher temperatures did not produce high rank hydrochars, they significantly affected the yield of organic phosphorus in the hydrochar. Acid leaching was employed on hydrochar o at 180 °C, followed by precipitation with calcium oxide (CaO) to recover phosphorus. The research concludes that hydrothermal carbonization can effectively valorize sewage sludge digestate by recovering phosphorus and generating energy. The process not only enhances phosphorus recovery but also produces a solid fuel with promising energy content (Marin-Batista et al. 2020).

The research by Atienza-Martínez et al. (2014) explores the regaining of phosphorus from char ash derived from the pyrolysis of sewage mud. The research highlights the impact of different temperatures during the pyrolysis process on phosphorus recovery efficiency. The study concludes that phosphorus recovery is feasible and effective from char ash, suggesting that optimized thermal treatments can enhance recovery rates and contribute to sustainable waste management (Atienza–Martínez et al. 2014).

The report by Thabo Falayi, investigates the phosphorus (P) regaining from sewage mud using an alkaline drain process. The research is significant because it addresses two critical issues: the efficient recovery of phosphorus, a vital nutrient in agriculture, and the stabilization of the remaining sewage sludge, which can pose environmental challenges if not managed properly. Sewage sludge is treated with an alkaline solution, typically sodium hydroxide (NaOH), to solubilize phosphorus, making it easier to recover. The study found that the alkaline treatment effectively liberated phosphorus from the sewage mud. Additionally, the action improved the stability of the sludge residue, reducing potential environmental risks associated with its disposal (Falayi 2019).

The study by C. Adam et al. investigates a thermochemical treatment process for regaining phosphorus from sewage slurry ashes (SSA). The process involves two main steps: mono-incineration of municipal sewage mud taken by thermochemical action of the resulting ashes. The initial incineration step completely destroys organic pollutants in the sewage sludge, leaving behind SSA that contains phosphorus but also heavy metals, which are undesirable in agricultural contexts. The principal motto of thermochemical action was to enhance the bioavailability of phosphorus and to reduce heavy metal content. The treatment resulted in the making of new mineral states (such as farringtonite, chlorapatite, and stanfieldite) which significantly improved the bioactivities of phosphorus in the ashes from about 20–46% to 76–100% in the treated samples. The process also effectively decrease the concentration of heavy metals in the final outcome, making them safer for potential agricultural use (Adam et al. 2009).

The report by Paula Guedes and colleagues attentive on the regaining of phosphorus from sewage slurry ash using an electrodialytic (ED) process. Two different sewage sludge ash (SSA) samples were analyzed: one taken right after ignite (SA) and the other referenced from an open storage site (SB). These two samples underwent ED treatment as mixed adjournment in H_2SO_4 for varying durations of 3, 7, and 14 days. After 14 days, a significant amount of phosphorus was directed towards the anode (about 60 % from SA and 70 % from SB), while heavy metals predominantly migrated to the cathode. The resulting analyte comprised 98 % phosphorus, primarily in the form of orthophosphate, with only 2 % heavy metals present. Overall, the electrodialytic process proved effective for phosphorus recovery from SSA by successfully separating it from heavy metal contaminants was observed (Guedes et al. 2014).

The study by Lee et al. (2018) investigates phosphorus regaining from sewage mud make use of calcium silicate hydrates (CSH). The procedure involved the use of acid and base leaching solutions to elute phosphorus, followed by comparing the recovery efficiency of CSH with other calcium compounds. The findings indicated that about 55% of phosphorus in dried sewage mud could be regained effectively with CSH, suggesting its potential as a sustainable fertilizer source (Lee et al. 2018).

The review article by Kwapinski et al. (2021) probes on various thermal treatment technologies for sewage sludge, emphasizing their effectiveness in phosphorus recovery. The review outlines different thermal processes, including pyrolysis, incineration, gasification, and hydrothermal carbonization. Each method involves specific conditions for heat application, temperature ranges, and time durations, influencing the resultant product's characteristics and phosphorus availability. The processes typically involve subjecting dewatered sewage sludge to high temperatures in controlled environments to make easy the crash of organic matter and the regaining of valuable nutrients, particularly phosphorus (Kwapinski et al. 2021).

The study titled Phosphorous regaining from sewage discharge by vanadium-exchanged composite integrated via go back recrystallization path by Willbrynner Pereira Marques et al. focuses on synthesizing a composite material aimed at recovering phosphorus from sewage effluent. The researchers used a reverse co-precipitation method to synthesize a vanadium-substituted composite. This method involves the simultaneous precipitation of vanadium with other metal ions from the effluent, forming a composite material capable of adsorbing phosphorus. Optimal adsorption occurred at a specific pH, and the study identified the kinetics and isotherm models that best described the adsorption process, indicating a strong interaction between phosphorus and the composite material (Pereira Marques et al. 2024).

The Volpe M. et al. looks into the capability of hydrothermal carbonization (HTC) as a method for operating sewage mud to regain phosphorus and produce a solid biofuel (hydrochar). The researchers used anaerobically digested sewage slurry from a wastewater action plant in Trento, Italy, with a total solid content of 21 wt%. The sludge was subjected to hydrothermal carbonization at 190 and 210 °C temperatures for durations of 1 and 3 h. The resulting hydrochar was filtered and then leached with a 4 M hydrochloric acid (HCl) solution at room temperature to extract phosphorus and other inorganic component. Phosphorus was recovered by precipitating phosphate salts through alkalization of the acidic leachate by taking a solution of 5 M sodium hydroxide (NaOH) to raise the pH up to 9 (Volpe et. al. 2020).

The report by Yi Xiao et al. investigates how phosphorus transformations occur during the ignition of sewage mud, focusing specifically on the effects of moisture and mineral additives. The researchers conducted incineration experiments of sewage mud samples at various temperatures. They introduced steam and various minerals, such as magnesium chloride ($MgCl_2$) and calcium chloride ($CaCl_2$), to observe their effects on phosphorus transformations. After incineration, the residual ash was collected and analyzed (Xiao et al. 2021).

The research by Akira Sano et al. focuses on developing an alkali removal process for phosphorus regaining from raw sewage sludge that contains iron phosphate ($FePO_4$). The researchers optimized the conditions for alkali extraction by examining the extraction characteristics of phosphorus alongside organic and inorganic matter. The results indicated that unheated alkali extraction proved to be more effective for extracting phosphorus from $FePO_4$-containing sewage sludge than previously employed techniques using incinerated ash (Sano et al. 2012).

The report by Yu et al. (2021) focuses on an creative approach to regain phosphorus from sewage mud ash. The analysis utilized a specific adsorbent designed to selectively capture phosphorus from the ignited sewage slurry ash (ISSA). The adsorbent demonstrated high efficiency in capturing phosphorus from the ISSA under optimized conditions (Yu et al. 2021).

7.5.1.2 Nitrogen Removal

The Katarzyna Chojnacka et al. reports the innovative proposal to regain nitrogen from sewage mud. The authors describes the various methods to recover nitrogen from wastewater treatment processes, particularly focusing on technologies that can transform nitrogen in sewage sludge into reusable fertilizer forms (Chojnacka et al. 2023).

The report by Beckinghausen et al. (2020) explores different nitrogen recovery techniques, emphasizing the need for a shift from traditional nitrogen removal methods to more sustainable recovery approaches. Authors examined various methodologies including gas permeable membranes, vacuum membrane distillation, and ion exchange processes. Report provide that Gas Permeable Membrane Technology is a good method for ammonia regain, offering low energy consumption and potential for revenue through product sale (Beckinghausen et al. 2020).

The research by Van der Hoek et al. investigates methods to regain nitrogen from sewage, addressing the growing need for sustainable agricultural practices. The research focuses on the wastewater operation plant (WWOP) Amsterdam-West and evaluates various nitrogen recovery technologies based on sustainability, practicality, and the concentration of nitrogen in the wastewater. The Haber–Bosch process produces Nitrogen-based fertilizers by the industrial obsession with of N_2 into NH_3. Sustainability regarding energy utilize and dinitrogen emissions and ability to regain nitrogen in usable forms (Van der Hoek et al. 2018).

Gong et al. (2017) focuses on improving the efficiency of energy and nutrient regaining from waste water. The research explored a novel method of sewage treatment using a membrane-based pre-concentration (MPC) process combined with ion exchange (IE). The first step was pre-concentrating the sewage to recover organic compounds. This step helps increase the chemical oxygen demand (COD) mass, that can be utilized for energy regaining, particularly via anaerobic digestion (AD). Following MPC, the wastewater was treated using ion exchange to regain nitrogen in the shape of ammonium ($NH_4{}^+$-N). Batch investigations were executed to control the efficiency of ammonium recovery. The ion exchange process recovered 37.5 % of ammonium $\left(NH_4^+\text{-}N\right)$, although it faced challenges with selectivity for ammonium over other ions like hard ions (e.g., calcium, magnesium). Energy Efficiency: The overall process could recover 0.38 kWh/m^3 of energy, with 0.26 kWh/m^3 from the digestion of organic matter and 0.12 kWh/m^3 from nitrogen reuse (Gong et al. 2017).

The study by Mika Horttanainen, Ivan Deviatkin, and Jouni Havukainen focuses on the release of nitrogen from mechanically evaporated sewage mud throughout thermal drying and its capability for regaining. The research aims to analyze the nitrogen emissions that occur through the thermal drying of sewage mud and to evaluate the potential for nitrogen recovery from these emissions. The researchers conducted a series of thermal drying experiments using mechanically dewatered sewage sludge. Throughout the time of the drying action, they monitored the release of nitrogen compounds, primarily ammonia, into the drying fumes. The study also investigated different nitrogen recovery techniques, including air stripping and scrubbing processes, which aimed to capture and convert the nitrogen released during the drying phase into useful products like ammonium sulfate fertilizers. This research highlights the importance of integrating nitrogen recovery processes into sludge management practices to both improve sustainability and reduce nitrogen emissions (Horttanainen et al. 2017).

Munasinghe-Arachchige et al. (2020a, b) explores an inventive method for nitrogen recovery from sewage sludge. This research pivots on capturing ammoniacal nitrogen from the centrate (a liquid byproduct of anaerobic digestion) using a gas-permeable membrane system. The study employs a non-pressurized membrane reactor where ammoniacal nitrogen in the centrate is selectively transferred across a hydrophobic membrane. The membrane's opposite side, sulfuric acid absorbs the NH_3, leading to the production of ammonium sulfate, a widely used nitrogen fertilizer. The efficiency of the membrane in capturing nitrogen was tested, and the resulting ammonium sulfate was analyzed for its purity and safety for use in agriculture. The system demonstrated a nitrogen recovery efficiency of up to 98%, making it highly effective for converting waste nitrogen into a usable form. This approach offers a promising solution for recycling nitrogen from wastewater, reducing environmental impact, and creating valuable fertilizers for crop production (Munasinghe-Arachchige et al. 2020b).

The study by Koskue et al. (2021a) astonished the depletion and regaining of nitrogen from actual digested sewage mud using an electro concentration process. The research primarily aimed to address the high ammonium $\left(NH_4^+\right)$ concentration in reject water, which is a byproduct of sewage operation, and to regain it for potential reutilize. The experiment utilized a three-chamber cell isolated by anion and cation exchange membranes. The cathode chamber received the reject water, where ammonium was changed to ammonia gas (NH_3) because of the greater pH induced by the electrochemical reactions. This ammonia was stripped from the water and collected in a stripping column. The remaining liquid then moved through the anode chamber, where further electrochemical reactions facilitated additional recovery of ammonium ions. The concentrate was collected for further analysis. The system achieved up to 94 % nitrogen removal efficiency and an 87% nitrogen recovery rate. The influent, which started with about 983 mg/L of ammonium nitrogen, was reduced significantly to 57 mg/L in the treated water (Koskue et al. 2021a).

In the study by Thant Zin and Kim (2021), the researchers explored the concurrently regaining of nitrogen (N) and phosphorus (P) from sewage slurry ash (SSA) and food wastewater (FW) by utilizing magnesium (Mg)-enhanced biochar. This method aimed to address the dual challenges of nutrient recovery and waste recycling. The study used two types of waste biomass (palm tree trunk and ground coffee bean) to produce Mg-biochar, which was then applied to SSA and FW. The Mg-biochar facilitated the regaining of nutrients through a fusion of precipitation, adsorption, and cation-conversation processes, forming struvite (a slow-release fertilizer). This process targeted phosphate $\left(PO_4^{3-}\text{-P}\right)$ and ammonium $\left(NH_4^+\text{-N}\right)$, which are key nutrients in agricultural fertilizers. Palm tree trunk-derived Mg-biochar demonstrated a higher recovery efficiency, capturing 54.8 % of NH_4^+-N and 92.2 % of PO_4^{3-}-P. Ground coffee bean-derived Mg-biochar also performed well, recovering 79.5 % of PO_4^{3-}-P and 38.6 % of NH_4^+-N. The study concluded that this process supports both nutrient recovery and pollution mitigation while offering an economical way to produce struvite (Thant Zin and Kim 2021).

In the study by Munasinghe-Arachchige et al. (2020a, b), the researchers developed an innovative algal track for nutrient regaining from urban waste. The focus was on using algae to recover nitrogen (N) and phosphorus (P), which are often lost during traditional sewage treatment processes. The study utilized an algal-based sewage treatment and resource recovery (STaRR) system. The process leveraged algae's natural ability to assimilate nutrients like P and N from discharge waste. Unlike traditional bacterial processes, which struggle due to carbon limitations, the algal system efficiently captured nutrients, helping to reduce energy and carbon inputs in treatment. After nutrient accumulation by algae, downstream biomass processing technologies were applied to regain the nutrients in the configuration of $(NH_4)_2SO_4$ and struvite. The STaRR system demonstrated the ability to recover approximately 31 % of N and 74 % of P from the waste water. From 100 m^3 of settled sewage, about 1.7 kg of $(NH_4)_2SO_4$ and 2.5 kg of struvite were produced, both valuable fertilizers. The study highlighted the potential of algal systems to not only improve nutrient recovery but also decrease the environmental effect and energy utilisation of sewage treatment (Munasinghe-Arachchige et al. 2020a).

The study by Abeysiriwardana-Arachchige et al. (2020) compares nutrient elimination and regaining from municipal sewage using conventional and algal approaches. The researchers used both an algal-based system and conventional sewage treatment methods. The algal system was designed to capture nitrogen and phosphorus more efficiently, while the conventional system relied on established treatment technologies. The algal approach demonstrated better nutrient recovery efficiency, especially for phosphorus, which is critical for producing biofertilizers. The study supports algal systems as a sustainable alternative to traditional methods (Abeysiriwardana-Arachchige et al. 2020).

The study by He et al. (2015) focuses on a new process that combines hydrothermal deamination and air clearing to regain nitrogen from dried sewage sludge (DSS). The system uses accelerated hydrothermal deamination to break down nitrogenous compounds, followed by air stripping to recover the released ammonia. Three distinct deamination regimes were identified to optimize nitrogen release. The process achieved effective nitrogen removal and recovery, offering a more sustainable approach to handling sewage sludge compared to traditional methods. Nitrogen recovery significantly reduced emissions of nitrogen-containing pollutants, providing environmental and economic benefits (He et al. 2015).

The report by Rahimi et al. (2020) provides an extensive review of current biological system for nitrogen elimination from sewerage. The authors reviewed several biological nitrogen elimination technologies, including: Nitrification and Denitrification: Traditional biological techniques that change nitrogen composite into harmless N_2 gas. Anaerobic Ammonium Oxidation (Anammox): A more energy-efficient process compared to conventional nitrification–denitrification, especially suitable for nitrogen-rich wastewater. Bio-electrochemical Systems: These systems can directly recover nitrogen in the form of ammonia without needing additional chemicals or aeration, making them highly energy-efficient. Microalgal Growth: The use of microalgae for nitrogen recovery, which has the dual benefit of reducing nitrogen while generating valuable biomass for biofertilizers and biofuels. The study highlights that each technology has its own set of advantages and challenges, especially when it comes to efficiency under different operational conditions like temperature and wastewater composition. While traditional methods are energy-intensive, newer methods such as Anammox and bio-electrochemical systems offer more sustainable options, albeit with challenges related to stability and large-scale implementation. Systems like bio-electrochemical processes show promise for recovering nitrogen as ammonia, which can be reused in fertilizers, thus reducing the need for synthetic nitrogen production through energy-intensive methods like the Haber–Bosch process (Rahimi et al. 2020).

7.5.1.3 Potassium

The study concurrent regaining of P and K as magnesium potassium phosphate from artificial sewage mud wastewater by Nakao et al. (2016) investigates the regaining of P and K as magnesium potassium phosphate (MPP) from artificial sewage mud wastewater. Bench-scale experiments were conducted to decide the optimal circumstances for MPP precipitation. The optimal pH for MPP formation was identified as 11.5. The phosphorus concentration needed to be at minimal 3 mM, and the molar ratio of K had to be above 3 to promote the crystallization of MPP (Nakao et al. 2017).

The research by Takahashi et al. (2017) investigates a method for recovering phosphorus, particularly potassium phosphate, among the ash of burned sewage mud using an acidic treatment process. Phosphorus was extracted from the ash using sulfuric acid. This step resulted in a phosphorus-containing solution primarily composed of aluminum phosphate. To remove the aluminum component, potassium sulfate and sulfuric acid were added to the extracted solution. This led to the crystallization of alum (aluminum potassium sulfate), which was separated by filtration. The phosphorus in the form of potassium phosphate was regained after the addition of potassium hydroxide to the filtrate. This converts the aluminum phosphate into a usable form of potassium phosphate, which was confirmed through X-ray analysis. The process successfully recovered phosphorus as a mixture of potassium phosphate and potassium sulfate, which can be used as fertilizers. However, some phosphorus remained in the residue after the crystallization, indicating that further optimization is needed to enhance recovery efficiency (Takahashi et al. 2017).

The study by Wilsenach et al. (2007) explores the recovery of potassium and phosphate from origin-isolated urine using struvite crystallization. This process is a method for recovering valuable nutrients, such as magnesium ammonium phosphate (MAP) and potassium magnesium phosphate (KMP), from urine, which has a high phosphate concentration compared to other wastewater streams. The team used a stirred tank reactor to facilitate the precipitation of struvite (a crystal composed of magnesium, ammonium, and phosphate). Fluidized bed reactors were tested for the precipitation process. The higher concentration of phosphate in urine made the precipitation simpler and less costly compared to conventional processes. Liquid–solid separation was integrated within the reactor design to enhance the recovery efficiency of struvite. Both MAP (ammonium-based struvite) and KMP (potassium-based struvite) were successfully precipitated from urine, though MAP had better compaction and crystal formation. Approximately 95% of phosphate from the influent was removed during the process, with only slight differences in efficiency between ammonium and potassium struvite formations (Wilsenach et al. 2007).

Lucy Marta Schellin and Sandro José Froehner had investigated the potential to recover potassium from domestic sewage. The research focuses on the formation of sylvite (KCl) through precipitation processes. The procedure involves extracting potassium from sewage effluent and exploring the chemical conditions that facilitate the formation of sylvite. The study addresses the importance of resource recovery in circular wastewater treatment systems. The study found that adjusting the concentrations of potassium and other ions in the sewage effluent could lead to the efficient formation of sylvite. The process offers a dual advantage by reducing the environmental impact of potassium disposal and providing a valuable resource that can be repurposed as fertilizer. The research demonstrates that this method could be integrated into existing sewage treatment processes with minor modifications (Schellin and Froehner 2024).

7.5.2 Metal Removal

Metal recovery from sewage mud is the main process because of the presence of worthy metals like copper, zinc, nickel, and even rare earth elements. Several methods are employed to extract and recover these metals like Thermal Treatment (Incineration and Ash Leaching): Incineration reduces sludge volume, converting organic material to ash, where metals become concentrated. This ash then undergo acid leaching or alkaline leaching to extract metals like zinc, copper, and others. The ash can also be processed through hydrometallurgical techniques, which involve dissolving metals and separating them through chemical precipitation or solvent extraction. Bioleaching involves using microorganisms (such as *Acidithiobacillus ferrooxidans*) to oxidize metal sulfides in the sludge, leading to metal dissolution (Pathak et al. 2009). This method is particularly effective for recovering metals like copper, zinc, and nickel without the need for high energy inputs. This process can be performed under acidic conditions, and is environmentally friendly as it produces fewer toxic emissions than thermal processes. In some advanced methods, supercritical carbon dioxide is used to extract metals directly from sludge. This method has shown promise due to its ability to recover both metals and lipids from sludge under specific conditions. It is energy-efficient and offers an alternative to chemical leaching. Electrokinetic methods utilize an electric field to drive metal ions out of the sludge into a recovery solution. This technique can target metals like lead, cadmium, and chromium, which are often tightly bound to the sludge's solid matrix. Electrokinetic systems are particularly useful in recovering heavy metals while reducing the toxicity of the treated sludge (Wang et al. 2022).

The study deportation and regaining of heavy metals from sewage mud via a three-stage integrated exercise by Yesil et al. (2021) addresses the challenge of heavy metal contamination in sewage mud, which limits its use for land application. Ultrasonication Pretreatment was used to enhance the breakdown of sludge, making it more susceptible to the subsequent treatment stages. The sludge underwent anaerobic fermentation in an alkaline environment, which facilitated the hydrolysis of organic matter and partial metal leaching. Three pH conditions were tested: free pH, acidic, and alkaline. Microbial bioleaching was employed to further extract metals from the sludge. This process used the sludge's own microbial community to leach metals into solution. The leachate produced was processed using a hold up liquid membrane fertilized with a basic carrier, allowing for the particular regaining of metals from the solution. The mixing of alkaline fermentation and ultrasonication was the most effective, achieving 33.7% hydrolysis, 10.5% acidification, and 11–33% metal leaching. Bioavailability of poisonous heavy metals was reduced above 25%. The membrane-based recovery system showed an efficiency of 39–68% in reclaiming metals like copper, zinc, and nickel from the bioleaching effluent (Yesil et al. 2021).

The article by Mulchandani and Westerhoff (2016) discusses regaining chances for energy and metals from sewage muds. The authors research various thermal processes such as hydrothermal liquefaction and incineration, which lyse cells and release metals. They also cover liquid extraction methods, including conventional lipid extraction, supercritical CO_2 for both lipids and metals, and the use of acids and green solvents. The study highlights sustainability assessments, exploring the potential to recover valuable resources like metals while addressing environmental challenges (Mulchandani and Westerhoff 2016).

The review article by Pathak et al. (2009) explores the potential of bioleaching as an effective method for recovering heavy metals from sewage mud. The review discusses the role of specific microorganisms, particularly sulfur-oxidizing bacteria like *Acidithiobacillus ferrooxidans*, in the bioleaching process. These bacteria oxidize sulfide minerals, releasing metals in a soluble form. The review compiles various experimental setups and results from previous studies, including batch tests and continuous leaching systems, to compare the effectiveness of different microbial strains and conditions. The review highlights that bioleaching can effectively remove significant amounts of heavy metals (e.g., zinc, copper, cadmium) from sewage sludge, with efficiencies often exceeding 80% under optimal conditions (Pathak et al. 2009).

The article by Gu et al. (2018), reviews recent advancements in bioleaching technology and its applications for metal recovery from various waste materials, including sewage sludge and fuel ash. The authors detail various bioleaching mechanisms, highlighting the role of specific microorganisms (e.g., bacteria and fungi) that can solubilize metals through biochemical processes. Microorganisms oxidize metals, converting them into soluble forms. Bacteria such as *Acidithiobacillus ferrooxidans* produce organic acids that facilitate metal solubilization. The review highlights successful recovery rates of metals such as copper, zinc, and nickel from sewage sludge using bioleaching, with efficiencies often reaching above 90% under optimized conditions (Gu et al. 2018).

The article by Jacklin et al. (2021), investigates the effectiveness of using native plant species in biofilters to upgrade the standard of urban wastewater runoff, particularly regarding nutrient and metal removal. The researchers conducted a field study using biofilters populated with indigenous South African plants. These biofilters were installed in urban areas to capture stormwater runoff. The study focused on various indigenous plant species known for their ability to tolerate and uptake nutrients and metals from water. Specific plant types were chosen based on their ecological suitability and potential for effective biofiltration. Wastewater runoff samples were collected before and after passing through the biofilters. The efficiency of the biofilters was assessed by comparing the concentration of nutrients and metals in the stormwater before and after treatment. The study found that the indigenous plant biofilters significantly reduced concentrations of phosphorus and nitrogen in the wastewater runoff. Elimination ability was often more than 70% for these nutrients. The biofilters also successfully decreased gathering of heavy metals, including zinc and copper, with removal rates varying based on the specific metal and the plant species (Emmerich et al. 1982).

The article by Veeken and Hamelers (1999) investigates a method for eliminating heavy metals from sewage mud using carbonic acids as extractants. The researchers focused on the removal of heavy metals from dewatered sewage mud using various carbonic acids, together with acetic acid, citric acid, and formic acid. The extraction process involved mixing the sludge with the organic acid solutions at different concentrations. Several exercises were managed to judges the ability of metal extraction under varying conditions. The gathering of heavy metals in the sludge prior and behind treatment were measured using appropriate analytical techniques to determine the effectiveness of the extraction process. The study found that organic acids could effectively extract significant amounts of heavy metals from sewage mud. Citric acid was particularly noted for its high extraction efficiency, outperforming the other acids tested. The optimal conditions for metal extraction were identified, with specific concentrations of organic acids and pH levels that maximized removal rates. For instance, higher concentrations of citric acid led to increased extraction efficiencies. The extraction process allowed for the recovery of heavy metals in a soluble form, which could potentially be further processed for recycling or safe disposal (Veeken and Hamelers 1999).

The article by Otero et al. (2009), explores the utilize of sewage mud-obtained substances as sorbents for the extraction of heavy metals from hydrated mixtures. The researchers make ready sorbents from sewage sludge by applying thermal treatment and chemical exhilarating methods. This aimed to enhance the surface effects and adsorption capacity of the sludge-derived materials. A sequence of batch adsorption observations were supervised to assess the effectiveness of the sorbents in extracting various heavy metals from hydrated mixtures, counting copper (Cu), zinc (Zn) cadmium (Cd), and lead (Pb). The study assessed competitive effects by analyzing how the nearness of multiple heavy metals in the solution influenced the adsorption efficiency of each metal. Various initial concentrations of heavy metals were tested. The concentrations of heavy metals in the solution before and after treatment were measured using atomic absorption spectrometry (AAS) to determine the removal efficiencies. The study found that sewage sludge-based sorbents could effectively remove significant amounts of heavy metals from aqueous solutions. The removal efficiencies differing basing on the kind of metal, with lead and copper existence removed more effectively compared to cadmium and zinc (Otero et al. 2009).

7.6 Advantages and Disadvantages of Used Processes

Recovering nutrients and metals from sewage sludge involves various processes, everyone with its allow set of superiority and drawback. Here's a summary based on multiple sources:

7.6.1 Advantages

Nutrients like nitrogen, potassium, and phosphorus can be recovered and reused as fertilizers, lowering the necessity for synthetic fertilizers and upgrading circular economy practices (Mulchandani and Westerhoff 2016) Metals like zinc, lead, and copper can be recovered and recycled, reducing the environmental impact of mining new ores (Yunta et al. 2024).

Reduces the volume of waste that requires to be landfilled or reduce to ashes, minimizing environmental pollution and greenhouse gas emissions (Koskue et al. 2021b). Enhances the quality of treated water, reducing eutrophication and heavy metal contamination in waterways (Jacklin et al. 2015).

By recovering valuable resources, treatment facilities can create additional revenue streams, potentially lowering treatment costs. Promotes job creation in the recycling and resource recovery sectors (Mulchandani and Westerhoff 2016).

Supports sustainable wastewater management practices and helps meet environmental regulations by reducing contaminants in effluent (Horttanainen et al. 2020).

7.6.2 *Disadvantages*

Recovery processes is technically complex, requiring sophisticated equipment and processes, which may increase operational costs (Munasinghe-Arachchige et al. 2020a).

The efficiency of nutrient and metal recovery may vary significantly basing on the type of sludge, the processes used, and the operational circumstances (Gu et al., 2018; Otero et al., 2009).

There is a risk of mobilizing hazardous substances during recovery, particularly heavy metals, which may pose health and environmental risks if not governed properly (Yunta et al., 2024; Thant Zin and Kim 2021).

The recovery and reuse of materials from sewage sludge may face public skepticism regarding safety and potential contamination, which can hinder implementation (Mulchandani and Westerhoff, 2016).

Compliance with environmental regulations and safety standards can complicate recovery processes and limit the use of recovered materials (Horttanainen et al. 2020).

While the regaining of nutrients and metals from sewage sludge presents significant opportunities for resource conservation and environmental protection, it also poses challenges that need to be addressed to maximize benefits and minimize risks.

7.7 Future Scope and Development Needed.

The regaining of nutrients and metals from sewage mud presents numerous future research opportunities and potential advancements in sustainable wastewater management. Here are several key areas for future study:

Research focusing on optimizing existing recovery processes to increase the ability of nutrient and metal removal. This includes exploring various kinds of sorbents, bioleaching techniques, and extraction methods. Innovative technologies, such as electrochemical methods and membrane filtration, can be developed or refined to enhance recovery rates and reduce costs (Munasinghe-Arachchige et al. 2020a).

Integrating nutrient and metal recovery processes with energy recovery from sewage sludge, such as anaerobic digestion, can create synergistic systems that improve overall sustainability and reduce operational costs (Horttanainen et al. 2020; Munasinghe-Arachchige et al. 2020a).

Conducting comprehensive lifecycle assessments (LCA) of recovery processes can provide insights into the environmental impacts of nutrient and metal recovery compared to traditional waste management practices (Koskue et al. 2021b). Further research into the potential risks of contaminant mobilization during recovery processes, especially concerning heavy metals, is essential for safe implementation (Yunta et al. 2024).

Understanding public perception and societal acceptance of recovered materials can guide effective communication strategies and foster public trust in recovered products (Mulchandani and Westerhoff 2016). Development of guidelines and regulations for the safe use of recovered nutrients and metals can facilitate broader implementation in agriculture and industry.

Exploring the application of recovered nutrients in sustainable agricultural practices, including organic farming and precision agriculture, can maximize the benefits of nutrient recycling (Munasinghe-Arachchige et al. 2020a). Investigating the industrial applications of recovered metals, including their use in manufacturing and construction, can enhance resource recovery's economic viability (Pathak et al. 2009).

Research into advanced characterization techniques for analyzing the chemical composition and bioavailability of recovered nutrients and metals can inform their safe and effective use.

The future scope of research in the regaining of nutrients and metals from sewage sludge is broad and multifaceted. Emphasizing sustainable practices, technological innovation, and regulatory support will be crucial for maximizing the benefits of these recovery processes.

7.8 Conclusion

The recovery of nutrients and metals from sewage sludge plays a decisive role in sustainable waste management by reducing dependency on natural resources and minimizing environmental impacts. As natural ecosystems experience the depletion of essential nutrients like phosphorus and nitrogen, their recovery from sewage addresses these shortages while simultaneously mitigating the environmental hazards posed by heavy metal contamination. The process of metal recovery help to reduce the environmental risks associated with pollutants such as cadmium, lead, and copper, supporting cleaner water bodies and ecosystems. These recovery techniques reduce dependence on virgin raw materials while fostering the developement of a circular economy, where waste is transformed into valuable products, thereby closing the loop on resource utilization. This transition is crucial given global resource scarcity and the growing demand for sustainable agricultural practices. Furthermore, developing efficient technologies for nutrient and metal recovery contributes to energy savings and reduces greenhouse gas emissions, aligning with environmental sustainability goals.

Advanced methods such as struvite precipitation, membrane filtration, ion-exchange resins, and biochar application have shown great potential in recovering nutrients. For example, ion-exchange resins can achieve efficiencies of 85–98% in adsorbing targeted metal ions from wastewater, while reverse osmosis, a form of membrane filtration, can reach 95–99 % removal of metals and substantial nutrient recovery. Electrocoagulation involves passing an electric current through sewage to form metal ion flocs, can achieve recovery efficiencies of 90–95 %. Moreover, certain aquatic plants, like water hyacinth and aquatic grasses, naturally absorb metals and nutrients, with recovery efficiencies ranging from 50 to 80 %, depending on the plant species and environmental conditions. Microbial systems that break down organic matter can also remove nutrients such as nitrogen and phosphorus, with recovery rates of 70–90 %. In addition, hydrothermal carbonization (HTC) can convert sewage sludge into hydrochar while recovering metals at a rate of 80–90 %. These methods not only improve recovery rates but also enhance the environmental footprint of urban waste management by integrating these technologies into existing wastewater treatment systems. Ultimately, these advancements in resource recovery are vital for fostering sustainability, supporting circular economy models, and reducing the ecological impact of waste. This chapter provides a unified platform for understanding the significance of sewage treatment and nutrient/metal recovery as key components of a sustainable waste management strategy.

References

Abeysiriwardana-Arachchige ISA, Munasinghe-Arachchige SP, Delanka-Pedige HMK, Nirmalakhandan N (2020) Removal and recovery of nutrients from municipal sewage: algal vs. conventional approaches. Water Res 175:115709. https://doi.org/10.1016/j.watres.2020.115709

Adam C, Peplinski B, Michaelis M et al (2009) Thermochemical treatment of sewage sludge ashes for phosphorus recovery. Waste Manag 29:1122–1128. https://doi.org/10.1016/j.wasman.2008.09.011

Aiyuk S, Forrez I, Lieven DK et al (2006) Anaerobic and complementary treatment of domestic sewage in regions with hot climates—a review. Bioresour Technol 97:2225–2241. https://doi.org/10.1016/j.biortech.2005.05.015

Atienza–Martínez M, Gea G, Arauzo J et al (2014) Phosphorus recovery from sewage sludge char ash. Biomass Bioenergy 65:42–50. https://doi.org/10.1016/j.biombioe.2014.03.058

Beckinghausen A, Odlare M, Thorin E, Schwede S (2020) From removal to recovery: an evaluation of nitrogen recovery techniques from wastewater. Appl Energy 263:114616. https://doi.org/10.1016/j.apenergy.2020.114616

Camargo FP, Sérgio Tonello P, dos Santos ACA, Duarte ICS (2016) Removal of toxic metals from sewage sludge through chemical, physical, and biological treatments—a review. Water Air Soil Pollut 227:433. https://doi.org/10.1007/s11270-016-3141-3

Chojnacka K, Skrzypczak D, Szopa D et al (2023) Management of biological sewage sludge: fertilizer nitrogen recovery as the solution to fertilizer crisis. J Environ Manage 326:116602. https://doi.org/10.1016/j.jenvman.2022.116602

Cieślik B, Konieczka P (2017) A review of phosphorus recovery methods at various steps of wastewater treatment and sewage sludge management. The concept of no solid waste generation and analytical methods. J Clean Prod 142:1728–1740. https://doi.org/10.1016/j.jclepro.2016.11.116

Cloete T (2001) The role of extracellular exopolymers in the removal of phosphorus from activated sludge. Water Res 35:3595–3598. https://doi.org/10.1016/S0043-1354(01)00093-8

Cloete TE, Bosch M (1994) Acinetobacter cell biomass, growth stage and phosphorus uptake from activated sludge mixed liquor. Water Sci Technol 30:219–230. https://doi.org/10.2166/wst.1994.0562

Emmerich WE, Lund LJ, Page AL, Chang AC (1982) Movement of heavy metals in sewage sludge-treated soils. J Environ Qual 11:174–178. https://doi.org/10.2134/jeq1982.00472425001100020006x

Falayi T (2019) Alkaline recovery of phosphorous from sewage sludge and stabilisation of sewage sludge residue. Waste Manag 84:166–172. https://doi.org/10.1016/j.wasman.2018.11.041

Griffith JW, Raymond DH (2002) The first commercial supercritical water oxidation sludge processing plant. Waste Manag 22:453–459. https://doi.org/10.1016/S0956-053X(02)00029-6

Gong H, Wang Z, Zhang X et al (2017) Organics and nitrogen recovery from sewage via membrane-based pre-concentration combined with ion exchange process. Chem Eng J 311:13–19. https://doi.org/10.1016/j.cej.2016.11.068

Gu T, Rastegar SO, Mousavi SM et al (2018) Advances in bioleaching for recovery of metals and bioremediation of fuel ash and sewage sludge. Bioresour Technol 261:428–440. https://doi.org/10.1016/j.biortech.2018.04.033

Guedes P, Couto N, Ottosen LM, Ribeiro AB (2014) Phosphorus recovery from sewage sludge ash through an electrodialytic process. Waste Manag 34:886–892. https://doi.org/10.1016/j.wasman.2014.02.021

Havukainen J, Nguyen MT, Hermann L et al (2016) Potential of phosphorus recovery from sewage sludge and manure ash by thermochemical treatment. Waste Manag 49:221–229. https://doi.org/10.1016/j.wasman.2016.01.020

He C, Wang K, Yang Y et al (2015) Effective nitrogen removal and recovery from dewatered sewage sludge using a novel integrated system of accelerated hydrothermal deamination and air stripping. Environ Sci Technol 49:6872–6880. https://doi.org/10.1021/acs.est.5b00652

Heberer T, Reddersen K, Mechlinski A (2002) From municipal sewage to drinking water: fate and removal of pharmaceutical residues in the aquatic environment in urban areas. Water Sci Technol 46:81–88. https://doi.org/10.2166/wst.2002.0060

Horttanainen M, Deviatkin I, Havukainen J (2017) Nitrogen release from mechanically dewatered sewage sludge during thermal drying and potential for recovery. J Clean Prod 142:1819–1826. https://doi.org/10.1016/j.jclepro.2016.11.102

Jacklin DM, Brink IC, Jacobs SM (2021) Exploring the use of indigenous Western Cape plants as potential water and soil pollutant phytoremediators with a focus on green infrastructure. Water SA 47(3):317–325. https://doi.org/10.17159/wsa/2021.v47.i3.11860

Jacklin K, Pace JE, Warry W (2015) Informal dementia caregiving among indigenous communities in Ontario, Canada. Care Manag Journals 16:106–120. https://doi.org/10.1891/1521-0987.16.2.106

Jiabao Q, Zhan Q, Hongyu C et al (2023) Pyrometallurgy treatment of electroplating sludge, emulsion mud and coal ash: $ZnAlFeO_4$ spinel separation and stabilization in calcium metasilicate glass. J Environ Manage 329:117101. https://doi.org/10.1016/j.jenvman.2022.117101

Jung IS, Lovitt RW (2011) Leaching techniques to remove metals and potentially hazardous nutrients from trout farm sludge. Water Res 45:5977–5986. https://doi.org/10.1016/j.watres.2011.08.062

Khan MN, Mohammad F (2014) Eutrophication: challenges and solutions. In: Eutrophication: causes, consequences and control. Springer Netherlands, pp 1–15

Koskue V, Freguia S, Ledezma P, Kokko M (2021a) Efficient nitrogen removal and recovery from real digested sewage sludge reject water through electroconcentration. J Environ Chem Eng 9:106286. https://doi.org/10.1016/j.jece.2021.106286

Koskue V, Rinta-Kanto JM, Freguia S et al (2021b) Optimising nitrogen recovery from reject water in a 3-chamber bioelectroconcentration cell. Sep Purif Technol 264:118428. https://doi.org/10.1016/j.seppur.2021.118428

Kwapinski W, Kolinovic I, Leahy JJ (2021) Sewage sludge thermal treatment technologies with a focus on phosphorus recovery: a review. Waste and Biomass Valorization 12:5837–5852. https://doi.org/10.1007/s12649-020-01280-2

Lau P, Tam NF, Wong Y (1995) Effect of algal density on nutrient removal from primary settled wastewater. Environ Pollut 89:59–66. https://doi.org/10.1016/0269-7491(94)00044-E

Lee C-G, Alvarez PJJ, Kim H-G et al (2018) Phosphorous recovery from sewage sludge using calcium silicate hydrates. Chemosphere 193:1087–1093. https://doi.org/10.1016/j.chemosphere.2017.11.129

Mahmoud IY, Al-Bahry SN, Al-Musharafi SK (2013) Fresh water habitat pollution by treated sewage effluent in relation to multiple-antibiotic-resistant bacteria. APCBEE Proc 5:363–367. https://doi.org/10.1016/j.apcbee.2013.05.062

Margonwar AS, Rathod PB, Singh MP (2024) Microbial populations, function, and impact on environmental changes. In: Microbial metagenomics in effluent treatment plant. Elsevier, Amsterdam, pp 141–166

Marin-Batista JD, Mohedano AF, Rodríguez JJ, de la Rubia MA (2020) Energy and phosphorous recovery through hydrothermal carbonization of digested sewage sludge. Waste Manag 105:566–574. https://doi.org/10.1016/j.wasman.2020.03.004

Takahashi M, Takemoto Y, Yuki E, Seno K (2017) Potassium phosphate recovery from incinerated ash of sewage sludge. J Mater Sci Eng B 7. https://doi.org/10.17265/2161-6221/2017.7-8.003

Maske PD, Rathod PB, Singh MP (2025) Rhizoremediation: a biotechnology approach for water and soil remediation. In: Biotechnologies for wastewater treatment and resource recovery. Elsevier, Amsterdam, pp 353–365

Volpe M, Fiorib L, Merzarib F, Messineo A, Andreottolab G (2020) Hydrothermal carbonization as an efficient tool for sewage sludge valorization and phosphorous recovery. Chem Eng Trans 80. https://doi.org/10.3303/CET2080034

Meng X, Huang Q, Xu J et al (2019) A review of phosphorus recovery from different thermal treatment products of sewage sludge. Waste Dispos Sustain Energy 1:99–115. https://doi.org/10.1007/s42768-019-00007-x

Horttanainen M, Liikanen M, Havukainen J, Deviatkin (2020) Sustainable solid waste management using life cycle modeling for environmental impact assessment. In: Information modellingand knowledge bases XXXI, vol 321. IOS Press. https://doi.org/10.3233/FAIA200003

Mulchandani A, Westerhoff P (2016) Recovery opportunities for metals and energy from sewage sludges. Bioresour Technol 215:215–226. https://doi.org/10.1016/j.biortech.2016.03.075

Munasinghe-Arachchige SP, Abeysiriwardana-Arachchige ISA, Delanka-Pedige HMK, Nirmalakhandan N (2020a) Algal pathway for nutrient recovery from urban sewage. Algal Res 51:102023. https://doi.org/10.1016/j.algal.2020.102023

Munasinghe-Arachchige SP, Cooke P, Nirmalakhandan N (2020b) Recovery of nitrogen-fertilizer from centrate of anaerobically digested sewage sludge via gas-permeable membranes. J Water Process Eng 38:101630. https://doi.org/10.1016/j.jwpe.2020.101630

Munir MT, Li B, Mardon I et al (2019) Integrating wet oxidation and struvite precipitation for sewage sludge treatment and phosphorus recovery. J Clean Prod 232:1043–1052. https://doi.org/10.1016/j.jclepro.2019.06.007

Nakao S, Nishio T, Kanjo Y (2017) Simultaneous recovery of phosphorus and potassium as magnesium potassium phosphate from synthetic sewage sludge effluent. Environ Technol 38:2416–2426. https://doi.org/10.1080/09593330.2016.1264485

Otero M, Rozada F, Morán A et al (2009) Removal of heavy metals from aqueous solution by sewage sludge based sorbents: competitive effects. Desalination 239:46–57. https://doi.org/10.1016/j.desal.2008.03.005

Otoo M, Drechsel P, Hanjra MA (2015) Business models and economic approaches for nutrient recovery from wastewater and fecal sludge. Wastewater. Springer, Netherlands, pp 247–268

Pasalari H, Farzadkia M, Khosravani F et al (2024) Phosphorous recovery from sewage sludge via chemical and thermal technologies. Chem Eng J 496:153869. https://doi.org/10.1016/j.cej.2024.153869

Pathak A, Dastidar MG, Sreekrishnan TR (2009) Bioleaching of heavy metals from sewage sludge: a review. J Environ Manage 90:2343–2353. https://doi.org/10.1016/j.jenvman.2008.11.005

Pereira Marques W, Barros de Souza J, da Silva Júnior AH et al (2024) Phosphorous recovery from sewage effluent by vanadium-substituted composite synthesized via reverse coprecipitation route. J Clean Prod 469:143184. https://doi.org/10.1016/j.jclepro.2024.143184

Rahimi S, Modin O, Mijakovic I (2020) Technologies for biological removal and recovery of nitrogen from wastewater. Biotechnol Adv 43:107570. https://doi.org/10.1016/j.biotechadv.2020.107570

Rathod PB, AjishKumar KS, Chappa S et al (2018a) Pd^{2+}-loaded magnetic nanoassembly formed by magnetite nanoparticles crosslinked with poly(acrylic acid) via amide bonds for catalyzing mizoroki-heck coupling reaction. ChemistrySelect 3:8151–8158. https://doi.org/10.1002/slct.201801682

Rathod PB, Chappa S, Ajish Kumar KS et al (2018b) Poly(ethylenimine) functionalized magnetic nanoparticles for sorption of Pb, Cu, and Ni: potential application in catalysis. Sep Sci Technol 00:1–11. https://doi.org/10.1080/01496395.2018.1520731

Rathod PB, Chappa S, Kumar KSA et al (2019) Cadmium(II)-Loaded Fe_3O_4 @MPTS nanoparticles: Preparation and application as catalyst for C-N coupling reactions. Chem Select 4:11796–11800. https://doi.org/10.1002/slct.201902432

Rathod PB, Kalel RA, Singh Tomar MP et al (2023) Reformed metal-organic frameworks (MOFs) for abstraction of water contaminants—heavy-metal ions. Met Org Fram Wastewater Contam Remov 227–250. https://doi.org/10.1002/9783527841523.ch10

Rathod PB, Singh MP, Markad US (2025) Approach of microbial consortia in the field of industrial and pharmaceutical wastewater remediation, pp 109–140

Rathod PB, Pandey AK, Meena SS, Athawale AA (2016) Quaternary ammonium bearing hyper-crosslinked polymer encapsulation on Fe_3O_4 nanoparticles. RSC Adv 6:21317–21325. https://doi.org/10.1039/c6ra01543c

Rathod PB, Singh MP, Taware AS et al (2024) Comprehensive insights into water remediation: chemical, biotechnological, and nanotechnological perspectives. Environ Pollut Bioavailab 36. https://doi.org/10.1080/26395940.2024.2329660

Sano A, Kanomata M, Inoue H et al (2012) Extraction of raw sewage sludge containing iron phosphate for phosphorus recovery. Chemosphere 89:1243–1247. https://doi.org/10.1016/j.chemosphere.2012.07.043

Schellin LM, Froehner SJ (2024) Precipitation of potassium from domestic sewage as sylvite, pp 43–47

Schweitzer L, Noblet J (2018) Water contamination and pollution. In: Green chemistry. Elsevier, Amstesrsdasm, pp 261–290

Sharma M, Kant R, Sharma AK, Sharma AK (2024) Exploring the impact of heavy metals toxicity in the aquatic ecosystem. Int J Energy Water Resour. https://doi.org/10.1007/s42108-024-00284-1

Singh MP, Rathod PB, Margonwar AS (2024) Industrial wastewater remediation by using microbial communities

Singh MP, Rathod PB, Sakhare SK (2022) Photocatalytic degradation of dye from various metal/metal oxides derived from diverse plants. In: Phytonanotechnology. Springer, Singapore, pp 299–318

Sugurbekova G, Nagyzbekkyzy E, Sarsenova A et al (2023) Sewage sludge management and application in the form of sustainable fertilizer. Sustainability 15:6112. https://doi.org/10.3390/su15076112

Taware AS, Rathod PB, Katariya AP et al (2024) +Technological advancement in the development of nano fertilizers for sustainable agriculture. J Soil Sci Plant Nutr. https://doi.org/10.1007/s42729-024-01674-4

Tognacchini A, Rosenkranz T, van der Ent A et al (2020) Nickel phytomining from industrial wastes: growing nickel hyperaccumulator plants on galvanic sludges. J Environ Manage 254:109798. https://doi.org/10.1016/j.jenvman.2019.109798

Thant Zin MM, Kim D-J (2021) Simultaneous recovery of phosphorus and nitrogen from sewage sludge ash and food wastewater as struvite by Mg-biochar. J Hazard Mater 403:123704. https://doi.org/10.1016/j.jhazmat.2020.123704

Van der Hoek J, Duijff R, Reinstra O (2018) Nitrogen recovery from wastewater: possibilities, competition with other resources, and adaptation pathways. Sustainability 10:4605. https://doi.org/10.3390/su10124605

Veeken A, Hamelers B (1999) Effect of temperature on hydrolysis rates of selected biowaste components. Bioresour Technol 69:249–254. https://doi.org/10.1016/S0960-8524(98)00188-6

Wang X, Cui X, Fang C et al (2022) Agent-assisted electrokinetic treatment of sewage sludge: heavy metal removal effectiveness and nutrient content characteristics. Water Res 224:119016. https://doi.org/10.1016/j.watres.2022.119016

Wei SP, van Rossum F, van de Pol GJ, Winkler M-KH (2018) Recovery of phosphorus and nitrogen from human urine by struvite precipitation, air stripping and acid scrubbing: a pilot study. Chemosphere 212:1030–1037. https://doi.org/10.1016/j.chemosphere.2018.08.154

Wilsenach JA, Schuurbiers CAH, van Loosdrecht MCM (2007) Phosphate and potassium recovery from source separated urine through struvite precipitation. Water Res 41:458–466. https://doi.org/10.1016/j.watres.2006.10.014

Xiao Y, Zhao R, Xiong Z et al (2021) Transformation of phosphorous during incineration of sewage sludge: influence of steam and mineral. Fuel 303:121307. https://doi.org/10.1016/j.fuel.2021.121307

Yakkala K, Chappa S, Rathod PB et al (2021) Silver nanoparticles embedded cation-exchange membrane for remediation of Hg species and application as the dip catalyst in organic transformation. Mater Today Chem 22:100547. https://doi.org/10.1016/j.mtchem.2021.100547

Yesil H, Molaey R, Calli B, Tugtas AE (2021) Removal and recovery of heavy metals from sewage sludge via three-stage integrated process. Chemosphere 280:130650. https://doi.org/10.1016/j.chemosphere.2021.130650

Yu X, Nakamura Y, Otsuka M et al (2021) Development of a novel phosphorus recovery system using incinerated sewage sludge ash (ISSA) and phosphorus-selective adsorbent. Waste Manag 120:41–49. https://doi.org/10.1016/j.wasman.2020.11.017

Yunta F, Schillaci C, Panagos P et al (2024) Ecological risk assessment of heavy metals from application of sewage sludge on agricultural soils in Europe. Eur J Soil Sci 75. https://doi.org/10.1111/ejss.13562

Zhang J, Chen Z, Liu Y et al (2022a) Phosphorus recovery from wastewater and sewage sludge as vivianite. J Clean Prod 370:133439. https://doi.org/10.1016/j.jclepro.2022.133439

Zhang X, Chen S, Ai F et al (2022b) Identification of industrial sewage sludge based on heavy metal profiles: a case study of printing and dyeing industry. Environ Sci Pollut Res 29:12377–12386. https://doi.org/10.1007/s11356-021-16569-5

Chapter 8
Aerobic Treatment of Municipal Sewage and Wastewater

Ambika Kumar, Anshu Kumar, Neetu Yadav, Atul Samiran, and Rajneesh Kumar

Abstract Currently, the treatment of sewage water, whether it comes from industrial or residential sources, has become a progressively important procedure. Rapid industrialization and modern urbanization are the key factors of numerous harmful compounds released into natural water sources, which lead to the severe contamination of freshwater supplies. Because of its increasing importance and widespread influence, environmentalists from all over the globe have been focussing more on this issue as of late. Many methods have been suggested for improving wastewater management; however, most of these have drawbacks including low mineralization, high operational costs, insufficient capacity, and the generation of waste byproducts. In this context, aerobic treatment is a feasible option for cleaning concentrated industrial effluent, as it offers a quicker alternative to the more time-consuming anaerobic treatment methods. The use of aerobic wastewater treatment as a cost-effective approach with minimal operational and maintenance expenses is becoming more common. To handle increasing wastewater loadings, it is necessary to increase biomass and dissolved oxygen levels. This chapter has evaluated the potential of pure oxygen aeration and its constraints. By using pure oxygen, it can efficiently handle larger amounts of municipal sewage while maintaining the quality of the discharged

A. Kumar (✉)
Department of Chemistry, Bhagalpur National College, Bhagalpur (A Constituent Unit of T. M. B. U. Bhagalpur), Bhagalpur 812007, Bihar, India
e-mail: kumarambika.1115@gmail.com

A. Kumar
Department of Chemistry, T. N. B. College, Bhagalpur (A Constituent Unit of T. M. B. U. Bhagalpur), Bhagalpur 812007, Bihar, India

N. Yadav
Department of Chemistry, Indian Institute of Roorkee, Roorkee 247667, Uttarakhand, India

A. Samiran
Department of Zoology, Tilka Manjhi Bhagalpur University, Bhagalpur 812004, Bihar, India

R. Kumar
Department of Chemistry, Patna University, Patna 800005, Bihar, India

S. Madhav et al. (eds.), *Municipal Sewage and Sludge: Treatment and Disposal Strategies*, Environmental Science and Engineering,
https://doi.org/10.1007/978-3-032-07597-0_8

effluent. Even at lower flow rates, aeration using pure oxygen can produce a better decontamination result than mixed air.

Keywords Aerobic treatment · Industrial wastewater · Domestic sewage · Aeration · Pure oxygen

8.1 Introduction

Water is essential for ecological systems as it supports plant growth, provides habitats, facilitates nutrient cycling, regulates climate, maintains biodiversity, offers ecosystem services, and is crucial for human survival (Smith et al. 2013). Over the last century, rapid growth in population followed by urbanization and industrialization have significantly degraded vital ecosystems necessary for human survival. A major contributor to the pollution of oceans and rivers is the discharge of untreated wastewater from various sources (Preisner 2020; Singh et al. 2018; 2020; Gupta 2018). These wastewaters may initially have high pollutant concentrations, but their impact on ecosystems might be severe due to high Chemical Oxygen Demand (COD), Total Suspended Solid (TSS), and Biochemical Oxygen Demand (BOD) (Gaspar et al. 2025). Moreover, various industrial processes release hazardous compounds into the environment, including pesticides, dyes, phenols, chlorophenols, disinfectants, and other complex aromatic substances (Table 8.1). This contamination can result from leaks, spills, improper disposal, or transportation accidents in oil and gas, paint manufacturing, and chemical production. The high toxicity of these substances poses serious health risks to both humans and animals, further exacerbated by their tendency to bioaccumulate in the food chain (Roy 2024). The escalating pressures on our aquatic ecosystems are being tackled by various governmental bodies through the establishment of more stringent regulations concerning pollution discharge, primarily aimed at minimizing waste. In numerous situations, the remediation methods developed by the industry are identified as a requirement mandated by regulatory bodies. The implementation of these systems not only leads to unfavorable economic outcomes but also contributes to a rise in both capital and operational expenditures across the organization as a whole. The adherence to laws concerning mental health and environmental matters need not invariably lead to increased costs; instead, it ought to be viewed as a prospective avenue for generating supplementary revenue. In this regard, wastewater can be treated through an array of various physicochemical methodologies (Gupta and Gupta 2016; 2018; Kumar et al. 2022). Nonetheless, each of these approaches encounters constraints that are inherently technological and economic. Among the array of procedures presently accessible, biological methods or bioremediation processes emerge as alternatives that hold promise for economic viability, energy efficiency, and environmental feasibility (Srivastava 2015; Saeed et al. 2022). Among various biological processes for sewage and wastewater treatment, aerobic, anaerobic, and anoxic methods are reported as the most effective for achieving optimal pollutant removal and resource recovery (Meena et al. 2019).

The process of anaerobic digestion yields methane gas, which possesses significant renewable energy potential; consequently, biological treatment emerges as a viable strategy for wastewater management. Biological methods can effectively address almost all biodegradable wastewaters exhibiting a BOD/COD ratio ranging from 0.5 or higher, contingent upon the execution of necessary analyses and the maintenance of environmental monitoring (Młyńska et al. 2024). When evaluated alongside alternative wastewater treatment methods, it demonstrates a lack of additional contamination and offers the further benefit of reduced treatment expenses. There exist feasible alternatives that encompass both aerobic and anaerobic mechanisms. Microorganisms, often referred to as aerobic organisms, engage in the transformation of organic waste into biomass and carbon dioxide via aerobic processes (Reineke 2001). Available or dissolved oxygen has the capability to produce this transformation. Wastewater management encompasses the aerobic treatment of sewage, commonly known as Aerobic digestion is a microbiological treatment that transpires in the presence of oxygen. Organic material undergoes oxidation in aerobic circumstances, yielding compounds such as nitrate, phosphate, and carbon dioxide (Martin et al. 2015). The aerobic treatment of sewage necessitates using oxygen-consuming microorganisms, primarily bacteria, to promote the biological decomposition of organic matter in wastewater. This process not only diminishes contaminants but also produces energy in the form of biogas, presenting a considerable advantage over conventional physicochemical treatment methods. This approach, therefore, reflects a greater sense of environmental responsibility and enhanced energy efficiency (Gikas 2016; Patel et al. 2021).

Table 8.1 Various contaminants in wastewater and their adverse impact

S. No	Contaminants	Effects
1	Biodegradable organics (proteins, carbohydrates, fats)	Reduces the concentration of dissolved oxygen
2	Dissolved inorganics	Contamination of the domestic water
3	Heavy metals	Toxic material, which cannot be removed by conventional methods
4	Nutrients (Nitrogen, Phosphorus, etc.)	Can cause eutrophication in water reservoirs
5	Oil and floating materials	This can lead to unsightly conditions
6	Pathogens	Cause diseases in humans and animals
7	Priority pollutants	Can lead to several adverse impacts, e.g., carcinogenic, mutagenic, and teratogenic effects, and can have great toxicity
8	Refractory organics (surfactants, phenolics, and pesticides)	Hazardous materials
9	Suspended solids	Can produce sludge deposits, which affect the aquatic environment

Table 8.2 Comparative features of anaerobic and aerobic assessment

S. No	Feature	Anaerobic	Aerobic
1	Alkalinity requirement	High for some wastes	Low
2	Bioenergy and nutrient recovery	Yes	No
3	Sewage quality	Moderate to poor	Excellent
4	Energy requirement	Lesser	Higher
5	Nutrient requirement	Lesser	Higher
5	Odor	Extremely high	Less odor
6	Organic removal efficiency	High	High
7	Sludge production	Lesser	Higher
9	Setup time	High (2–4 month)	Less (2–4 weeks)
10	Temperature sensitivity	Higher Sensitivity	Lesser Sensitivity

The fundamental objective of aerobic sewage treatment process is to eliminate organic pollutants, pathogenic microorganism and suspended solids present in the municipal sewage. This ensures that the wastewater is rendered safe for environmental disposal or potential reuse. (Lier 2008; Raman et al. 2016). The accessibility of pure oxygen facilitates the proliferation of aerobic bacteria, which adeptly decompose organic compounds, thereby achieving this process. Several physical and biological methods associated with oxygen requirements work together to accomplish the goal. Aerobic and anaerobic methods are widely used to treat wastewater with several advantages as outlined in Table 8.2, and both demonstrate the capacity to achieve a significant degree of organic elimination efficiency.

In the current chapter, we delve deeper into the various factors and processes governing aerobic treatment. It becomes evident that optimizing these processes is essential for achieving high treatment efficiency and compliance with environmental regulations. Understanding the intricacies of aerobic treatment units will empower practitioners to implement effective solutions in decentralized wastewater management systems.

8.2 Biological Treatment Methods

Biological treatment methods of municipal sewage or wastewater mainly utilize microbes, such as bacteria and protozoa, to decompose organic contaminants (Kumar et al. 2023). There are three primary approaches classified as biological methods for wastewater treatment. The methods include aerobic treatment, anaerobic treatment, and anoxic water treatment processes (Sutton 2006). The selectivity of the suitable treatment methods is based on the qualitative and quantitative features of municipal sewage. Wastewater with reduced biological oxygen demand (BOD) and chemical oxygen demand (COD) can be efficiently treated using aerobic methods, while anaerobic treatment is favored compared to aerobic treatment in scenarios involving elevated concentrations of organic contaminants in wastewater (Chan et al. 2009). Combining all three treatment methods is a common method for reducing sludge yield. Biological treatment methods are classified under various criteria. This classification approach of the biological wastewater treatment systems enables the optimization of the most suitable category for specific wastewater treatment requirements. Primarily, the basis of classifications is on the type of microbial growth, which may be either suspended, attached. Another basis of classification is the operational condition, which can be aerobic, anoxic, or anaerobic. Suspended growth reactors are further classified into Activated Sludge Process (ASP), Sequencing Batch Reactor (SBR), Anaerobic SBR (ASBR), Up-flow Anaerobic Sludge Blanket Reactor (UASB), and Membrane Biological Reactor (MBR), which operates through maintaining microorganisms in suspension within the wastewater. On the other hand, Attached Growth Reactors employ fixed media to form the biofilms and to include technologies such as Trickling Filters (TF), and Anaerobic Filters (AF), whereas, Hybrid reactors combine Moving Bed Biofilm Reactors (MBBR), Rotating Biological Contactors (RBC), and Submerged Aerobic Filters (SAF) (Ghangrekar 2022).

8.2.1 Bacterial Metabolism in Wastewater Treatment

The process in which each microbial cell undergoes a series of metabolic reactions that can supply energy for growth is known as metabolism (Deng et al. 2021). According to Sidhu et al. (2017), there are primarily two types of metabolism namely, catabolism and anabolism. The energy-producing metabolic processes of cells are known as catabolism, whereas the energy-using biosynthetic metabolism of microbes is known as anabolism (Liu and Tay 2000). The release of chemical energy from an organic substance occurs in the catabolic reaction; this energy is then used in the anabolic process to synthesize additional components of the cell (McKnight et al. 2020). When anabolism and catabolism work in harmony, cells, and organisms expand in all directions.

8.3 Aerobic Treatment System

The aerobic method of sludge treatment is often considered more effective among biological methods for several reasons, particularly for wastewater with lower COD and BOD values (Chan et al. 2009). Aerobic treatment is an oxidative process in which microorganisms break down organic materials and other contaminants in the presence of oxygen (Reineke 2001). Aerobic decomposition of organic molecules frequently includes oxidation (or respiration), fermentation, endogenous respiration, and biosynthesis as the governing mechanism. In these cases, an aerobe is a component that relies on oxygen for survival and growth and uses aerobic respiration to generate energy. The reaction produced CO_2, ammonia, water, energy, and other byproducts. Here C (carbon) O (oxygen) H (Hydrogen) N (Nitrogen) and S (Sulphur) represent the combination of a waste organic compound.

In general, the process can be represented in the following steps:

Step I: Production of energy

$$\text{COHNS} + O_2 + \text{aerobes} \rightarrow CO_2 + H_2O + NH_3 + \text{other products} + \text{energy}$$

By recycling a portion of the energy produced during the oxidation process, it is possible to transform certain waste materials into operational cells.

Step II: Consumption of the produced energy

$$\text{COHNS} + O_2 + \text{aerobes} + \text{energy} \rightarrow C_5H_7NO_2(\text{new cells or sludge})$$

Generally, bio-decomposable materials are organic substances with larger energy content, upon oxidation these compounds generate lower-energy chemicals, which are associated with the release of energy. In this regard, microbes utilize this released energy to decompose the organic waste material into new cells (Pagliano et al. 2017). Decomposing bacteria generally favor aerobic environments over anaerobic ones because they rely on dissolved oxygen during aerobic metabolism (Zainudin et al. 2022). When the oxygen supply rate during aeration is less than or equal to the consumption rate, the concentration of dissolved oxygen will eventually fall below the requisite level to sustain an aerobic process (Bennett and Lash 1980). In wastewater treatment, which employs biochemical oxidation of wastewater treatment, supplementary oxygen is provided to aerobic bacteria, allowing them to utilize the reaction substrate to augment their metabolism. Thus, the final result is associated with the transformation of organic contaminants into inorganic substances, along with the production of new microbial cells.

The aerobic method is deemed appropriate for wastewater with reduced COD values (COD < 1000 mg/L) (Chan et al. 2009). Anaerobic treatments are favored when wastewater contains a significant concentration of organic contaminants. Thus, while managing the severly contaminated municipal sewage, aerobic treatment is typically favored following an anaerobic decontamination process. The efficacy of aerobic treatment as a decontamination procedure is contingent upon the quantitative feature of the organic contaminants present in the wastewater. The aerobic treatment generates odorless sludge post-treatment, rendering it suitable for fertilizer application. The often-utilized aerobic methods for wastewater treatment are the Activated Sludge Process (ASP), Sequencing Batch Reactor (SBR), Aerated Lagoon, Trickling Filters, Biological Active Filter (BAF), and Fluidised Bed Bioreactor (FBB).

8.3.1 Activated Sludge Process (ASP)

Activated sludge, also known as active microbial sludge, is comprised of bacteria and various microbes, such as protozoa and fungi (Seviour and Nielsen 2010). The suspension of the mixture is maintained using either mechanical or diffused aeration. The process comprises the partial oxidation of dissolved and colloidal organic compounds in wastewater that do not settle, resulting in their conversion to carbon dioxide. Concurrently, the residual material is converted into settleable cellular biomass, which facilitates the elimination of these organic substances from the municipal wastewater. The success of the procedure is contingent upon the aerobic bacteria utilized, which are kept in interaction with the organic materials found in the contaminated water. The characteristics and actions displayed by microorganisms in an ASP are influenced by the biochemical conditions in their environment. The choice of an aeration method in the activated sludge process can significantly impact the efficiency of wastewater treatment (Ozturk et al. 2016). Aeration occurs primarily through three methods namely, mechanical and diffused air aeration, and a combination of these aeration techniques.

8.3.1.1 Diffused Air Aeration

Diffused air aeration enhances oxygen transfer efficiency and improves mixing in the activated sludge process, leading to more effective biological treatment of wastewater (Pittoors et al. 2014). This method supports aerobic microorganisms, reduces odors, and is cost-effective for long-term operation. In this method of aeration system, pre-filtered air is channeled through assembled diffusers located within the aeration tank. In the activated sludge process, the Primary Settling Tank (PST) allows heavy solids to settle out of raw sewage, while the Secondary Settling Tank (SST) separates treated water from biological flocs after aeration (Li et al. 2024). Diffused air aeration enhances this process by supplying oxygen and promoting mixing, which optimizes microbial activity for effective wastewater treatment (Fig. 8.1). The air is filtered through an air filter before reaching the diffusers, ensuring the elimination of dust and dirt. The air compressors operate to produce the requisite pressure to offset the water column's depth and the corresponding friction loss. The volume of air supplied for sewage treatment depends on the biochemical oxygen demand (BOD) of the sewage and the required treatment level.

8.3.1.2 Mechanical Aeration

Mechanical aeration provides effective mixing in the activated sludge process, enhancing the breakdown of organic matter (Wang et al., 2009). This method also reduces maintenance requirements, making it ideal for large-scale wastewater treatment facilities. The mechanical aeration process is performed to disperse wastewater

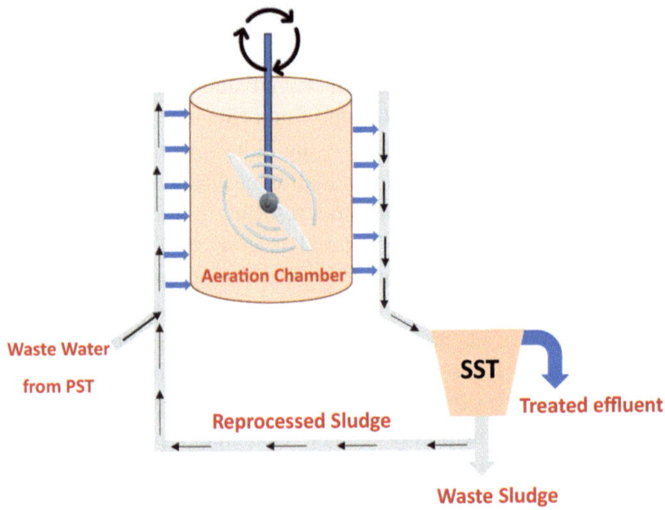

Fig. 8.1 A schematic diagram of diffused air aeration

droplets into the air, ensuring that fresh surfaces of the wastewater come into contact with the air continuously (Gresch et al. 2011). This approach prevents stagnation at the air–water interface and enhances the rate of oxygen transfer in the contaminated water. In dispersed air aeration, only a fraction of the total compressed air is effectively utilized for the oxidation of organic materials, with the majority being primarily allocated for mixing purposes. To address this limitation, the mechanical aeration system proves to be more effective. This system can use either fixed or floating surface aerators. To maximize use, rectangular aeration tanks are segmented into various parts, each equipped with a mechanical aerator or mixer. The impellers are designed to draw wastewater from the center and propel it as a fine spray over the surface when the electric motors are activated. This spraying action increases the surface area of the wastewater exposed to air, thereby accelerating oxygen mass transfer.

8.3.1.3 Combined Mechanical and Diffused Air Aeration

Combined mechanical and diffused air aeration systems in the activated sludge process enhance oxygen transfer and mixing, improving wastewater treatment efficiency (Wagner et al. 2014). This integration allows for optimal performance across varying tank depths, reduces energy consumption, and minimizes maintenance needs while ensuring uniform treatment throughout the system. This design typically features submerged paddles that rotate in opposition to rising air bubbles from the diffusers. The integration allows for enhanced mixing throughout the tank, reducing dead zones where sludge might accumulate. The mechanical component often operates at a lower speed to balance mixing and energy efficiency. The combination allows for more effective oxygen transfer due to the synergistic effects of both methods. The rising bubbles from diffusers create turbulence that aids in mixing while also delivering oxygen directly to microorganisms. Also, by leveraging both systems, facilities can achieve better aeration with potentially lower energy costs than using either method alone. Studies indicate that switching to combined systems can lead to energy savings of 30–40% (Jamaludin et al. 2024). The hyphenation facilitates the adaption of different tank sizes and wastewater properties, making it suitable for an extensive range of applications in municipal and industrial contaminated water treatment.

8.3.2 Sequencing Batch Reactor (SBR)

In a sequencing batch reactor (SBR), treatment occurs in distinct batches, with a series of operations carried out within a single tank (Gerardi 2011). The SBR system integrates primary sedimentation, biochemical oxidation of organic waste, and the settling of generated bacterial cells into a singular aeration tank. The Sequencing Batch Reactor system is beneficial for wastewater treatment due to its efficient integration of multiple treatment processes within a single tank, which optimizes space and reduces operational complexity (Wang and Li 2009). This flexibility effectively removes organic matter, nutrients, and suspended solids, making it suitable for varying flow conditions and contributing to high-quality effluent with low pollutant levels. Moreover, SBR systems are cost-effective, require fewer mechanical components, and offer lower power consumption compared to traditional treatment methods. The standard sequence comprises five stages, including the filling cycle, the reaction cycle, the settling cycle, the decant cycle, and an optional idle phase (Radetic 2024). This procedure is preferred because all treatment stages occur within a single reactor basin. The reaction cycle involves the completion of biochemical reactions facilitated by microbes, with continuous aeration maintained throughout the process. In the settling phase, aeration ceases, allowing for the effective separation of particles from liquids (Fig. 8.2).

The concluding phase in this method is referred to as the idle phase, which primarily aims to synchronize the operational cycles of multiple reactors (Casellas et al. 2006). During this inactive period, aeration may be activated or deactivated. Turning off aeration promotes the denitrification process of nitrified wastewater in the SBR. The idle phase often proves unnecessary in an SBR system, as the timing can be effectively coordinated to align the filling stage of one reactor with the reaction, settling, and decanting phases of others. Within the SBR, an anoxic zone is created to aid denitrification, and the wastewater is circulated internally between the two zones to make it even more effective (Zhao et al. 2016). The SBR's batch operation guarantees that all wastewater is treated uniformly, which keeps the process

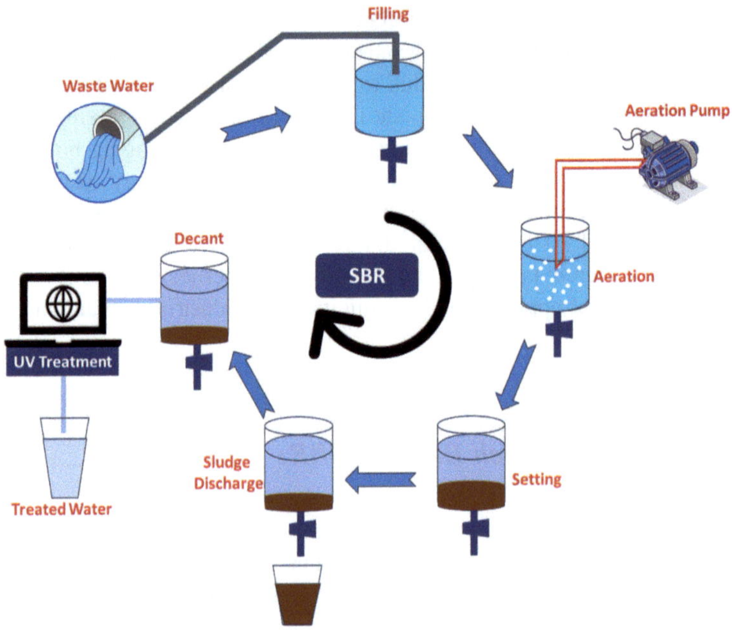

Fig. 8.2 Different cycles of operations associated with sequential batch reactor (SBR)

performance consistent. Since all of the sludge is contained in a single tank, performance is unaffected by variations in the Sludge's Volume Index (SVI). Automation and control are essential for ensuring dependable operation because manual process control is impractical due to the operation of many SBRs in various phases within the treatment plant.

The primary advantage of this technology over traditional activated sludge processes is its reduced spatial requirements due to modular construction, the elimination of a separate settling tank, minimal aeration needs, and automated operations (Mace and Mata-Alvarez 2002). To keep the volume level of reactor's container at the required necessary level, the accumulated waste is regularly removed from the container. Because SBR is a suspended development process, the form and classification of bio-aggregates are crucial to the system's proper functioning. Settling is a key phase in SBR technology, making the morphology of bio-aggregates that facilitate this process vital. Previous studies indicate that flocculant particles demonstrate inadequate settling properties, while granular bio-aggregates promote effective settling mechanisms. The accumulation of biomass in both aerobic and anaerobic SBRs is influenced by various parameters. For instance, calcium present in influent wastewater has been shown to aid in granule formation. The secretion and function of extracellular polymeric substances (EPS) are also important, along with inorganic cementing fillers (Flemming et al. 2007a). EPS consists of hydrated polymers made up of proteins, polysaccharides, and DNA (Flemming et al. 2007b) and is often released in response to environmental stimuli. The attainment of a detailed configuration for bio-aggregates is a complicated process that remains inadequately comprehended. Maintaining a low Food-to-microorganism (F/M) ratio and high relative shear promotes sludge granulation, leading to smaller and denser granules (Beun et al. 1999). In addition to sustaining the F/M ratio, several factors influence the size of aerobic granular biomass produced in SBR systems (Liu et al. 2005).

8.3.3 Aerated Lagoon

Aerated lagoons represent a technique for aerobic suspended growth biological wastewater treatment (Godini et al. 2021; Bartsch and Randall 1971). This system facilitates continuous treatment of wastewater, incorporating the option for recycling solids from the secondary sedimentation tank. Unlike oxidation ponds that rely on photosynthesis for oxygen, aerated lagoons typically use surface aerators to supply oxygen (Kim and Kim 2014). Floating aerators are commonly used due to their adaptability to changes in water levels caused by seasonal evaporation. Aerators installed on platforms or diffused air aeration devices may be utilized in smaller lagoons. The agitation generated by mechanical aerators or the rising air bubbles from diffusers keeps the biomass suspended within the lagoon. The depth of the lagoons is generally determined by the type of aerators used and their effective influence on depth. The contents of an aerobic lagoon can be either partially or fully mixed. The effluent typically contains about one-third to one-half of the incoming biochemical oxygen demand (BOD) as cell biomass, depending on the selected detention time. Therefore, before discharging wastewater, particulates must be removed through sedimentation. Reintroducing solids to the lagoon from the settling tank results in a minimal difference compared to an extended aeration activated sludge process. A facultative stabilisation pond can be used for final clarification and the removal of bacterial cells from the effluent, as an alternative to using a settling tank after the aerated lagoon. This facultative lagoon maintains dissolved oxygen in its upper layer, creating an aerobic zone, while the suspended sediments at the bottom undergo anaerobic decomposition due to inadequate mixing. The sediment accumulating at the bottom should be removed every few years, typically within a 5 to 10-year timeframe. The lack of a final clarifier after the facultative lagoon presents a risk of particle instability in the final effluent. Aerated lagoons require less land than stabilization ponds and have lower construction and operational costs, when compared with the activated sludge processes (ASP), making them a popular choice for treating biodegradable wastewaters. However, they do require more land than activated sludge processes, and their construction costs are higher than those of stabilization ponds due to the need for aerator installation (Shammas et al. 2009).

8.3.4 Trickling Filter

During the early to mid-twentieth century, trickling filters were popular wastewater decontamination system in the United States (Bressani-Ribeiro et al. 2018). Although this technology was present in the U.S. and Great Britain since the 1890s, with large tanks filled with boulders, the first significant public treatment system was established in Madison and Wisconsin in 1912 (Metcalf and Eddy 2003). This method is classified as an unsubmerged attached growth process, where contaminated water is introduced from the upper part of the filter media and flows downward, resulting in the designation "trickling filter" (Lessard and Bihan 2003). The process involves initial absorption, succeeded by oxidation, as wastewater moves through a biofilm that develops on the filtration bed. A trickling filter is composed of three key components: (a) a filtering media that allows contaminated water to trickle over its surface, (b) a rotary distribution system that ensures even distribution of wastewater across the filtration bed, and (c) an under-drainage system that efficiently removes the filtered waste from the system (Fig. 8.3).

After undergoing preliminary treatment, the effluent is transferred by the nozzle of the rotating system to the trickling filter bed. The process primarily relies on the biological slime layer present on the filter medium, while additional processes such as flocculation and sedimentation within the bed material's interstices greatly improve the overall effectiveness of the treatment (Wang et al. 1986).

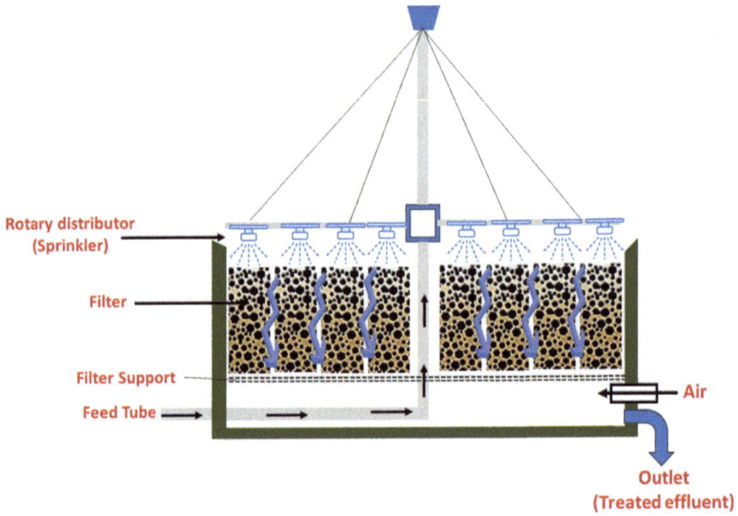

Fig. 8.3 A sectional elevation of trickling filter

8.3.4.1 Benefits and Drawbacks of a Trickling Filter

The trickling filter maintains efficient performance over diverse organic and hydraulic loading ranges and functions as a fundamental biological wastewater treatment approach while demonstrating resistance to sudden changes in loading (Särner 1981). The trickling filter system achieves good nitrification with low organic loading rates while delivering better effluent quality for BOD and suspended particles and uses less energy and land than stabilization ponds (Forbis-Stokes et al. 2018). However, there filters require more capital investment while needing extra treatment stages to meet strict discharge rules and specialized construction knowledge (Harrison et al. 1984). A continuous wastewater flow together with regular monitoring remains a fundamental requirement. The functioning of trickling filters presents operators with multiple problems, including vector control and odor issues, which require pre-treatment of wastewater to operate while demanding treatment for sloughed biomass in secondary settling tanks and face significant clogging risks.

8.3.5 Biological Active Filter (BAF)

The Biological Active Filter (BAF) system simultaneously conducts biological oxidation of wastewater's organic matter while filtering through a used filter medium to extract organic contaminants (Tang et al. 2021). BAFs function as an effective solution for removing both suspended and soluble organic debris from water (Mendoza-Espinosa and Stephenson 1999). Filtered wastewater descends through the filter media which performs dual functions of removing suspended particles and allowing bacterial biomass to adhere to its surface. The biological active filter promotes vital aeration at its foundation, allowing for the aerobic breakdown of organic substances found in wastewater. This process gathers treated wastewater from its upper section following the completion of the filtration process. In BAFs, different filter media types including granular activated carbon along with anthracite, sand, and bituminous coal determine removal efficiency while responding to wastewater influent characteristics (Spanjers 2017). The filter media must undergo regular backwashing with water and air to keep free from blockages while preserving filtration speed, removing inactive biomass, and retaining active media. The BAF system maintains effective performance at moderate volumetric and organic loadings in small reactors which leads to reduced physical space requirements. BAFs require aerobic decomposition of organic matter which drives up both initial investment and ongoing operation expenses because of aeration demands. The main application of BAFs occurs when treated water reuse is needed because they perform effectively better after tertiary treatment processes particularly when organic matter levels are low (Mendoza-Espinosa and Stephenson 1999). The brief mechanism inside Biologically Active Filter is illustrated in Fig. 8.4

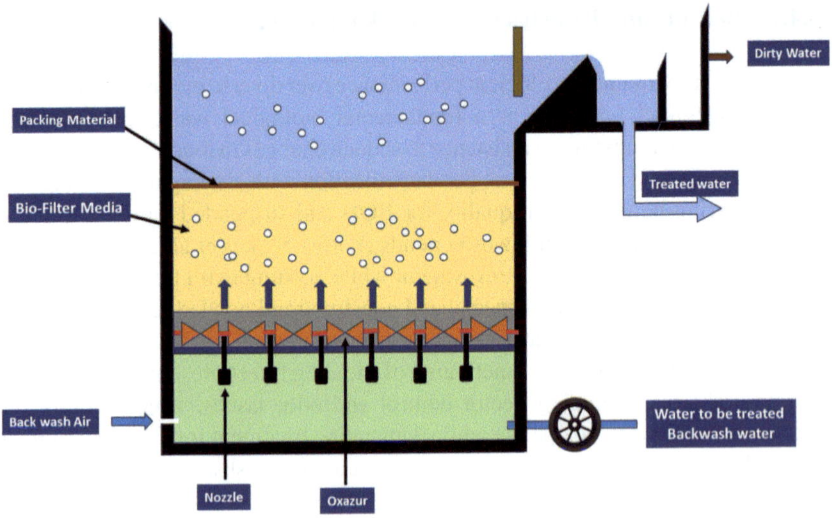

Fig. 8.4 Representation of biologically active filter

8.3.6 Fluidized Bed Bioreactor (FBB)

Fluidized Bed Bioreactors (FBBs) utilize a fluid to suspend solid particles, enhancing mass transfer and mixing for efficient microbial growth. They are widely used in wastewater treatment and biotechnology for producing bioproducts (Shieh and Keenan 2005). Fluidization is an impressive process in which particles remain suspended due to the upward flow velocity of liquid within a reactor (Yates and Lettieri 2016). This fluidization presents numerous benefits, such as efficient mixing, larger specific surface area, improved mass transfer rates, homogeneous particle distribution, and uniform temperature profiles throughout the reactor. Liquid–solid fluidization is utilized in the majority of fluidized bed bioreactors (FBBRs) (Nelson et al. 2017). Small fluidized media are utilized in a fluidized bed bioreactor (FBBR), a modern wastewater treatment system, to immobilize and retain bacterial species. The formation of biofilm on fluidised media increases biomass concentration within the Fluidised Bed Bioreactor (FBBR). This enhancement contributes to improved stability and performance, while simultaneously decreasing the risk of excessive cell washout. Aerobic FBBRs are utilized for the treatment of groundwater severely polluted by harmful wastes, whereas anaerobic FBBRs are primarily employed for the treatment of poorly polluted wastewater. Pre-dissolving oxygen is required for microbial development, the effluent in aerobic FBBRs must be recirculated or directed through an oxygenation tank. In an aerobic fluidised bed bioreactor, activated carbon serves as the packing medium for the degradation of organic pollutants (Fig. 8.5).

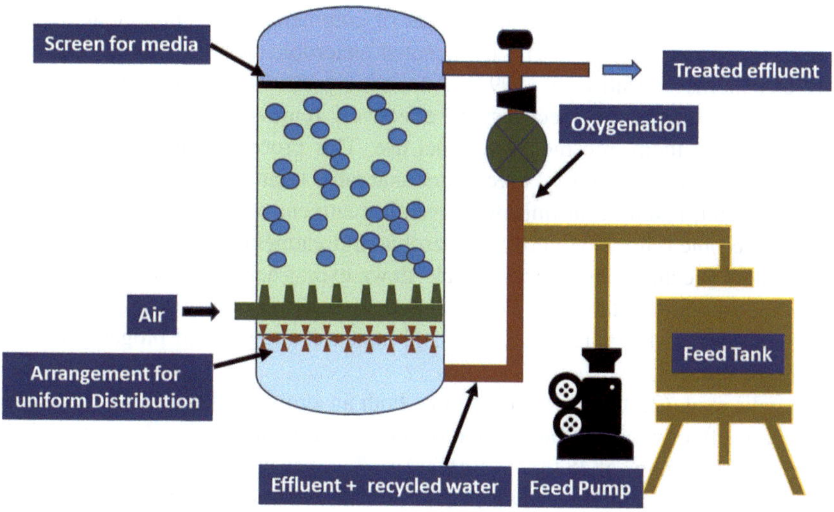

Fig. 8.5 Fluidized bed reactor

8.4 Benefits of Aerobic Treatment

Aerobic systems have quickly emerged as an effective solution for treating commercial and municipal wastewater, focusing primarily on the removal of nutrients (Carey and Migliaccio 2009). One of the primary benefits of aerobic treatment is its ability to produce high-quality secondary effluent suitable for safe discharge into the environment (Demirbas et al. 2017). This process is particularly effective in polishing industrial wastewater that has already undergone anaerobic treatment, ensuring complete degradation of contaminants and compliance with stringent environmental regulations (Jonstrup et al. 2011). Aerobic systems are known for their reliability and efficiency. Also, aerobic treatments generate less odor than anaerobic systems, as the former do not produce hydrogen sulfide or methane, making them more suitable for populated areas. Moreover, the sludge produced from the aerobic treatment is often odor-free and can be repurposed as a high-quality agricultural fertilizer (Demirbas et al. 2017). The versatility of aerobic treatment systems is another advantage, as they can be tailored to various industrial applications, including food and beverage processing, chemical manufacturing, and municipal wastewater management (Austermann-Haun et al. 1999). Conventional technologies exemplify the diverse approaches available for aerobic treatment. However, the aeration process in aerobic treatment necessitates a substantial amount of energy, but their effectiveness in nutrient removal and overall pollutant degradation makes them a valuable option in modern wastewater management strategies. By ensuring high-quality effluent and promoting sustainable practices, aerobic treatment plays a vital role in preserving water resources and the ecosystem. Conventional anaerobic–aerobic treatment plants have also been utilized to handle the bulk of commercial and municipal wastewater (Sikosana et al. 2019). The hyphenated anaerobic–aerobic process can have several benefits (Garuti et al. 1992). Anaerobic pretreatment presents significant opportunities for resource recovery by effectively removing organic contaminants and converting them into biogas for fuel use. This process boasts high treatment efficiency, as the addition of oxygen after anaerobic treatment enhances the quality of the effluent, resulting in improved overall treatment outcomes and stabilization of the anaerobic output. Moreover, anaerobic pretreatment leads to reduced sludge production. The anaerobic process breaks down excess aerobic sludge, which minimizes the amount of stabilized sludge that requires disposal (Wei et al. 2003). This reduction lowers sludge removal costs and increases the yield of biogas generated during the process.

Aggelis et al. (2001) demonstrated that both anaerobic and aerobic technologies fell short in effectively treating green olive debittering wastewater on their own, which presents a COD range of 25,000 to 100,000 mg/L (Chan et al. 2009). Neither aerobic nor anaerobic treatment on its own yields effluents that comply with discharge regulations for high organic strength industrial wastewaters. Utilizing both anaerobic and aerobic processes can slash operational costs by up to eight times compared to relying solely on aerobic treatment (Shim et al. 2002). At the same time, these processes remove a lot of organic matter, make less aerobic sludge, and the pH remains unaltered. Energy efficiency is another notable advantage. Anaerobic pretreatment functions as an influent equalization tank, managing daily fluctuations in oxygen demand effectively. This regulation helps reduce the necessary maximum aeration capacity, leading to lower energy consumption. Moreover, during anaerobic treatment, volatile organics are broken down efficiently, preventing their volatilization in subsequent aerobic treatment phases (Smet et al. 1999). This aspect is crucial for maintaining environmental standards and improving overall treatment effectiveness. When the combination of anaerobic and aerobic techniques is applied to highly contaminated industrial wastewater, it facilitates an economical and practical alternative method for the wastewater treatment. This integrated approach harnesses the benefits of aerobic digestion, which significantly reduces chemical oxygen demand and volatile suspended solids, while also capitalizing on anaerobic digestion's capability for biogas production. In summary, anaerobic pretreatment emerges as a powerful method for wastewater treatment, offering substantial environmental and economic advantages.

The modified Aerobic granular sludge (AGS) technology can also offer several benefits for wastewater treatment, including enhanced settling properties that allow for effective biomass and water separation, which leads to improved treatment efficiency (Nancharaiah and Sarvajith 2019). AGS can simultaneously remove nitrogen, phosphorus, and other pollutants in a single reactor, resulting in significant energy savings of up to 50% compared to conventional methods (Nancharaiah and Reddy 2018). Aerobic granulation also creates a denser and stronger aggregate structure, helps solids settle and separate from wastewater, increases biomass concentration, and makes the material stronger against shock loads (Liu et al. 2009). Because granules are compact, they keep biomass from being lost in waste and can handle high-strength organic loads as well as changes in organic loads. One more benefit of this method is that the same granules can be used over and over again for a long time. Aerobic granulation can also be used to get rid of harmful metal ions in different types of trash from businesses and cities (Jagaba et al. 2024). It looks like aerobic grains could be very useful for removing harmful chemicals from the environment because they have a lot of surface area, are porous, and are easy to settle (Adav et al. 2009). Nancharaiah et al. (2006) have also demonstrated that aerobic grainy biomass serves as an effective biosorbent for the extraction of uranium from diluted nuclear waste.

8.5 Challenges Encountered in Aerobic Treatment

Aerobic treatment procedures provide proper treatment with diluted Domestic Wastewater (DWW) but require a huge amount of energy for aeration Also, a high organic load of DWW (more than 40,000 mg/L) and an extremely low BOD: COD ratio (between 0.3 and 0.4) is not suitable for aerobic treatment (Johnson et al. 2019). High COD and recalcitrant compounds like melanoidins and polyphenols can constrain the growth of aerobic bacteria. Moreover, scaling up of aerobic reactors is challenging since mass transfer limitations are high and this may lead to a lesser efficiency in pilot scale reactors than laboratory reactors (Noorman 2011). Also, aerobic treatment is not favorable in in cyanobacterial treatment, which uses chemical methods like algaecides and antibiotics, to control harmful algal blooms and mitigate their ecological and health impacts, as their oxygen production during the day and consumption at night disrupts the oxygen balance needed for effective aerobic processes.

Despite several challenges in aerobic treatment, the process can be modified and hyphenated with modern technologies, such as membrane bioreactors, anaerobic digestion, and advanced oxidation processes, to enhance efficiency and effectiveness (Tabrizi and Mehrvar 2004). By integrating these innovative approaches, wastewater treatment systems can better address issues like nutrient removal and toxin degradation, ultimately providing improved opportunities for sustainable water management. Furthermore, the hyphenation with other technologies can optimize resource recovery and minimize operational costs, making wastewater treatment more resilient and adaptable to varying conditions.

8.6 Case Study

8.6.1 *Successful Implementation of Aerobic Treatment Systems in Urban Areas*

The city of Akron, Ohio, has successfully implemented a comprehensive aerobic treatment system for its municipal wastewater (Ju et al. 2005; Arunachalam et al. 2004). The city's wastewater treatment plant utilizes a combination of advanced technologies, including enhanced primary solids removal, efficient aeration systems, and advanced sludge management strategies. The enhanced primary solids removal process, which involves microsieving and continuous backwash upflow media filtration, has significantly reduced the overall energy demand of the aeration process, leading to substantial cost savings and a lower carbon footprint. The efficient aeration system, designed to optimize oxygen transfer and minimize energy consumption, has further improved the performance of the aerobic treatment process, ensuring effective removal of organic matter and pathogens. (Sivakkumar et al. 2019).

The city has launched a comprehensive sludge management initiative that utilises dewatering and drying techniques to minimize the water quantity of the sludge. This has enabled the city to explore beneficial uses for the sludge, such as land application for agricultural purposes or use as a fuel source in energy recovery processes. By adopting these innovative approaches, the city of Akron has successfully demonstrated the viability and benefits of aerobic treatment of sewage, serving as a model for other urban areas to emulate in their efforts to achieve sustainable wastewater management (Vanitha 2020; Sivakkumar et al. 2019; Gikas 2016; Smirnova et al. 2020).

8.6.2 *Challenges Faced During the Upgrade of Existing Treatment Plants*

The city of Malmö Sweden has highlighted an essential case study, which demonstrates the difficulties encountered by conventional sewage treatment methods (Bautista et al. 2018). The sewer system of the city faced difficulties because the infrastructure was built several decades ago yet needed to adapt to population growth alongside changing precipitation patterns caused by climate change. The severe accumulation of solids and sludge within the sewer system created major issues for existing treatment facilities because the changing population and lifestyle choices of residents made this situation worse. (Smirnova et al. 2020).

The city initiated a detailed transformation of its wastewater treatment facilities by implementing diverse technological approaches to resolve its infrastructure problems. Malmö integrated basic technological components including sewer system improvements with existing asset reuse and a versatile treatment process setup to address its challenges. The city utilized this approach to create a dependable solution which proved both affordable and operationally efficient for addressing rising permit needs regarding effluent standards while boosting the complete treatment system performance. Malmö city employed strategic asset management together with technological innovations in order to upgrade its sewage treatment capabilities thus becoming a model for other wastewater infrastructure adaptation projects (Barton 2016).

8.7 Perspectives on the Future of Aerobic Treatment

The future of aerobic wastewater treatment is mainly affected by associated advancing technologies, regulatory demands, and an increasing focus on sustainability (Capodaglio et al. 2017). With the rising demand for effective and efficient wastewater treatment solutions, various trends are emerging that underscore possible breakthroughs in aerobic treatment methods. Incorporation of sophisticated technologies, aerobic treatment approaches are progressively improving their efficiency and efficacy. In this context, Aerobic Granular Sludge (AGS) systems are becoming increasingly favored due to their exceptional settling properties and lower energy usage when stacked against conventional activated sludge methods (Liu et al. 2023). AGS can attain elevated biomass concentrations and enhanced pollutant removal efficiencies, rendering them a sustainable alternative for future wastewater treatment plants. Moreover, future wastewater treatment facilities are anticipated to emphasise resource recovery in conjunction with contaminant elimination. Anaerobic digestion technologies combined with aerobic systems can enhance biogas output and reduce total energy usage. This transition to a circular economy model will boost the sustainability of wastewater treatment and contribute to energy neutrality in treatment plants. The current advancements in aerobic treatment involve the implementation of Advanced Oxidation Processes (AOPs), which employ potent oxidants to decompose intricate contaminants that conventional approaches may inadequately eliminate (Ponnusami et al. 2023). These methods are extremely beneficial for the treatment of pharmaceuticals and other developing pollutants, therefore enhancing overall water quality and safety. The amalgamation of AOPs with current aerobic systems may yield more thorough treatment solutions that comply with rigorous regulatory standards. Enhancing aerobic treatment procedures can be greatly achieved through the integration of cognitive technologies like artificial intelligence and machine learning (Singh et al. 2023). Predictive analytics enables operators to modify aeration rates, chemical dosing, and other parameters in real-time, hence improving treatment efficiency and minimising energy use. This data-centric methodology facilitates proactive oversight of wastewater treatment processes, resulting in enhanced performance and reduced operational expenses (Sanni et al. 2022). Also, as climate change increasingly affects global water resources, aerobic treatment methods must adjust to evolving environmental conditions. Future systems may integrate characteristics that bolster robustness against severe weather occurrences or variations in influent quality. The emphasis on attaining carbon neutrality will stimulate developments designed to diminish greenhouse gas emissions linked to wastewater treatment operations (Kumar et al. 2021). As laws around water quality and public health intensify, aerobic treatment methods must adapt to achieve compliance. The capacity to generate high-quality effluent that complies with or beyond regulatory criteria will be essential for the future sustainability of these technologies. Ongoing enhancements in process efficiency and pollution removal efficacy will be essential to tackle the issues presented by urbanization and industrialization. In summary, the future of aerobic wastewater treatment is optimistic, incorporating technological innovations, resource recovery strategies, intelligent automation, and an emphasis on sustainability. These advancements will improve the efficacy of municipal sewage treatment while also making substantial contributions to environmental preservation and public health. As these tendencies develop, aerobic treatment procedures will become increasingly essential in tackling global water issues while safeguarding water resources for future generations.

8.8 Conclusion

Wastewater treatment through aerobic treatment exists as a powerful and effective method, which mainly depends on aerobic bacterial processes to break down organic substances and remove the concerned pollutants. Through this method, the elimination of elements like nitrogen and phosphorus occurs with a simultaneous reduction of both biochemical oxygen demand (BOD) as well as the total suspended solids (TSS). Thus, the decontaminated effluent can be safely discharged or recycled for the irrigation process. Aerobic treatment systems have the benefit of generating high-quality effluent with little odor, which is especially advantageous in urban and rural areas with strict environmental restrictions. These systems can adapt to various applications, which include decentralized rural treatment and upgrading existing malfunctioning anaerobic systems. Aeration energy requirements, together with excess sludge management, remain as ongoing challenges for these systems. The advancement of technology, along with process optimization, will prove crucial for enhancing both efficiency and sustainability standards of aerobic treatment systems. The global attention on environmental concerns will drive the further importance of aerobic treatment for sustainable wastewater management, which will improve water quality and create a healthier ecosystem.

Acknowledgements Ambika Kumar expresses sincere gratitude to the Principal, B. N. College, Bhagalpur, and the faculty members of the Department of Chemistry, B. N. College, Bhagalpur, for their invaluable support and a conducive environment for his work in the completion of this chapter. Dr. Kumar would also like to dedicate this work to the memory of his late father, Prof. Krishna Deo Roy, who was a dedicated college teacher in the Department of Chemistry, L. K. V. D. College, Tajpur, Samastipur. Throughout Dr. Ambika Kumar's academic and professional pursuits, the scientific zeal and educational commitment of his father, Prof. Krishna Deo Roy, have been a continual source of inspiration. Dr. Ambika Kumar is also grateful to his wife, Mrs. Rubi Roy, for her unwavering support and encouragement throughout this journey.

References

Adav SS, Lee DJ, Lai JY (2009) Treating chemical industries influent using aerobic granular sludge: recent development. J Taiwan Inst Chem Eng 40(3):333–336

Aggelis GG, Gavala HN, Lyberatos G (2001) SE—Structures and environment: combined and separate aerobic and anaerobic biotreatment of green olive debittering wastewater. J Agric Eng Res 80(3):283–292

Arunachalam RS, Shah HK, Ju LK (2004) Aerobic sludge digestion under low dissolved oxygen concentrations. Water Environ Res 76(5):453–462

Austermann-Haun U, Meyer H, Seyfried CF, Rosenwinkel KH (1999) Full scale experiences with anaerobic/aerobic treatment plants in the food and beverage industry. Water Sci Technol 40(1):305–312

Barton M (2016) Nature-based solutions in urban contexts: a case study of Malmö, Sweden. IIIEE Masters Thesis

Bartsch EH, Randall CW (1971) Aerated lagoons: a report on the state of the art. J (Water Pollut Control Federat) 699–708

Bautista Angeli JR, Morales A, LeFloc'h T, Lakel A, Andres Y (2018) Anaerobic digestion and integration at urban scale: feedback and comparative case study. Energy Sustain Soc 8(1):1–23

Bennett GF, Lash L (1980) Oxygen transfer rates, mechanisms, and applications in biological waste water treatment. Crit Rev Environ Sci Technol 9(4):301–392

Beun JJ, Hendriks, A, Van Loosdrecht MCM, Morgenroth E, Wilderer PA, Heijnen JJ (1999) Aerobic granulation in a sequencing batch reactor. Water research, 33(10):2283–2290

Bressani-Ribeiro T, Almeida PGS, Volcke EIP, Chernicharo CAL (2018) Trickling filters following anaerobic sewage treatment: state of the art and perspectives. Environ Sci: Water Res Technol 4(11):1721–1738

Capodaglio AG, Callegari A, Cecconet D, Molognoni D (2017) Sustainability of decentralized wastewater treatment technologies. Water Pract Technol 12(2):463–477

Carey RO, Migliaccio KW (2009) Contribution of wastewater treatment plant effluents to nutrient dynamics in aquatic systems: a review. Environ Manage 44:205–217

Casellas M, Dagot C, Baudu M (2006) Set up and assessment of a control strategy in a SBR in order to enhance nitrogen and phosphorus removal. Process Biochem 41(9):1994–2001

Chan YJ, Chong MF, Law CL, Hassell DG (2009) A review on anaerobic–aerobic treatment of industrial and municipal wastewater. Chem Eng J 155(1–2):1–18

Demirbas A, Edris G, Alalayah WM (2017) Sludge production from municipal wastewater treatment in sewage treatment plant. Energy Sources, Part a: Recovery, Utilization Environ Effects 39(10):999–1006

Deng Y, Beahm DR, Ionov S, Sarpeshkar R (2021) Measuring and modeling energy and power consumption in living microbial cells with a synthetic ATP reporter. BMC Biol 19(1):101

Flemming HC, Neu TR, Wozniak DJ (2007a) The EPS matrix: the "house of biofilm cells". J bacteriology, 189(22):pp.7945–7947

Flemming HC, Strathmann M, Morales CFL (2007b) Microbial effects. In Sediment dynamics and pollutant mobility in rivers: An interdisciplinary approach. Berlin, Heidelberg: Springer Berlin Heidelberg. pp. 343–378

Forbis-Stokes AA, Rocha-Melogno L, Deshusses MA (2018) Nitrifying trickling filters and denitrifying bioreactors for nitrogen management of high-strength anaerobic digestion effluent. Chemosphere 204:119–129

Garuti G, Dohanyos M, Tilche A (1992) Anaerobic-aerobic combined process for the treatment of sewage with nutrient removal: the ananox® process. Water Sci Technol 25(7):383–394
Gaspar E, Irimia O, Stanciu M, Barsan N, Mosnegutu E (2025) Strategies for a sustainable economy: optimizing processes for BOD, COD and TSS removal from wastewater. Water 17(3):318
Gerardi MH (2011) Troubleshooting the sequencing batch reactor. John Wiley & Sons
Ghangrekar MM (2022) Wastewater to water
Gikas P (2016) Towards energy positive wastewater treatment plants. J Environ Manage 203:621. https://doi.org/10.1016/j.jenvman.2016.05.061
Godini K, Azarian G, Kimiaei A, Dragoi EN, Curteanu S (2021) Modeling of a real industrial wastewater treatment plant based on aerated lagoon using a neuro-evolutive technique. Process Saf Environ Prot 148:114–124
Gresch M, Armbruster M, Braun D, Gujer W (2011) Effects of aeration patterns on the flow field in wastewater aeration tanks. Water Res 45(2):810–818
Gupta H, Gupta B (2016) Adsorption of polycyclic aromatic hydrocarbons on banana peel activated carbon. Desalin Water Treat 57(20):9498–9509
Gupta H, Gupta B (2018) Vehicular tire as potential adsorbent for the removal of polycyclic aromatic hydrocarbons. Polycyclic Aromat Compd 38(4):354–368
Gupta H (2018) PAH determination in effluent and sludge samples of paper industry. Environ Technol Innov 9:115–121. https://doi.org/10.1016/j.eti.2017.11.009
Harrison JR, Daigger GT, Filbert JW (1984) A survey of combined trickling filter and activated sludge processes. J (Water Pollut Control Federation 1073–1079
Jagaba AH, Lawal DU, Abdulazeez I, Lawal IM, Mu'azu ND, Birniwa AH, Usman AK, Hagar HS, Yaro NSA, Noor A, Abioye KJ (2024) Advancements and prospects in the utilization of metal ions and polymers for enhancing aerobic granulation during industrial effluent treatment and resource recovery. J Water Process Eng 66:105972
Jamaludin M, Tsai YC, Lin HT, Huang CY, Choi W, Chen JG, Sean WY (2024) Modeling and control strategies for energy management in a wastewater center: a review on aeration. Energies 17(13):3162
Johnson I, Ali MAS, Kumar M (2019) Cyanobacteria/microalgae for distillery wastewater treatment-past, present and the future. In: Microbial wastewater treatment, Elsevier, pp 195–236
Jonstrup M, Punzi M, Mattiasson B (2011) Comparison of anaerobic pre-treatment and aerobic post-treatment coupled to photo-Fenton oxidation for degradation of azo dyes. J Photochem Photobiol A 224(1):55–61
Ju LK, Shah HK, Porteous J (2005) Phosphorus release in aerobic sludge digestion. Water Environ Res 77(5):553–559
Kim HD, Kim KH (2014) Water purification using aeration techniques in the lagoon. Appl Mech Mater 675:602–607
Kumar A, Thanki A, Padhiyar H, Singh NK, Pandey S, Yadav M, Yu ZG (2021) Greenhouse gases emission control in WWTS via potential operational strategies: a critical review. Chemosphere 273:129694
Kumar L, Chugh M, Kumar S, Kumar K, Sharma J, Bharadvaja N (2022) Remediation of petrorefinery wastewater contaminants: a review on physicochemical and bioremediation strategies. Process Saf Environ Prot 159:362–375
Kumar V, Sehgal R, Gupta R (2023) Microbes and wastewater treatment. In: Development in wastewater treatment research and processes. Elsevier, pp 239–255
Lessard P, Bihan YL (2003) Fixed film processes. In: Handbook of water and wastewater microbiology, pp 317–336
Li Y, Hu Z, Liu X, Dong Y, Wang Y, Zhang S, Xu Z, Yang Q (2024) Characteristics of bioaerosol emissions from a municipal wastewater treatment plant: health risk assessment and microbial composition. Sci Total Environ 934:173096
Liu Y, Tay JH (2000) Interaction between catabolism and anabolism in the oxidative assimilation of dissolved organic carbon. Biotech Lett 22(19):1521–1525

Liu Y, Wang ZW, Qin L, Liu YQ, Tay JH (2005) Selection pressure-driven aerobic granulation in a sequencing batch reactor. Appl Microbiol Biotechnol 67:26–32
Liu S, Zhou M, Daigger GT, Huang J, Song G (2023) Granule formation mechanism, key influencing factors, and resource recycling in aerobic granular sludge (AGS) wastewater treatment: a review. J Environ Manage 338:117771
Liu XW, Yu HQ, Ni BJ, Sheng GP (2009) Characterization, modeling and application of aerobic granular sludge for wastewater treatment. In: Biotechnology in China I: from bioreaction to bioseparation and bioremediation, pp 275–303
Mace S, Mata-Alvarez J (2002) Utilization of SBR technology for wastewater treatment: an overview. Ind Eng Chem Res 41(23):5539–5553
Martin J, Santos JL, Aparicio I, Alonso E (2015) Pharmaceutically active compounds in sludge stabilization treatments: anaerobic and aerobic digestion, wastewater stabilization ponds and composting. Sci Total Environ 503:97–104
McKnight CL, Griffard JH, Bhat SG, Bollig RW, Lawson CM, Long CA, Daley BJ (2020) Introduction to metabolism. In: Surgical metabolism: the metabolic care of the surgical patient, pp 3–16
Meena RAA, Kannah RY, Sindhu J, Ragavi J, Kumar G, Gunasekaran M, Banu JR (2019) Trends and resource recovery in biological wastewater treatment system. Bioresource Technol Reports 7:100235
Mendoza-Espinosa L, Stephenson TOM (1999) A review of biological aerated filters (BAFs) for wastewater treatment. Environ Eng Sci 16(3):201–216
Metcalf and Eddy (2003) In: Wastewater engineering treatment and reuse, 4th ed, McGraw Hill
Młyńska A, Halecki W, Chmielowski K (2024) Efficient biological treatment: achieving exceptional reductions in pollutants and ensuring environmental compliance. In: Desalination and water treatment, pp 100552
Nancharaiah YV, Joshi HM, Mohan TVK, Venugopalan VP, Narasimhan SV (2006) Aerobic granular biomass: a novel biomaterial for efficient uranium removal. Current Sci 503–509
Nancharaiah YV, Reddy GKK (2018) Aerobic granular sludge technology: mechanisms of granulation and biotechnological applications. Biores Technol 247:1128–1143
Nancharaiah YV, Sarvajith M (2019) Aerobic granular sludge process: a fast growing biological treatment for sustainable wastewater treatment. Current Opinion Environ Sci Health 12:57–65
Nelson MJ, Nakhla G, Zhu J (2017) Fluidized-bed bioreactor applications for biological wastewater treatment: a review of research and developments. Engineering 3(3):330–342
Noorman H (2011) An industrial perspective on bioreactor scale-down: what we can learn from combined large-scale bioprocess and model fluid studies. Biotechnol J 6(8):934–943
Ozturk MC, Serrat FM, Teymour F (2016) Optimization of aeration profiles in the activated sludge process. Chem Eng Sci 139:1–14
Pagliano G, Ventorino V, Panico A, Pepe O (2017) Integrated systems for biopolymers and bioenergy production from organic waste and by-products: a review of microbial processes. Biotechnol Biofuels 10:1–24
Patel K, Patel N, Vaghamshi N, Shah K, Duggirala SM, Dudhagara P (2021) Trends and strategies in the effluent treatment of pulp and paper industries: a review highlighting reactor options. Current Res Microbial Sci 2:100077. https://doi.org/10.1016/j.crmicr.2021.100077
Pittoors E, Guo Y, Van Hulle SW (2014) Oxygen transfer model development based on activated sludge and clean water in diffused aerated cylindrical tanks. Chem Eng J 243:51–59
Ponnusami AB, Sinha S, Ashokan H, Paul MV, Hariharan SP, Arun J, Gopinath KP, Le QH, Pugazhendhi A (2023) Advanced oxidation process (AOP) combined biological process for wastewater treatment: a review on advancements, feasibility and practicability of combined techniques. Environ Res 237:116944
Preisner M (2020) Surface water pollution by untreated municipal wastewater discharge due to a sewer failure. Environ Processes 7(3):767–780
Radetic B (2024) Sequencing batch reactors in municipal used water purification. Handbook of water and used water purification. Springer International Publishing, Cham, pp 537–555

Raman AAA, Asghar A, Buthiyappan A, Daud WMAW (2016) Treatment of recalcitrant waste. In: Elsevier eBooks, Elsevier BV, pp 409. https://doi.org/10.1016/b978-0-444-63665-2.00016-3
Reineke W (2001) Aerobic and anaerobic biodegradation potentials of microorganisms. In: Biodegradation and persistence, pp 1–161
Roy T (2024) Aquatic ecosystem toxicity and food chain. In: Toxicity of aquatic system and remediation, CRC Press, pp 25–36
Saeed MU, Hussain N, Sumrin A, Shahbaz A, Noor S, Bilal M, Aleya L, Iqbal HM (2022) Microbial bioremediation strategies with wastewater treatment potentialities–A review. Sci Total Environ 818:151754
Sanni SE, Okoro EE, Sadiku ER, Oni BA (2022) Data-centric intelligent systems for water quality monitoring, assessment, and control. In: Current trends and advances in computer-aided intelligent environmental data engineering. Academic Press, pp 129–160
Särner E (1981) Removal of dissolved and particulate organic matter in high-rate trickling filters. Water Res 15(6):671–678
Seviour R, Nielsen PH (eds) (2010) In: Microbial ecology of activated sludge. IWA Publishing
Shammas NK, Wang LK, Wu (2009) Waste stabilization ponds and lagoons. In Biological treatment processes. Totowa, NJ: Humana Press. pp. 315–370
Shieh WK, Keenan JD (2005) Fluidized bed biofilm reactor for wastewater treatment. In: Bioproducts, Berlin, Heidelberg, Springer, pp 131–169
Shim H, Shin EB, Yang ST (2002) A continuous fibrous-bed bioreactor for BTEX biodegradation by a co-culture of pseudomonas putida and pseudomonas fluorescens. Adv Environ Res 7:203–216
Sidhu C, Vikram S, Pinnaka AK (2017) Unraveling the microbial interactions and metabolic potentials in pre-and post-treated sludge from a wastewater treatment plant using metagenomic studies. Front Microbiol 8:1382
Sikosana ML, Sikhwivhilu K, Moutloali R, Madyira DM (2019) Municipal wastewater treatment technologies: a review. Proc Manufact 35:1018–1024
Singh S, Kumar A, Gupta H (2020) Activated banana peel carbon: a potential adsorbent for Rhodamine B decontamination from aqueous system. Appl Water Sci 10(8):185. https://doi.org/10.1007/s13201-020-01274-4
Singh NK, Yadav M, Singh V, Padhiyar H, Kumar V, Bhatia SK, Show PL (2023) Artificial intelligence and machine learning-based monitoring and design of biological wastewater treatment systems. Biores Technol 369:128486
Singh S, Parveen N, Gupta H (2018) Adsorptive decontamination of rhodamine-B from water using banana peel powder: a biosorbent. Environ Technol Innov 12:189–195
Sivakkumar SN, Periysamy K, Pandiyan B, Natarajan N, Vasudevan M (2019) Examine design considerations for sustainable management of an overloaded sewage treatment plant in a residential campus. Int J Recent Technol Eng (IJRTE) 8:104. https://doi.org/10.35940/ijrte.d1025.1284s219
Smet E, Van Langenhove H, De Bo I (1999) The emission of volatile compounds during the aerobic and the combined anaerobic/aerobic composting of biowaste. Atmos Environ 33(8):1295–1303
Smirnova E, Larionova Y, Larionov A (2020) Sewer system optimization in housing and communal services. In: E3S web of conferences, vol 157. pp 2002. https://doi.org/10.1051/e3sconf/202015702002
Smith P, Ashmore MR, Black HI, Burgess PJ, Evans CD, Quine TA, Thomson AM, Hicks K, Orr HG (2013) The role of ecosystems and their management in regulating climate, and soil, water and air quality. J Appl Ecol 50(4):812–829
Spanjers MG (2017) Biologically active filtration media properties: practical and mechanistic implications
Srivastava S (2015) Bioremediation technology: a greener and sustainable approach for restoration of environmental pollution. In: Applied environmental biotechnology: present scenario and future trends, pp 1–18

Sutton PM (2006) Membrane bioreactors for industrial wastewater treatment: Applicability and selection of optimal system configuration. In WEFTEC 2006. Water Environment Federation. pp. 3233–3248

Tabrizi GB, Mehrvar M (2004) Integration of advanced oxidation technologies and biological processes: recent developments, trends, and advances. J Environ Sci Health Part A 39(11–12):3029–3081

Tang P, Xie W, Tiraferri A, Zhang Y, Zhu J, Li J, Lin D, Crittenden JC, Liu B (2021) Organics removal from shale gas wastewater by pre-oxidation combined with biologically active filtration. Water Res 196:117041

van Lier JB (2008) High-rate anaerobic wastewater treatment: diversifying from end-of-the-pipe treatment to resource-oriented conversion techniques. Water Sci Technol 57:1137. https://doi.org/10.2166/wst.2008.040

Vanitha S (2020) Performance of different types of constructed wetlands for treating wastewater in short time horizon. In: IOP Conference series: materials science and engineering, November, vol 955(1). IOP Publishing, pp 012095

Wagner M, Stenstrom MK, Jenkins D, Wanner J (2014) Aeration and mixing. In: Activated sludge-100 years and counting. IWA publishing, pp 131–154

Wang LK, Wang MHS, Poon CP (1986) Trickling filters. Biological treatment processes. Humana Press, Totowa, NJ, pp 361–425

Wang LK, Li Y (2009) Sequencing batch reactors. In: Biological treatment processes, pp 459–511

Wang LK, Wu Z, Shammas NK (2009) Activated sludge processes. Biological treatment processes, pp 207–281

Wei Y, Van Houten RT, Borger AR, Eikelboom DH, Fan Y (2003) Minimization of excess sludge production for biological wastewater treatment. Water Res 37(18):4453–4467

Yates JG, Lettieri P (2016) Fluidized-bed reactors: processes and operating conditions, vol 26. Basel, Switzerland: Springer International Publishing

Zainudin MHM, Singam JT, Sazili AQ, Shirai Y, Hassan MA (2022) Indigenous cellulolytic aerobic and facultative anaerobic bacterial community enhanced the composting of rice straw and chicken manure with biochar addition. Sci Rep 12(1):5930

Zhao W, Zhang Y, Lv D, Wang M, Peng Y, Li B (2016) Advanced nitrogen and phosphorus removal in the pre-denitrification anaerobic/anoxic/aerobic nitrification sequence batch reactor (pre-A2NSBR) treating low carbon/nitrogen (C/N) wastewater. Chem Eng J 302:296–304

[illegible]

Chapter 9
Anaerobic Treatment of Sewages in Cities: A Better Perspective and Usage of Technologies

Rahul Maddheshiya, K. P. Singh, Vinod, Daya Nidhi Urmaliya, and Himanshu Gupta

Abstract Sewage waste water is a major area of concern in highly urbanized localities. The elimination of the contaminants dissolved in waste water are a heated topic. The untreated sewage gets directly discharged into the main rivers resulting in the eutrophication and biological toxicity to living ones. With the advent of the improved fermentation technologies the anaerobic sewage treatment plants are flourishing. Biochemical, microbiological, and digester technological advances in anaerobic technology has shown some promising results in the sewage stabilization. One of the most significant aims of using anaerobic process is that low maintenance, very minimal requirement of micro–macro nutrients and production of relatively low sludge. Anaerobic digestion is done if BOD/COD is 0.5 or higher; high temperature of effluents; presence of cyanides, phenols, heavy metals, long chain fatty acids, ammonia and sulphides sometimes inhibit the process. Different types of the digesters are installed for the anaerobic treatment of industrial effluents such as UASB, Anaerobic baffled reactor, Anaerobic sequencing batch reactor, anaerobic filters, expanded and fluidized bed reactor etc. But much work in this field is needed because of lack of awareness. Therefore, this chapter is aimed to focus on the different anaerobic microbes and their biochemistry associated with anaerobic digestion. This chapter

R. Maddheshiya
IFTM University, Moradabad, Uttar Pradesh, India

K. P. Singh
Department of Zoology, University of Allahabad, Prayagraj, India

Vinod
Department of Chemistry, University of Lucknow, Lucknow, Uttar Pradesh, India

D. N. Urmaliya
Environmental Engineering Divisions, CSIR—National Environmental Engineering Research Institute, Nagpur, India

H. Gupta (✉)
Department of Chemistry, School of Sciences, IFTM University, Moradabad, Uttar Pradesh, India
e-mail: hims.research@gmail.com

S. Madhav et al. (eds.), *Municipal Sewage and Sludge: Treatment and Disposal Strategies*, Environmental Science and Engineering,
https://doi.org/10.1007/978-3-032-07597-0_9

also outlines the various reactor deployed for anaerobic digestion of the sewage waste water and the factors affecting the working of digesters.

Keywords Sewage · Anaerobic · Effluents · Sequencing batch reactor · Anaerobic filters

9.1 Introduction

Rapid industrialisation and urbanisation have caused significant damage to Mother Earth by releasing pollutants directly into the environmental segments, particularly, water and soil (Gupta 2016; Gupta and Gupta 2016; Kumar et al. 2014, 2017; Gupta and Dhiman 2024). The unchecked raw industrial effluents and sewage water has severely harmed to fauna and flora along with human being, resulting in a low floral and faunal density and diseases in humans. Along with the waste management, the energy crisis is also a major concern of today (Abdelrahman et al. 2023; Wang and Zhou 2023; Gupta 2018). These pollutants also affect the natural healing ability of the environment by inhibiting the natural remedial process for example, gentisic acid and gallic acid from vinasse inhibit the anaerobic microbial activity involved in degradation of effluents (Strong and Burgess 2007; Melamane et al. 2007). In sewage waste water, sludge is the major contaminant/component involved, it is due to the high ratio of household wastewater is directly connected to the central treatment plants.

Nowadays, anaerobic treatment of sewage has been widely deployed over aerobic microbial methods because the former one requires lower consumption of energy concerning that of later one and some pollutants such as cyanides and phenols cause inhibitory effects on the later microbes (aerobic autotrophs) in the process of oxidation of ammonia. Similarly, heterotrophs also face the same problem in the reactor (Iaconi et al. 2009; Lofrano et al. 2010). Sulphate accumulation due to mining, for example sulphuric acid formed as a result of oxidation of pyrite (H_2S) oxidation in the presence of oxygen and H_2O, causes dissolution of many heavy minerals as a result acid mine drainage (AMD) is overloaded with heavy minerals and cause contamination (Akcil and Koldas 2006; Oberholster et al. 2013). The conventional treatment methods using metal hydroxide precipitation etc. proved very costly and less efficient in sulphate removal and sludge formation (Garcia et al. 2001). Several studies have shown the path of cost effectiveness and improved efficiency by anaerobic method (Van Lier et al. 2001; Bekmezci et al., 2011; Rodriguez et al. 2012; Kieu et al. 2011; Xingyu et al. 2013). Therefore, researchers have switched to the anaerobic mode of detoxification of sewages and industrial effluents.

Anaerobic digesters are a gold technology for the treatment of sewage resulting in the formation and processing of the high amount of sludge without any need of pretreatment. These digesters has several positive factors such as formation of biogas, increased stabilization of sludge, less odor, reduced pathogens and low sludge dried mas (Ward et al. 2008; Hanum et al. 2019). In anaerobic process of treatment less sludge is produced and more biogases (methane) are produced which can be further stored and used for late usage in different purposes as cooking. Phenolic removal from sewage water is carried by up flow anaerobic sludge blanket reactors (UASB) and in expanded granular sludge blanket (Fang et al. 2006; Scully et al. 2006). Another reactor used for the purification of the sewage and removal of impurities is anaerobic sequential reactor (ASBR). ASBR reactor works on for regular and continuous cycles such as feeding, reaction, settling, discharge, and idle time. It has various advantage over other mentioned earlier in that it has different work capacity such as in fed batch, batch or in both (Zaiat et al. 2001). ASBR using the maximum feeding time lowers the accumulation of toxic compounds and therefore decreases the possibility of inhibition by substrate.

Separation of various chemical reactions such as acidogenesis, and methanogenesis lets the favorable condition for each microbial colony to flourish and restricts the formation of intermediary byproducts which could harm the reactor (Bouallagui et al. 2004; Donoso-Bravo et al. 2009).

Anaerobically treatment of sewage water is obtained in 4 steps: (1) hydrolysis: organic complex components are broken into much simpler water soluble forms such as carbohydrates and amino acids; (2) acidogenesis: the products of hydrolysis is further degraded into smaller fatty acids, hydrogen (H_2), carbon dioxide (CO_2) and phenols; (3) acetogenesis: higher molecular weight organic acids are consumed by acetogenic microbes and degraded to carbon dioxide, acetic acid, and hydrogen; (4) methanogenesis: the methanogens finally breakdown the acetic acid into methane (CH_4 and CO_2) (Sompong et al. 2016; Stamatelatou et al. 2011; Muller et al. 2017). The usage of biogas demand as a renewable energy is increasing dramatically because it has less impact on environment in comparison to conventional fuels available. Anaerobic breakdown of sewage pollutant can be a better option and it may address the global energy requirements of fuel (Tambon et al. 2010; Rehl and Muller 2011; Jiang et al. 2011).

Much technical emphasis should be given in the field, so that efficiency of anaerobic reactors could be improved. In the treatment of high strength sewage anaerobically, volatile fatty acid (VFA) affects the process negatively due to imbalance in the rate formation and consumption. Instability of reactors related to anaerobic sewage water management is mainly due to the hydraulic overloading. Therefore, in order to counter the mistakes and hydraulic overloading, a suitable and more appropriate hydraulic loading rate (HLR) is engaged. In tropical world, overheating problem of the reactors is naturally avoided because of lower environmental temperature and this low temperature is favorable for the reactors.

Therefore, this chapter is confined to cover the important aspects and technologies and some case studies associated with the anaerobic processing of sewage or industrial effluents.

9.2 Anaerobic Microbial Community in Sewage Treatment

9.2.1 *Methanogens*

In the absence of electron digesters, different groups of bacteria from various trophic levels aid in anaerobic digestion of the sewage substrate such as monomeric and polymerics. Breakdown of proteins, suspended organic, long chain fatty acids, and polysaccharide components in to their less complex forms by means of enzymatic hydrolysis (facultative or obligate bacterial fermentation). Through the initial examination, it is observed that formic acid, acetate, carbon dioxide, methanol and H_2 are produced as a result of hydrolytic and non-hydrolytic fermentation by bacteria. The initiation process is called as acidogenesis (Fang et al. 2002).

Furthermore, acidogenesis is followed by acetogenesis, resulting in breakdown of end product of acidogenesis and oxidative release of carbon dioxide, acetate and hydrogen. The whole process is mediated by acetogens that produce hydrogen in obligate manner. The conversion of propionate, ethanol, or butyric acid to hydrogen, formic acid, carbon dioxide, or acetic acid is an endergonic process. But when hydrogen partial pressure is low such as during the during formate usage the reaction becomes exergonic. Therefore authors are view that acetogens must necessarily grow in syntropy with hydrogen using methanogens and sulphate using bacteria (SRB). The reaction and involvement of acetogens are considered as link between basic fermenting and methanogenic step (Schink 2002).

The last and final step to anaerobic sewage water treatment is methanogenesis. This group of bacteria is strictly anaerobic such as *Archaea. Archaea* is further subdivided into two groups, (a) *Acetoclastic* methanogens are responsible for the production of methane by decarboxylation of acetate and (b) Hydrogenotrophic species are responsible for methane formation by reduction of carbon dioxide and hydrogen. In comparison to both groups Acetoclastic group is more potent accounting to nearly 70% of total methane production during anaerobic digestion (O'Flaherty et al. 2006).

9.2.2 *Sulphate Reducing Bacteria (SRB)*

In an anaerobic condition, SRB grows well (Litho and heterotropically) in the sulphate, sulphate, and thiosulphate rich sewage substrate. SRB generally are specialized in using sulphate ions as a whole sole electron acceptor for molecular hydrogen oxidation. During the low sulphate ions concentration, SRB uses acetogenic pathway in anaerobic condition in sewage. It has the ability to involve different substrates in the presence of abundant sulphate also. This feature of SRB increases competitive ability of them to that of other types of bacterial group and therefore, resulting in the formation of hydrogen sulphide (H_2S) rather than methane. The competition between sulphate reducers to that of other groups of bacteria show some demerits also as, low methane yield, H_2S formed has inhibitory effect on other group of bacteria, corrosion of pipes etc. (O'Flaherty and Colleran 2000).

9.2.3 Acetogenic Bacteria

Homoacetogenic group is an important auto-heterotrophic bacterial group which results in the formation of acetate from carbon dioxide and hydrogen or other complex carbon compounds. *Desulphobacter, Eubacterium,* and *Acetobacterium,* are autotrophs and grow on CO_2 and H_2, but heterotrophs homoacetogens are unable to grow on both CO_2 and H_2 (Ryan et al. 2004). On the basis of metabolism usually this group has been divided into two: (1) Obligate (2) Homoacetogens.

Obligate hydrogen producer acetogens (OHPA/Proton reducer) (*Syntrophobacter Wolinii* and *S. wolfei*) by Beta-oxidation pathway breakdown larger fatty acids such as butyric acid, valerate, stearate, myristate, isocalerate, and propionate, alcohols to produce CO_2, H_2, and acetic acids. These are also essential in breaking down of aromatic compounds. It thrives well in the low H_2 condition produced as result of metabolic product. Such condition may be only achieved in presence of methanogens thus showing a syntrophic relationship between two groups of anaerobic bacteria (Methanogens-sulphate reducers and OHPA). This relation can be disturbed by a very slight perturbation as methanogens are inhibited by OHPA substrate (fatty acid) and OHPA is inhibited by H_2 (Substrate for methanogens).

Other group is anaerobic homoacetogens help in the formation of acetate from CO_2, and H_2. They also show syntrophic association because they are actively engaged in the interspecies H_2 transfer required by the first group acetogens. Some examples of homoacetogens are *Clostridium, Acetogenium, Acetobacterium, Pelobacter, Butribacterium, and Acetoanaerobium* (Anderson et al. 2003).

9.2.4 Fatty Acid Synthesizer Bacteria

These groups of bacteria belong to genus *Clostridia* are an expert in producing fatty acids from ethanol or acetate but in low concentration in the presence of high hydrogen concentration. This is a reversing process of the reactions due to syntrophic bacteria. In 80's bacteria were isolated that were capable in the conversion of acetate to propionate (Hydrogen sink) by the help of hydrogenases (O'Flaherty et al. 2006).

9.2.5 Acidogenic Bacteria

This group of bacteria provides substrates for the methanogens and acetogens. They usually breakdown the carbohydrates and amino acids into more simpler forms such as hydrogen, acetates, and carbon dioxide. *Ruminococcus, Micrococcus, Lactobacillus, Bacteroides, Propionibacterium, Streptococcus, Butyribacterium, Clostridium* are some of the acidogenic bacteria. The facultative groups also aid in the covering of the oxygen sensitive methanogens by digesting traces of oxygen that enter the reactors via feed. The total count of acidogens may vary in the range of 10^6–10^8 per cell (Anderson et al. 2003).

9.3 Anaerobic Treatment Process of Sewage Waste Water

The anaerobic sewage treatment has certain profits as it gives energy for further use in form of biogas and has low sludge production. The sole performance of anaerobic digesters depends upon the efficiency (Kinetics) of the microbe population within the reactor. In a reactor around 70% of the fermentation product is acetate formed as a result of methanogenesis. There is less number of methanogens that utilize acetate route of methane formation alternatively CO_2/H_2 route is preferred. Second route offers low hydrogen pressure to the system. Thus hydrogenotrophs become more evident in the granular methane sludge formation comprising of high density because H_2 producer and consumer bacteria lie at very minimal distance providing a calculated measurement rate of degradation of more complex substrates to simpler form by H_2 transfer. By high affinity hydrogen utilization methanogens are capable of metabolizing substrate at very low partial pressure of 3–7 Pa, thus showing syntrophic relationship to acetogens (Poulsen 2003; Aiyuk et al. 2006).

These methanogens (Hydrogenotrophics) are widely used because of two reasons: (1) capability to act in presence of very low H_2 which is nearly not detectable; (2) Doubling time of hrydrogenotropic methanogens is nearly 6 h much lesser to that of acetoclastic ones (2.6 days). Therefore, hydrogenotrophics methanogens are widely used in stable digesters. In unstable digesters, a high hydrogen partial pressure is maintained resulting in high electron sink such as lactate, butyrate, ethanol, volatile fatty acids and propionate. The products formed further broken down by acetogenesis before being used by methanogens. In waste water from petroleum refineries that is rich in sulphur, a notable competition is observed between methanogens and sulphate reducers. This competition results in the sulphate reduction as the core electron acceptor in the process (Elferink et al. 1994). This process is also favored thermodynamically.

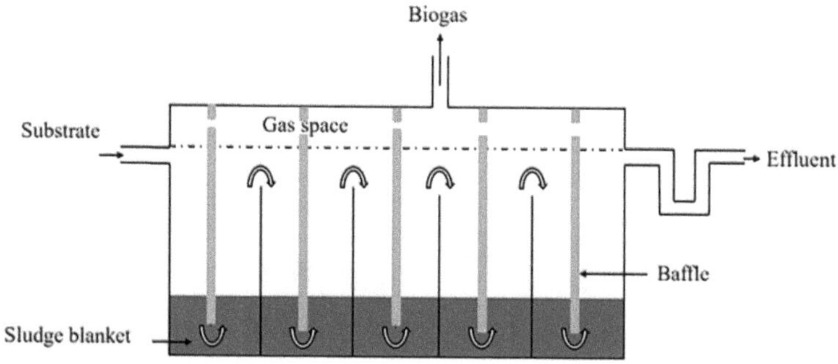

Fig. 9.1 Baffled reactor (anaerobic). Reprinted from original article by Khanal et al. (2017), after taking permission from Elsevier using RightsLink

9.4 Major Types of Anaerobic Digesters

9.4.1 Anaerobic Baffled Reactor (ABR)

In 1980s McCarty and Co-researchers of Stanford University developed ABR. This reactor consists of many compartments arranged in a series, baffled with the force of the incoming sewage water up through a lineage of blanked sludge. The baffles are aligned in a way such that they force the sewage water to flow above and below the baffles. Microbial colony gets accumulated in between the baffles and thus forming the granules of biomass along with time. Early during its introduction ABR was very rare in use but now a days its popularity is increasing tremendously (94–90). Systems stability is enhanced by the usage of granules. Therefore HRT can be separated from SRT, leading to better COD and solids removal, low sludge formation, and very small footprints. Bacteria in the ABR rise and settle at bottom due to flow traits and gas release in each separate compartment but show a slow and horizontal movement. This provides a rise in cell retention time (CRT-100 days) at HRT 20 h (Fig. 9.1).

ABR is preferable because it causes longitudinal separation of acidogenesis and methanogenesis in the reactor and allowing differential growth of bacterial groups. Furthermore, ABR is potent at high loading rates along with it extreme environmental factors and inhibitory substances have little effect. This reactor has shown higher level of tolerance against organic shock and hydraulic loads, biomass retention is for longer duration, and more importantly lower sludge formation. ABR has lesser risk of clogging and sludge bed expansion which choke other systems as UASB and AF.

9.4.2 Anaerobic Sequencing Batch Reactor (ASBR)

This technique is very promising in treating the high strength effluents containing medium solid concentration of TS 0.5–4%. Bio-flocculation and granulation are the

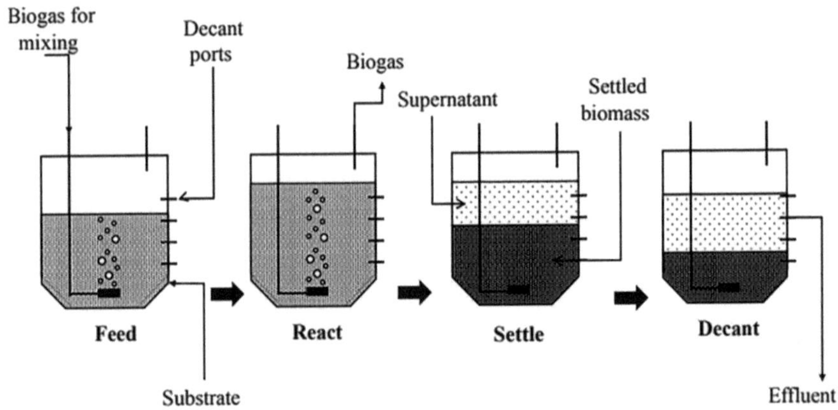

Fig. 9.2 Anaerobic sequencing batch reactor. Reprinted from original article by Khanal et al. (2017), after taking permission from Elsevier using RightsLink

two basic mechanism followed in ASBR. ASBR reactor follows sequential operation such that in many cyclic steps a batch mode system of operation is carried. In the process, during "feed cycle" the substrate is fed in the reactor thereafter with the help of anaerobic microbes the organic substrates are broken down in "react cycle" (Fig. 9.2). The content in the reactor is mixed either mechanically or with the help of biogas. During the "settle cycle" the microbial biomass is let to settle in the reactor after a desired time and decantation of supernatant is followed in "decant period" (Khanal et al. 2017).

9.4.3 Anaerobic Filters (AF)

Anaerobic filters are group of filter bed attached to the growth reactor and it is developed for the treatment of the sewage waste that is highly soluble (Khanal et al. 2017). It is configured into two components (1) Up flow AF, (2) Down flow AF. In case of up flow filter bed, the sewage flows upwards through the medium submerging the entire bed. Up flow bed is a fixed film reactor in which microbial colony is entrapped within the spaces between packing media and free colonies forms granular structure due to the gas bubbles. Gravels, ceramic tiles due to low porosity and lesser surface area were associated with clogging problems (40–50%). At present synthetic and light plastic media are employed which have better porosity and surface area (80–95% and 100 m^2/m^3 respectively). A quiet large quantity of biomass of the anaerobes microbes is retained this filter, thus achieving a long SRT irrespective of HRT. Excessive biomass is periodically removed from the bottom of the reactor to avoid short-circuiting and clogging (Khanal 2008).

9.4.4 Expanded-Bed Reactor (EBR)

An EBR is an attached growth reactor in which anaerobic colonies are rendered immobilized on certain down supportive media such as tire pieces, activated carbon/charcoal, or sand particles. The supportive bed may be subjected to expansion with the help of up flow fluid velocity of the sewage water rushing into it and effluent recirculation. The fluid flowing towards apex maintains the bed expansion rate upto 15–30%. More importantly, the clogging is minimal and better diffusion of the substrate is achieved within the biofilm. This reactor is ideally installed against the high strength factory sewage with high OLR (Khanal et al. 2017).

9.4.5 Fluidized-Bed Reactor (FBR)

It is similar to EBR but up flow velocity of liquid is maintained at 10–25 m/h so that the bed can expand by 25–30% of the total settled volume of the bed. Therefore, FBR is a fully fixed film reactor in which suspended solid particles and microbial colony is properly washed out of it (Khanal 2008). It is better preferred because of low clogging and short circuiting. The free moving bio-carriers in the bed is fully favored by liquid up flow velocity. FBR was 94% efficient in COD removal in a slaughter house (COD concentration = 4500 mg/L; OLR = 27 kg COD/m^3) (Borja et al. 1995).

9.4.6 Up Flow Anaerobic Sludge Blanket Reactor (UASB)

USAB reactor is employed for high strength sewage water in the moderate temperature range between 20 and 45 °C. The main configuration of this reactor is: (1) Higher cellular retention time due to high biomass concentration; (2) Granulation/sludge formation of multi-cellular aggregates; (3) Various group of microbes are involved that convert complex organic components into carbon dioxide and methane; (4) reduced requirement for nutrients and no overloading of sludge formation; (5) highly stable in relation to concentration and composition of influents; (6) High organic loading rates (OLR); (7) Lower installation rate and anaerobic process is self-sustained at adequate internal conditions; (8) Excessive sludge production is low. Due to the above sustainable points, UASB technique has attracted the researchers for its domestic installation. UASB has been widely used in countries such as Colombia, America, China, Cuba, Mexico, Brazil etc. There are certain differences between industrial and domestic sewages treated under UASB. Domestic sewage is although low strength but particulate COD is in very high along with proteins, carbohydrates, fat, and detergents. Excess of these products pose restriction to the UASB efficiency in terms of COD removal rate, low methane recovery, and maximum organic loading

(OLR). UASB sewage digester is used as first unit of the system installed for domestic waste water treatment at temperature more than 20 °C. At OLR less than 3 kg COD M^{-3} d^{-1}, COD removal capacity was up to 65–80% (HDT range = 6–10 h) (Chernicharo and Nascimento 2001; Rodriguez et al. 2001). In a full scale running reactor installed for domestic sewage treatment inoculation can be neglected (Passing et al. 2000). Granulation is observed showing that presence of carbohydrates pose less impact. AT lower OLR and higher HTD the COD of the filtered effluent lies in the range 80–120 mg l^{-1}. Therefore effluent COD value provides information about the USAB reactor performance.

9.5 Factors Affecting AD Digesters

Toxicity in anaerobic digestion may be caused due to different factors that inhibit the reaction process in digesters.

9.5.1 Ammonia Accumulation

In AD as a result of the anaerobic biological breakdown of nitrogen rich substrate such as urea and proteins, ammonia gets accumulated. Some of the process has been already proven in the estimation of degree to which ammonia is inhibiting a reaction such as specific enzyme inhibition reaction, increased in the maintenance energy requirement, and fluctuation in the pH (intracellular) (Whitemann et al. 1995). There are usually two types of ammonium nitrogen available in the reactor aqueous solution: Free ammonia (FA) and ammonium ion.

Free ammonia can easily pass through the filter membrane and therefore it is the main culprit in inhibition process by inhibiting the methanogens activity. Most affected methanogen species were *Methanospirillum hungatei* (most affected), *M. hungatei,* and *Methanobacterium formicium.* According to Liu and Sung (2002), since nitrogen is an integral part of the anaerobic digestion therefore, a concentration of ammonia below 200 mg/L in the reactor does not inhibit the reaction.

9.5.2 pH

Increased pH increases toxicity in the digester. It is just because of the fact that higher pH results in higher accumulation of free ammonia (ionized form). This ionized ammonia caused accumulation of volatile fatty acids (VFAs) and thus lowers pH which ultimately leads to low FA. It is a inhibited steady state where VFA, FA, and pH together cause the system to rum but with lower yield of total methane. Acidification of crab waste water is known to improve the working power of UASB reactors. A reduction in pH from 7.5 to 7 has been reported to increase the methane production in cow manure where thermophilic anaerobic process occurs. In pig form manure a adjustment of pH from 8 to 7.4 results in usage of excess VFA. This shows a optimum pH is required in the proper functioning of AD because both acidogenic and methanogenic microbes have very fixed optimal pH for efficient functioning (Chen et al. 2008).

9.5.3 *Temperature*

Temperature maintenance is the key to proper functioning of microbes in digesters. Increase in temperature in the digester positively impacts the microbe's metabolic rate but also results in the accumulation of higher concentration FA. Higher FA means more ammonia resulting in less stability at thermophilic temperature such as digestion with 50 °C of cow manure waste with TAN value of 3 g/L was showing difficulty in processing. Lower temperature (> 60 °C) with high concentration of ammonia was prot3ected by high FA concentration because of higher yield of biogas (Hasen et al. 1999).

9.5.4 *Presence of Ions*

Presence of some ions such as Mg^{2+}, Na^{+}, and Ca^{2+} was having antagonistic effect to NH_3 inhibition. It is based on simple phenomena in which presence of one ion decreases the toxic effect/inhibitory effect of other ion. NH_3 acts antagonistically to sodium and vice versa. For example addition of Na^{+} (0.002–0.005 M) to the media increases the methane production by 5% antagonistically addition of ammonia (0.15 M) to the reactor reduces the methane formation from acetic acid by 20%. Addition of both K^{+} and Na^{+} or Na^{+} and Mg^{2+} boosts the total yield of methane by 10% much higher than Na^{+} alone. It is also reported that the waste manure from poultry farms with $NH_4^{+}Cl$ with concentration as high as 30 g/L showed higher yield of biogas on the addition of phosphorite. This can be attributed to the fact that phosphorite prevent the biomass washout by immobilization on the mineral bed (Chen et al. 2008).

9.5.5 *Sulphide Inhibition*

Sulphide or sulphate is a common ingredient in much industrial sewage. Sulphate reducer microbes (SRM) convert sulphate into sulphide by two group of SRMs: (1) Incomplete oxidizer microbes that cause reduction of lactic acid into carbon dioxide and acetic acid, (2) Complete oxidizer microbes that convert lactate into acetic acid and further results in formation of carbon dioxide and HCO_3^- ion. In the primary inhibition a competition occurs for organic and inorganic substrates causing methane suppression and secondly, toxicity to microbes due to sulphide (Colleran et al. 1998; Colleran and Pender 2002).

9.5.6 N-Substituted *Aromatic Compounds*

This group of toxicant include nitrophenols, aromatic amines, nitrobenzenes, etc. these compounds usually hamper the metabolic activity by interfering with the function of the enzymes. According to Bhattacharya et al. (1996) these compounds show inhibitory concentration of 0.014–0.12 mM for methanogens. In contrast aromatic amines are less toxic due to hydrophobic nature ($IC_{50} = 3.2–67$ mM). amongst them all nitroanilines are most toxic to methanogens (Razo-Flores et al. 1997).

Addition or deletion of the nitro group may affect the toxicity of the *N-substituted* compounds such as an addition of amino to aniline results in the formation of phenylenediamines which is less toxic to parent compound. Among the nitrophenols studies the toxicity was in the order 4-nitrophenol > 2-nitrophenol > 3-nitrophenol for acetate abundant culture (Haghighi-Podeh and Bhattacharya 1996). Therefore acclimatization of anaerobes culture with *N-Substituted* aromatic chemicals decreases and helps in the enhancement of breakdown of the components in sewage.

9.5.7 Long Chain Fatty Acids (LCFAs)

Long chain fatty acids usually hamper the anaerobic process of sewage digestion. At lower concentration LCFAs has proven to be inhibitory in function for gram-positive microbes in the digesters but have nearly no effect on gram-negative micro-organisms. Methanogens cell wall shows similarity to that with gram-positive bacteria and therefore LCFA inhibitory is effective in them because they bind with the cell wall and hinder the important biological process. LCFAs sorption in light quantity in the biomass results in more sludge formation ultimately its washout from digester (Hwu et al. 1998).

Lauric acid and oleic acid almost exhibited the same inhibitory concentration of 4.3 mM. In presence of lauric acid, the toxicity of both myristic acid and capric acid was increased. The toxicity of LCFAs also dependent upon the physical traits of the sewage instead of its biological traits such as flocculent sludges face more inhibition because of more exposed surface are than to its granular form (Hwu et al. 1996). Mesophilic microbes were more affected by LCFAs than as compared to thermophiles (Hwu and Lettinga 1997). LCFA thus shows bactericidal effect and bacteria could not develop resistant to it. But current researches have shown that acclimatization of the bacteria in the oleic acid degradation fixed bed reactor has led to more resistive biofilm formation towards oleate and more biodegradable capacity compared to biofilm formation in lack of lipid in the digester (Alves et al. 2001a, b). LCFAs can be inhibited with the addition of Ca^{2+}, it may be due to the formation of insoluble salts.

9.6 Case Study in Full Scale Anaerobic Sewage Treatment Plant in Brazil

In the current study, 5 anaerobic sewage treatment plants (STP) has been discussed regarding the chemical oxygen demand (COD), settled effluents, and biological oxygen demand (BOD). All the reviewed sewage treatment plants fulfilled the desired role of organic load reduction except first STP. The median value of the BOD and COD ranged between 240–550 mg/L and 550–840 mg/L consecutively. The BOD and COD removal success rate was 65 (around 82%) and 60 (around 75%) respectively. HRT value for STP 2–4 was slightly more than expected (20%) and for STP 5 the HRT value was much higher than the designed reactor (80%). Therefore, one can say that STP was much more functional then expected (More HRT value the better hydrolysis power of the reactor). The higher HRT impairs the proper functioning of the reactor as it may cause less mixing of the fluids due to less up flow of the fluid velocity (< 1.5 m/h; For STP 5 up flow velocity was 0.3 m/h) (Bressani-Ribeiro et al. 2019).

The USAB reactors have unique trait in that they show high retention of the solid and therefore more sludge accumulation causes higher stabilization of sludges (Chernicharo et al. 2015, 2019). STP 1 with USAB reactor receives abnormal different waste from different sewage system, thus resulting in solid overload (more than its capacity) and their excessive washout with anaerobic effluents. This excessive solid washout consequently leads to much higher COD effluent concentration. This higher load has negative effect on the trickling filter of the post-treatment process. In STP 1 therefore average COD removal capacity moved to zero resulting in clogging of the trickling filters. The problems related with anaerobic reactor can be solved at a minimal cost (Bressani-Ribeiro et al. 2019).

9.7 Conclusion

The anaerobic reactors have been widely deployed for the treatment of high strength sewage water from houses and industries as well as for the bioenergy production. The reactors are used towards a pretreatment measure against the sewages in order to reduce the OLR. The reactors that are built on the basis of SRT and HRT separation especially for the liquid sewage are the major area of research. Few decoupling techniques engaged in these anaerobic reactors are mainly recycling of microbial colonial biomass, retaining of the biomass, granulation, and immobilization of biomass. This is achieved by using membrane system. The major motive of decoupling in the process is the prevention of slow developing methanogens from washout through the anaerobic bioreactor. It is dire need of time that some new innovative designs of the reactors should be incorporated that works efficiently, cheap, reliable, and capable of producing high quantity of effluents. With the emergence of advanced molecular tools, particular attention should be directed toward research and development activities that pertain to the integration of microbial information to the design and functioning of an anaerobic reactor. Moreover, with regard to the design of the bioreactor, process automation and control should be incorporated. As a result, process engineering as well as process microbiology are highly essential for the design and operation of a bioreactor.

References

Abdelrahman AM, Kosar S, Gulhan H, Cicekalan B, Gulin U, Ezgi A (2023) Impact of primary treatment methods on sludge characteristics and digestibility, and wastewater treatment plant-wide economics. Water Res 235:119920. https://doi.org/10.1016/j.watres.2023.119920

Aiyuk S, Forrez I, De Lieven K, van Haandel A, Verstraete W (2006) Anaerobic and complementary treatment of domestic sewage in regions with hot climates—a review. Bioresour Technol 97:2225–2241. https://doi.org/10.1016/j.biortech.2005.05.015

Akcil A, Koldas S (2006) Acid mine drainage (AMD): causes, treatment and case studies. J Clean Prod 14:1139–1145

Alves MM, Mota Vieira JA, Pereira A,´ lvares RM, Pereira MA, Mota M (2001a) Effects of lipids and oleic acid on biomass development in anaerobic fixed-bed reactors. Part II: oleic acid toxicity and biodegradability. Water Res 35(1):264–270

Alves MM, Mota Vieira JA, Pereira A, Ívares RM, Pereira MA, Mota M (2001b) Effects of lipids and oleic acid on biomass development in anaerobic fixed-bed reactors. Part I: biofilm growth and activity. Water Res 35(1):255–263.

Anderson K, Sallis P, Uyanik S (2003) Anaerobic treatment processes. In: The handbook of water and wastewater microbiology. ISBN 0-12-470100-0

Bekmezci OK, Ucar D, Kaksonen AH, Sahinkaya E (2011) Sulfidogenic biotreatment of synthetic acid mine drainage and sulfide oxidation in anaerobic baffled reactor. J Hazard Mater 189:670–676

Borja R, Banks CJ, Wang Z (1995) Effect of organic loading rate on anaerobic treatment of slaughter house wastewater in a fluidised-bed reactor. Bioresour Technol 52:157–162

Bouallagui H, Torrijos M, Godon J, Moletta R, Ben Cheikh R, Touhami Y, Delgenes J, Hamdi M (2004) Two-phase anaerobic digestion of fruit and vegetables wastes: bioreactors performance. Biochem Eng J 21:193–197

Bressani-Ribeiro T, Chamham-Silva LA, Chernicharo CAL (2019) Constraints, performance and perspectives of anaerobic sewage treatment: lessons from full-scale sewage treatment plants in Brazil. Water Sci Technol. https://iwaponline.com/wst/article-pdf/80/3/418/608870/wst080030418.pdf

Chen Y, Cheng Jay J, Creamer Kurt S (2008) Inhibition of anaerobic digestion process: a review. Bioresour Technol 99:4044–4064. https://doi.org/10.1016/j.biortech.2007.01.057

Chernicharo CAL, Bressani-Ribeiro T, Lobato LCS (2019) Operation of UASB reactors for sewage treatment. In: Chernicharo CAL, Bressani-Ribeiro T (eds) Anaerobic reactors for sewage treatment: design, construction and operation. IWA Publishing, London, pp 126–158

Chernicharo CAL, Nascimento CP (2001) Feasibility of a pilot-scale UASB/trickling filter system for domestic sewage treatment. Water Sci Technol 44(4):221–226
Chernicharo CAL, van Lier JB, Noyola A, Bressani Ribeiro T (2015) Anaerobic sewage treatment: state of the art, constraints and challenges. Rev Environ Sci Biotechnol 14(4):649–679
Colleran E, Pender S (2002) Mesophilic and thermophilic anaerobic digestion of sulphate-containing wastewaters. Water Sci Technol 45(10):231–235
Colleran E, Pender S, Phipott U, O'Flaherty V, Leahy B (1998) Full-scale and laboratory-scale anaerobic treatment of citric acid production wastewater. Biodegradation 9:233–245
Di Iaconi C, Ramadori R, Lopez A (2009) The effect of ozone on tannery wastewater biological treatment at demonstrative scale. Bioresour Technol 100:6121–6124
Donoso-Bravo A, Ruiz-Filippi G, Chamy R (2009) Anaerobic treatment of low-strength wastewater with a high fraction of particulate matter in an unconventional two-phase ASBRs system. Biochem Eng J 43:297–302
Elferink OSJWH, Visser A, Hulshoff LW, Stams AJM (1994) Sulphate reduction in methanogenic bioreactors. FEMS Microbiol Rev 15:119–136
Fang HHP, Liang DW, Zhang T, Liu Y (2006) Anaerobic treatment of phenol in wastewater under thermophilic condition. Water Res 40:427–434
Fang HHP, Liu H, Zhang (2002) Characterization of a hydrogen-producing granular sludge. Biotechnol Bioeng 78(1):44–52
Garcia C, Moreno DA, Ballester A, Blàzquez ML, Gonzàlez F (2001) Bioremediation of an industrial acid mine water by metal-tolerant sulphate-reducing bacteria. Miner Eng 14(9):997–1008
Gupta H (2016) Photocatalytic degradation of phenanthrene in the presence of akaganeite nanorods and the identification of degradation products. RSC Adv 6(114):112721–112727. https://doi.org/10.1039/C6RA24602H
Gupta H (2018) PAH determination in effluent and sludge samples of paper industry. Environ Technol Innov 9:115–121. https://doi.org/10.1016/j.eti.2017.11.009
Gupta H, Dhiman S (2024) Emerging contaminants in the environment introduction. In: Occurrence, distribution and toxic effects of emerging contaminants. CRC Press, pp 1–16. https://doi.org/10.1201/9781003335757
Gupta H, Gupta B (2016) Adsorption of polycyclic aromatic hydrocarbons on banana peel activated carbon. Desalin Water Treat 57(20):9498–9509
Haghighi-Podeh MR, Bhattacharya SK (1996) Fate and toxic effects of nitrophenols on anaerobic treatment systems. Water Sci Technol 34:345–350
Hansen KH, Angelidaki I, Ahring BK (1999) Improving thermophilic anaerobic digestion of swine manure. Water Res 33:1805–1810
Hanum F, Yuan LC, Kanahara H, Aziz HA, Atsuta Y, Yamada T, Daimon (2019) Treatment of sewage sludge using anaerobic digestion in Malaysia: current state and challenges. Front Energy Res 7:19. https://doi.org/10.3389/fenrg.2019.00019
Hwu CS, Donlon B, Lettinga G (1996) Comparative toxicity of long chain fatty acid to anaerobic kludges from various origins. Water Sci Technol 34(5–6):351–358
Hwu CS, Lettinga G (1997) Acute toxicity of oleate to acetate-utilizing methanogens in mesophilic and thermophilic anaerobic sludges. Enzyme Microb Technol 21:297–301
Hwu CS, Tseng SK, Yuan CY, Kulik Z, Lettinga G (1998) Biosorption of long-chain fatty acids in UASB treatment process. Water Res 32(5):1571–1579
Jiang X, Sommer SG, Christensen KV (2011) A review of the biogas industry in China. Energy Policy 39:6073–6081
Khanal S K (2008) Anaerobic biotechnology for bioenergy production. Wiley-Blackwell, Ames, IA
Khanal SK, Giri B, Nityavardhana S, and Gadhamshetty V (2017) Anaerobic bioreactors/digesters: design and development. In: Current developments in biotechnology and bioengineering: biological treatment of industrial effluents. https://doi.org/10.1016/B978-0-444-63665-2.00010-2

Kieu HTQ, Müller E, Horn H (2011) Heavy metal removal in anaerobic semicontinuous stirred tank reactors by a consortium of sulfate-reducing bacteria. Water Res 45(13):3863–3870

Kumar R, Gupta B, Gupta H, Rani M (2014) Distribution of persistent organic pollutants in urban aquatic systems. Int J Sci Res Environ Sci 2(7):233–243. https://doi.org/10.12983/ijsres-2014-p0233-0243

Kumar R, Rani M, Gupta H, Gupta B, Park D, Jeon BH (2017) Distribution of trace elements in flowing surface waters: effect of seasons and anthropogenic practices in India. Int J Environ Anal Chem 97(7):637–656. https://doi.org/10.1080/03067319.2017.1339035

Liu T, Sung S (2002) Ammonia inhibition on thermophilic acetoclastic methanogens. Water Sci Technol 45:113–120

Lofrano G, Meric S, Inglese M, Nikolaou AD, Belgiorno V (2010) Fenton oxidation treatment of tannery wastewater and tanning agents: synthetic tannin and nonylphenol ethoxylate-based degreasing agent. Desalin Water Treat 23

Melamane X, Tandlich R, Burgess J (2007) Anaerobic digestion of fungally pre-treated wine distillery wastewater. Afr J Biotechnol 6(17):1990–1993

Müller F, Maack G-C, Buescher W (2017) Effects of biogas substrate recirculation on methane yield and efficiency of a liquid-manure-based biogas plant. Energies 10:325

O' Flaherty V, Collins G, Mahony T (2006) The microbiology and biochemistry of anaerobic bioreactors with relevance to domestic sewage treatment. Rev Environ Sci Biotechnol 5:39–55. https://doi.org/10.1007/s11157-005-5478-8

Oberholster PJ, Genthe B, Hobbs P, Cheng PH, de Klerk AR, Botha AM (2013) An ecotoxicological screening tool to prioritise acid mine drainage impacted streams for future restoration. Environ Pollut 176:244–253

Passig FH, Villela LH, Ferreira OP (2000) Piracicamirim sewage treatment plant—conception utilizing anaerobic process followed by aerobic process—evaluation of operational conditions and compatibility of the processes. In: Preprints of the "VI Latin-American workshop and seminar on anaerobic digestion, vol 1, pp 53–59

Razo-Flores E, Donlon B, Lettinga G, Field JA (1997) Biotransformation and biodegradation of N-substituted aromatics in methanogenic granular sludge. FEMS Microbiol Rev 20:525–538

Rehl T, Müller J (2011) Life cycle assessment of biogas digestate processing technologies. Resour Conserv Recycl 56:92–104

Rodriguez JA, Peña MOR, Manzi V (2001) Application of an innovative methodology to improve the starting-up of UASB reactors treating domestic sewage. Water Sci Technol 44(4):295–303

Rodriguez RP, Oliveira GHD, Raimundi IM, Zaiat M (2012) Assessment of a UASB reactor for the removal of sulphate from acid mine water. Int Biodeter Biodegr 74:48–53

Ryan P, O'Reilly C, Golden T, Colleran E (2004) The role of proceedings of AD, Montreal, Canada, 29-8-2/9, pp 155–160

Schink B (2002) Synergistic interactions in the microbial world. Anton Leeuw Int J 81(1–4):257–261

Scully C, Collins G, O'Flaherty V (2006) Anaerobic biological treatment of phenol at 9.5–158 C in an expanded granular sludge bed (EGSB)-based reactor. Water Res 40:3737–3744

Sompong O, Suksong W, Promnuan K, Thipmunee M, Mamimin C, Prasertsan P (2016) Two-stage thermophilic fermentation and mesophilic methanogenic process for biohythane production from palm oil mill effluent with methanogenic effluent recirculation for pH control. Int J Hydrogen Energy 41

Stamatelatou K, Antonopoulou G, Lyberatos G (2011) Production of biogas via anaerobic digestion. Handb Biofuels Prod Process Technol 266–304

Strong PJ, Burgess JE (2007) Bioremediation of a wine distillery wastewater using white rot fungi and the subsequent production of laccase digestion. Water Sci Technol 56(2):179–186

Tambone F, Scaglia BD, Imporzano G, Schievano A, Orzi V, Salati S et al (2010) Assessing amendment and fertilizing properties of digestates from anaerobic digestion through a comparative study with digested sludge and compost. Chemosphere 81:577–583

Van Lier JB, Tilche A, Ahring BK, Macarie H, Moletta R, Dohanyos M, Hulshoff Pol LW, Lens P, Verstraete W (2001) New perspectives in anaerobic digestion. Water Sci Technol 43(1):1–18

Wang HU, Zhou Q (2023) Dominant factors analyses and challenges of anaerobic digestion under cold environments. J Environ Manage 348:119378. https://doi.org/10.1016/j.jenvman.2023.119378

Ward AJ, Hobbs PJ, Holliman PJ, Jones DL (2008) Optimization of the anaerobic digestion of agricultural resources. Bioresour Technol 99:7928–7940. https://doi.org/10.1016/j.biortech.2008.02.044

Whittmann C, Zeng AP, Deckwer WD (1995) Growth inhibition by ammonia and use of pH-controlled feeding strategy for the effective cultivation of *Mycobacterium chlorophenolicum*. Appl Microbiol Biotechnol 44:519–525

Xingyu L, Gang Z, Xiaoqiang W, Laichang Z, Jiankang W, Renman R, Dianzuo W (2013) A novel low pH sulfidogenic bioreactor using activated sludge as carbon source to treat acid mine drainage (AMD) and recovery metal sulfides: pilot scale study. Miner Eng 48:51–55

Zaiat M, Rodrigues JA, Ratusznei SM (2001) Anaerobic sequencing batch reactors for wastewater treatment: a developing technology. Appl Microbiol Biotechnol 55:29–35

Chapter 10
Law and Policies in Sewage Treatment and Sludge Management

Shazia Shifa, Madiha Hashmi, and Saweza Hashmi

Abstract This chapter comprehensively overviews the legal and policy frameworks governing sewage treatment and sludge management. Beginning with a historical perspective and tracing the evolution to modern management techniques, the chapter addresses the challenges posed by increasing urbanization and population growth, which have significantly escalated sewage generation rates. Key regulations and frameworks, including water policy legislation, national urban sanitation policy, and state sewerage management policies, are meticulously examined. The chapter explains various sewage treatment techniques, detailing their processes and efficacy. Emphasis is placed on international guidelines, permissible limits, standards, penalties, and compliance requirements essential for effective management. Additionally, the chapter highlights critical gaps in current policies and suggests measures for addressing these shortcomings to ensure robust and sustainable sewage management systems. Overall, this chapter function as a foundational asset for policymakers, researchers, and practitioners involved in environmental management, offering insights into the regulatory landscape and technological advancements crucial for mitigating the challenges associated with sewage treatment and sludge management in contemporary urban settings.

Keywords Laws and policies · Sewage · Treatment · Sludge · Environmental management

S. Shifa (✉)
Department of Environmental Engineering, Delhi Technological University, Delhi, India
e-mail: shaziashifa198@gmail.com

M. Hashmi
Faculty of Law, Integral University, Lucknow, India

S. Hashmi
Department of Environmental Science, Jamia Millia Islamia, Delhi, India

S. Madhav et al. (eds.), *Municipal Sewage and Sludge: Treatment and Disposal Strategies*, Environmental Science and Engineering,
https://doi.org/10.1007/978-3-032-07597-0_10

10.1 Introduction

10.1.1 Understanding of Water Pollution and Sewage Disposal

Pollution is any contamination in the environment that emerges from various anthropogenic activities, and distributed into different environmental matrices, and causes negative environmental effects, furthermore, those substances natural or man-made that can deteriorate the environment are referred to as environmental pollutants, they can be of many different types, such as heavy metals, pesticides, poisonous gases, toxic substances, etc. (Arzu Özkara 2018). The degradation of water due to natural as well as human-made sources is referred to as water pollution, it is the degradation of the natural state of existing water and, the deformation of its properties making it unfit for drinking and other activities. Sewage is a mixture of solid and liquid that sewers collect, treat, and then release into vast bodies of water (Zhang et al. 2016). According to Ahmed et al. (2017), in underdeveloped countries, 90% of sewage is untreated and discharged into watercourses, polluting the environment, this is the direct source of water pollution due to sewage. Sewage can pollute drinking water through septic tank infiltration, inadequately treated wastewater, sludge, and sewer line leakages (Gothwal and Shashidhar 2015). Sludge is the semi-solid residue of sewage treatment plants referred to as sewage sludge operated to treat the effluent (Demirbas et al. 2016). There are various methods employed for sewage disposal, but all have some negative complications, which cause pollution and degradation of water. The sewage disposal method in septic tanks is majorly adopted but responsible for the pollution of water and soil media (Nowak and Imperowicz 2016). This is further followed by sewage flow into stormwater, which can cause an elevated risk of flooding as well as air contamination (Palamuleni 2002). Discharging sewage into ponds is a major threat to vector-borne diseases as it serves as the breeding area for the growth of vectors (Petrie et al. 2015). The unregulated disposal of unprocessed waste into rivers increases the presence of toxic substances in aquatic plants as well as animals, further expanding their reach into the food via direct and indirect feeding on these species.

10.1.2 Significance of Sewage Treatment and Sludge Management

The management of sludge and sewage treatment is an emerging concern due to its harmful impacts on the environment, degradation of the natural quality of water, air, soil, etc., along with the harmful impact on the living species (Rorat et al. 2019), therefore, to maintain sustainable living and safeguard natural resources it is evident to protect them by regulating the discharge of sewage and other toxic waste directly in them. It is important to meet the standards of SDG 6, 14, and 15 to achieve environmental sustainability. The existing methods of sludge management include an energy recovery process by incineration, pyrolysis, or gasification. There are options for organic recycling as well that include direct utilization of sewage sludge as soil fertilizer or composting, and mechanical biological treatment is also opted, landfill is the least preferred method for disposal and management of sludge (Kacprzak et al. 2017). Effective sludge management is necessary to reduce the health risks associated with unsanitary conditions caused due to improper sludge disposal, which causes millions of deaths annually in developing countries (Jimenez et al. 2004). Proper sludge treatment is also essential to regulate agricultural reuse to harness its nutrient value, it also balances the benefits and potential risks of sludge (Sharma et al. 2022).

10.2 Regulatory Framework for Wastewater Management in India

There are different elements in the legislative framework of the Central government in India, for sewage treatment. It has a governance system that has different components including Constitution, Policies, Laws, and Developmental Plans. As reported in Breitenmoser et al. (2022) these elements related to "The Indian Constitution, 1950 are articles 246 & 248", the policies included are, "The National Urban Sanitation Policy 2008", "The National Water Policy 2012", "Tariff Policy 2016", followed by Laws governing legislative framework are, "The Water (Prevention & Control of Pollution) Act, 1974" and "The Environment (Protection) Act, 1986" and the developmental plans included in the framework are "The Ganga Action Plan I & II, Namami Gange Mission, Swachh Bharat Mission, and Smart City Initiative". The schematic flow of this framework is represented in Fig. 10.1.

10.3 Historical Development of Water Pollution Control in India

10.3.1 The Shore Nuisance Act, 1853

The history of Environmental change in India has had its course throughout. The wave of industrialization and urbanization that began in England extended to India in the nineteenth century when the British East India Company arrived (Kumar Gupta 1994). This ushered in the establishment of factories, machinery, and industries, but this industrial growth was intricately linked to significant environmental degradation. The policy of imperialism led to the overuse of the rich natural resources of our country. The growing port of Bombay saw demographic and economic changes under British colonial rule which led to a lot of strain on the growing population

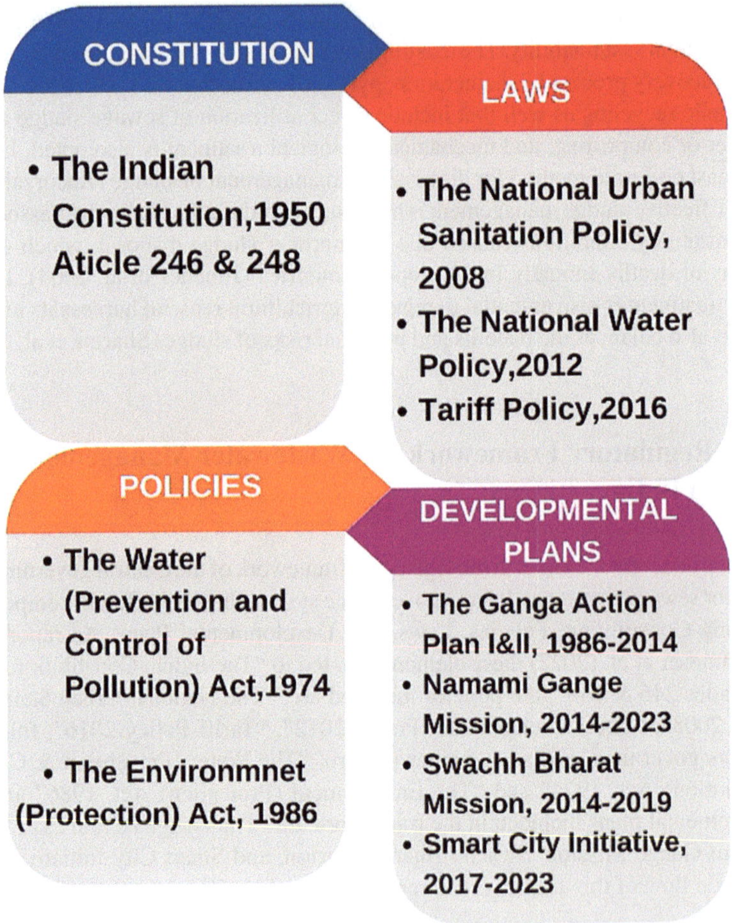

Fig. 10.1 Elements in sewage treatment legislative framework of Central Government in India

and pollution in the surrounding land and water bodies, like uncontrolled dumping of refuse, industrial waste, sewage, machinery wastes, and chemicals that led to the increase in coastal pollution and water bodies.

The Shore Nuisance which is also known as Bombay and Kolaba Act, was one of the initial measures taken to address water pollution control in the year 1853. The act states in the Preamble, "There is a large sea- shore in the islands of Bombay and Kolaba, and it is expedient, with a view to the safe navigation of the harbour of Bombay, and to the public interests generally, to facilitate the removal of nuisances, obstructions, and encroachments below high-water mark in the said harbour, or upon or about the shores of the said islands (India Code-Digital Repository 1853)". It was one of the foremost laws that aimed at addressing water pollution and was designed to regulate the discharge of waste materials from various industries situated near the coastal areas of Bombay and Kolaba. The Implementation of this Act brought about several key changes. These included improved regulation of the harbor area, which curbed unauthorized activity, making the harbor more orderly and functional, along with the Development of public infrastructure which allowed the construction of ports, roads, and public buildings contributing to the well-structured urban planning of the metropolis. Additionally, the act helped mitigate public health risks in crowded areas which laid the framework for improved public sanitation was another notable shift. Moreover, it set the legal framework for land management for local courts in land disputes and appeals for action by the collector of land revenue. Balancing government authority with individual rights played a significant role in shaping outcomes. Furthermore, it helped Bombay's evolution to Mumbai by turning it into India's largest and most important urban centre benefitting from structured port operations and land management policies initiated by laws like the Bombay Kolaba Act, and finally, an inspiration for future urban legislation by regulating the importance of public and coastal spaces and for later acts governing the areas for urban development.

10.4 The Evolution of Water Pollution Control Legislation in India

10.4.1 Growth of Environmental Governance in India

The evolution of legislation for water pollution controls started with the hey fining and movements dated back in the 90 s, some of which are briefed to make an understanding. Rachel Carson's "*Silent Spring*", in 1962 emphasized on hazards of the pesticide DDT and led the way for the environmental movement today, which highlighted the concept of humanity and its relationship with nature as how important is it for us to have self-control of our actions to maintaining harmony with nature as we are also a part of nature and if we harm or degrade nature we ultimately are harming ourselves (Rachel Carson 2015), International Environmental "Initiatives include the 42nd Amendment of the Constitution, introduced in 1976, which was inspired by the U.N. Conference held in Stockholm in 1972. This amendment introduced Article 48-A. This article states The State shall attempt to safeguard and strengthen the environment, forests, and wildlife of the Country (Paul 2024). Following this, Article 51-A (g) was added, which declares that preserving the country's environment including lakes, rivers, forests, and animals is the responsibility of every citizen.

The Stockholm Environmental Conference of 1972 had a considerable influence on India's environmental policies (Paul 2024).

10.4.2 Major Reforms and Legislative Developments

The 1974 Water Act serves as a central Indian law that aimed at mitigating and managing water pollution to preserve its sanitation and purity is maintained and controlled by central and state pollution control boards. It was first enacted in 1974 and later amended in 1988 and 2003. The Water Act creates SPCBs and CPCBs in each state, at the federal level, to monitor, prevent, and manage water pollution and define emission guidelines for streams and associated regions. This statute also incorporates the Water Cess Statute, 1977, which allows for the charge and collection of a cess on water used in industrial operations (MoEFCC 1974).

The Ganga Action Plan (GAP), which was implemented between 1986 and 2014, sought to enhance the quality standards of water in the Ganga River by reducing pollutants and ensuring that the river met acceptable water quality parameters. The fact that the central government contributed all of the funds for this project demonstrates that central resources were utilized to cover the whole cost. This plan laid the establishment of the National River Ganga Basin Authority and the designation of the river Ganga as a national asset. It was launched as a pioneering program by the Government in June 2014, in which the Namami Gange Programme is an integrated conservation campaign with a budget of Rs. 20,000 crores. Its main objective is pollution reduction, preservation, and revitalization of the Ganga National River. The National Mission for Clean Ganga envisions the long-term revival of sustainable populations of native and endangered species throughout the river, restoring them to their historic range and supporting the ecological balance of the Ganga River system (MoJS 2016). The 2014–2019 Swatch Bharat Abhiyan accelerated as a flagship initiative for water management in small cities, the way forward includes projects sanctioned under Swachh Bharat Mission Urban 2.0, with a total allocation of ₹11,785 crore for the establishment of sewage treatment plants with a capacity of 4900 MLD. The goal is to make sure that spent water in cities is gathered, cleaned, recycled, and used again for a variety of purposes. Sustainability is the focus of urban sanitation, and cities must protect the environment by not contaminating water bodies or by releasing waste and wasted water into them.

10.5 The Water (Prevention and Control of Pollution) Act, 1974

The Water Act of 1974 was the first comprehensive legislation aimed at controlling and preventing water contamination. Although water is a subject under the State List, Article 252(1) of the Constitution allows the Union to intervene. If two or more state legislatures agree to a central law, the government gains the authority to legislate on the matter. The Water Act of 1974 was enacted within this Constitutional system.

Fig. 10.2 Objectives of the water act

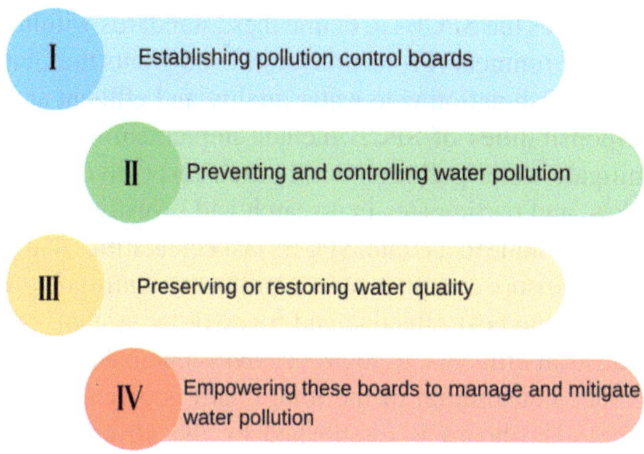

10.5.1 Key Provision and Objectives

The Water Act's key aims are mentioned in Fig. 10.2.

10.5.2 Institutional Framework and Compliance Mechanisms

The Water Act provides for the establishment of Central and State Pollution Control Boards to regulate and manage water pollution. Section 16 of the Act specifies the powers and responsibilities of the CPCB. The key role of CPCB is to maintain the cleanliness of wells and streams nationwide. Furthermore, it is authorized to conduct various essential functions to achieve this mandate (Abhijeet 2013). The major functions include advising the national government on matters about air and water pollution prevention and control, as well as air quality improvement tactics. It is responsible for creating and carrying out a nationwide program focused on avoiding and managing air and water pollution. The CPCB assists in resolving disputes and encourages collaboration among SPCBs. It also finances studies on preventing air and water pollution and provides State Boards with guidance and technical support. CPCB arranges and coordinates training for those involved in pollution prevention and control projects. It encourages a widespread public awareness campaign to inform the public about pollution prevention and control through the use of mass media. The CPCB collects, arranges, and disseminates statistical and technical data on air and water pollution, along with strategies for mitigating, preventing, and controlling it. It is also responsible for creating manuals and regulations for the management of industrial and sewage effluents as well as for the upkeep of air pollution control devices including ducts and stacks.

Section 17 of the Water Act outlines the roles and responsibilities of the SPCBs. While the Act does not set explicit standards for effluent discharge or water quality, it authorizes the SPCBs to define these standards within their authorities. Additionally, the Environment (Protection) Act, 1986 grants the Central Government the authority to establish nationwide water quality and effluent standards (Divan 2001). The key responsibilities of SPCB include implementing programs to prevent, control, and mitigate water pollution within their respective states. It actively promotes, undertakes, and participates in research and investigations to tackle water pollution challenges unique to its state. SPCBs inspect facilities that manage sewage and trade effluents to ensure compliance with pollution control measures. Its initiatives are focused on creating cost-effective and trustworthy systems for the treatment of sewage and industrial effluents, to improve environmental standards.

The constitution of the joint board is given under section 13 of the Water Act, 1974. The board is established by the government through an agreement. Joint Board Agreements can be formed by multiple state governments or between the Central Government and states bordering Union territories. The Agreements can be renewed for additional periods as specified in the agreement. These can be for all participating States or the Union territories. The Agreement Provisions for Joint Board provides for the apportionment of expenditure between participating States and Central Government. According to the Act, it establishes which State Governments would exercise and perform the State Government's duties and powers. The Act provides incidental and auxiliary measures as necessary or feasible to carry out the agreement and promotes interaction between the Central and participating State Governments. The official gazette of the participating states, as specified in clause (a) of sub-section (1), or the official gazette of the Union territories and states, as specified in clause (b) of the same sub-section, shall publish any agreement made under this section. According to Section 14 of the Water Act, 1974, the members of the joint board are listed in the Composition of the Joint Board. The members' information required is that a full-time chairperson who has expertise in managing institutions that deal with environmental protection issues or who has expertise or hands-on experience in ecological protection-related concerns would be appointed by the Central Government and the participating State Government will designate two officials from each participating State to represent that Government. Each participating State Government will appoint one individual from among the members of local authorities from the state. To represent the interests of industry, trade, agriculture, fisheries, or any other sector the State Government deems necessary, each participating State Government must appoint one non-official representative. The Central Government shall designate two individuals to represent the corporations or companies that the participating State Governments own, control, or manage. A full-time member-secretary with credentials, experience, and knowledge in science, engineering, or management who is selected by the Central Government. A Joint Board established through an agreement under clause (b) of section 13 and sub-section (1) will include a resolute chairperson possessing specialized knowledge or practical expertise in environmental protection or the administration of related institutions. Two representatives from participating Union Territories and States. A non-official member representing industries, commerce, agriculture, or fisheries inside the Union Territory or State, as well as a representation from local authorities that are working within the Union Territory or State. Two representatives of participating State Governments and the Central Government are connected to businesses or organizations that the Central Government owns, controls, or manages. A member-secretary who works full-time and has necessary training and expertise in engineering, science, or pollution control.

When a Joint Board is formed under clause (b) of section 13 and sub-section (1) the provisions of section 4 and sub-section (4) no longer apply to the Union territory concerned. Subject to subsection (3), The provisions of Sections 4–12 apply to both the Joint and State Boards and their Member Secretary. Any mention of the State Board in this Act should be interpreted unless the context suggests otherwise, to also encompass a Joint Board. The special provisions for issuing directions are outlined in Section 15 of the Water Act, 1974. This section states that, regardless of any other part of the Act, a Joint Board is established under Section 13. The State government for which the Joint Board is created has the authority to issue directions under this Act, but only when the matter falls exclusively within that State's territorial district. Under this Act, the Central Government has exclusive authority to issue directives when the matter involves the district of multiple States or a Union Territory. The provisions regarding new outlets and discharges of sewage and sludge incorporate restrictions under Section 24 of the Water Act, 1974. This clause prohibits the establishment of any business, operation, procedure, or system for the treatment and disposal of wastewater without prior consent from the State Board. The consent, granted applies to the type, configuration, thermal state, capacity, and speed of effluent discharge. The State Board may refuse consent for reasons recorded in writing and maintains a register documenting these conditions. Any establishment or use of new or modified outlets without consent may result in notices and compliance requirements issued by the State Board.

10.5.2.1 Refusal or Withdrawal of Consent by the State Board

Any sector, activity, method, processing, and waste management system, as well as the usage of a new or modified outlet, cannot be approved by a State Board unless it satisfies the necessary compliance standards. These conditions must allow the Board to exercise its right to collect effluent samples for monitoring and regulation purposes. A State Board may examine any condition imposed from time to time and issue a notice on the person to whom permission is granted modifying or withdrawing any such restriction, as well as make such orders as it deems appropriate MoEFCC (1974).

10.5.2.2 Appeal and Revision Mechanisms

Any individual who disagrees with an order from the State Board may file a request with the governing body within thirty days of receiving the order. However, if the supervisory authority deems that the appellant had a valid reason for missing the deadline, it may consider the appeal even after the thirty days have passed. The appellate authority may be composed of either one person or three persons, as determined by the State Government, and will be appointed by that Government. The fees associated with filing a request and the procedures to be followed by the appealing authority will be as specified. The appellate authority must manage the appeal promptly, ensuring that both the appellant and the State Board can present their cases. If the appeal board determines that any imposed requirement or any modification of a condition, is unreasonable, it may either annul the condition and replace it with a more reasonable alternative or adjust the condition to a more reasonable state while keeping it in effect. In the A.P. Pollution Control Board, vs the Appellate Authority under the Water and Air Act [ILR (2001) 2 AP 29 (34) (DB)] it has been held that the powers of the High Court under Art 226 do not include going the first time into the question as to whether an industry is a polluting industry, in the light of the decision of the Apex Court *in State of W.B.V. Nuruddin Mallick* (1998) 8 SCC 143.

10.5.2.3 Power to Take Samples and Procedure for Making Enquiry

State Boards or their authorized representatives hold the right to extract water samples from sewage discharge points for analysis. Upon sample collection, the occupier or their agent must be notified. The sample is securely sealed, labelled, and signed before being sent to a designated laboratory for examination. The analysis results are documented in triplicate: the first copy is provided to the occupier, the second is used for legal proceedings, and the third is retained by the Board. The cost of analysis is borne by the occupier and shall be enforced as unpaid property tac or public liability (MoEFCC 1974).

10.5.3 Penalties and Legal Procedures for Violation

Penalty for violating a law relating to the disposal of polluting materials: Whoever violates must be punished with imprisonment for at least two years and extending up to six years, and a fine. There is penalty for contravention of provision relating to a new outlet of discharge of sewage or existing discharge of sewage or effluent. Enhanced Penalty for Repeat Offenses, if a person previously convicted of an offense is found guilty of the same offense again, such an individual shall, upon their second and any subsequent conviction, be subject to imprisonment for a minimum term of two years, extendable up to seven years, in addition to a fine. However, if someone commits any offense then it won't be convicted for it if it happened more than 2 years ago. The penalty for breaking any rule is not based on any specific punishment instead, jail for 3 months and a fine up to Rs 10,100 or both, and still, if the violator violates the rule over and again, the fine increases to Rs 5000 per day (MoEFCC 1974). If any person commits the crime under this law again, then the court has full right to publish the name, address, offense, and punishment of the person in a newspaper or other public platforms and the cost of publication will be borne by the person himself as considered fine (MoEFCC 1974). Offenses by companies, if a company commits an offense under this Act, every individual responsible for overseeing and managing the company's business, along with the company itself, is considered guilty of the offense and may face prosecution and penalties. If an offense under this Act occurs, the company will be investigated. If found guilty of breaking the law, the director, manager, or any other officer involved will also face the same punishment (MoEFCC 1974). When any offense takes place and government departments, the Head of the department is also found to be involved and guilty, then they are also punished, but if in case they proved to be unaware of the offense and they tried to prevent it as well, then punishment can be revoked (MoEFCC 1974).

10.6 The Environmental Protection Act, 1986 and Its Relevance

The Environmental Protection Act of 1986 strengthened the enforcement mechanisms of the Water Act, of 1974, aiming to safeguard and improve environmental quality. It provided a comprehensive definition of the environment, including water, air, and land, and how are they interconnected with humans, living organisms, plants, and microorganisms. The Act also defined hazardous substances as materials that pose potential risks due to their physico-chemical properties as well as handling methods. It empowered the federal government to undertake necessary initiatives for preserving ecology, pollution prevention, and mitigation. Additionally, the legislation established standards for air, water, and soil quality, specified permissible pollutant concentration limits and outlined safety protocols for handling hazardous substances (Central Public Health & Environmental Engineering Organisation (CPHEEO) 2013).

10.6.1 Provisions Related to Sewage and Sludge Management

Key aspects of the Act relevant to sewerage and sanitation include that the Act grants the union government the authority to implement any appropriate actions needed to protect and enhance environmental quality, along with preventing, managing, and reducing environmental pollution. It also empowers the central government to set environmental standards, including permissible concentration limits for various pollutants (such as noise) across different areas. These standards are intended to maintain pollutant levels low enough to safeguard public health and welfare. The Act details specific penalties for violations. Furthermore, the central government is authorized to enter premises for inspection, equipment testing, and other related activities, as well as to collect air, water, soil, or other samples for analysis to ensure compliance with the Act. The Act also bans the release of pollutants exceeding prescribed limits and the handling of hazardous substances, except through approved regulatory procedures. It mandates that individuals responsible for discharges exceeding these limits take ongoing measures to control or reduce pollution and report such discharges to the appropriate government bodies at designated intervals (Central Public Health & Environmental Engineering Organisation (CPHEEO) 2013).

10.7 Regulations for Sewage Treatment and Sludge Management

10.7.1 Guidelines for Sewage Treatment Plants (STPs)

Sewage comprises around 99% water which is a mixture of industrial and domestic waste, produced by several commercial and industrial sources. It is a suspension of water as well as solid waste from different sources, and therefore to further remove this contamination from wastewater sewage treatment is required, which is the process of removing contamination from wastewater, by physical, chemical, and biological methods to produce environmentally safe treated wastewater (Demirbas et al. 2017).

10.7.2 Types of Treatment: Primary, Secondary, and Tertiary

10.7.2.1 Primary or Preliminary Treatment

This is the most basic and physical stage of treatment, where a sand or grit chamber, or a channel, is used to adjust the velocity of incoming sewage, allowing sand particles, stones, pebbles, broken glass, and similar materials to settle. Majorly all the coarse solid material is removed at this stage such as rags, sticks, plastic, etc. These solids are collected on a big mesh or a screen that stops them from moving onto further stages of the sewage treatment process. This process is further carried out in two stages: first is pre-aeration and second is sedimentation wherein, the settled and buoyant particles are removed while the residual liquid is sent for secondary treatment. In some cases, floating grease and oil can be recovered separately for saponification or production of biodiesel. In this stage, the sewage is passed through large tanks known as settling tanks in which sludge is settled, and oil or grease is skimmed off (Demirbas et al. 2017).

10.7.2.2 Secondary Treatment

This method utilizes microorganisms to remove biodegradable material in the sewage, which breaks down organic matter, leading to the formation of carbon dioxide and secondary products. This method also requires a separate process to degrade the microorganisms from treated water before it is sent to the tertiary treatment stage. This method focuses on breaking down the biological components of sewage into food, human waste, and further organic matter is broken down into carboxylic acid. The organic compounds in the sewage are converted into carboxylic acids, which are further transformed into carbon dioxide through aerobic fermentation, while methane is produced through anaerobic fermentation. The secondary treatment process utilizes a secondary settling tank to separate biological flocs from the secondary treatment bioreactor (Demirbas et al. 2017).

10.7.2.3 Tertiary Treatment

This is the final stage of the sewage treatment plant, which is specifically designed to improve the quality of wastewater before it is finally discharged into the environment or used further. It aims to remove the residual impurities even after the primary and secondary treatment operations, this stage ensures to meet the regulatory standards and norms for extended usage of wastewater. This includes advanced filtration methods, chlorination, nutrient removal, and chemical treatment by coagulation or flocculation and activated carbon adsorption. Other advanced tertiary treatment methods include ozonation and photo-Fenton (Arzate et al. 2019). The sequential steps involved in the sewage treatment process, outlining all major stages from pre-treatment to final disposal and resource recovery, are depicted in a flowchart presented in Fig. 10.3.

10.7.3 Design and Operation of STPs

Design criteria in STP are specific for all the individual components of the entire process, in the case of the screen chamber, the cross-sectional opening is to allow sewage to pass, at an average flow rate of 2.5–3 times hourly, and a velocity of 0.8–1.0 m/s. The design criteria for the equalization tank are based on the capacity to hold peak inflow volumes. In the case of a raw sewage lift pump, the designed criteria are based on the average daily operational capacity of STP, and are usually designed with a duplicate pump, i.e., dual pumps are used, which are fixed alongside each other, with only one operating at a time. Along with these during the design of STP, major key points considered are, the amount of sewage managed each day, types of pollutants specific to the source, and other factors such as BOD, COD, TSS, etc. in case of clarifier the designed criteria are based on the cross-sectional area which is typically considered between 12 and 18 $m^3/h/m^2$, whereas the depth

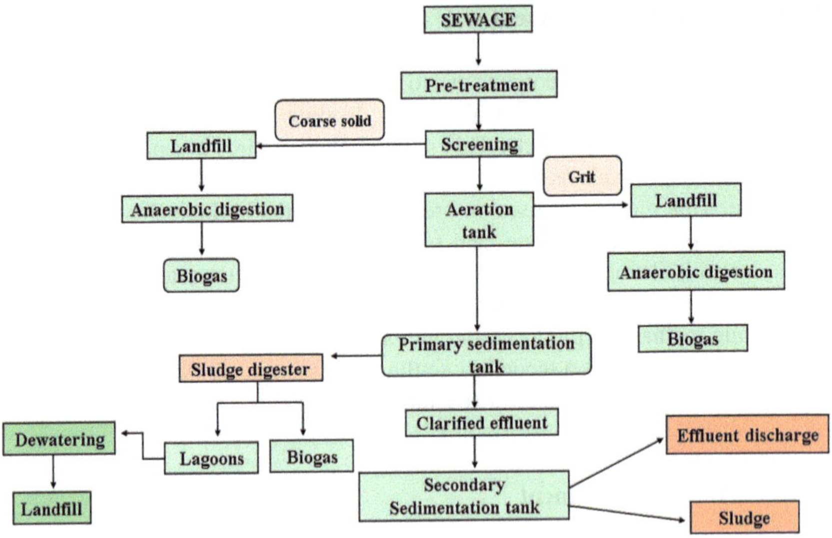

Fig. 10.3 Flowchart of sewage treatment

of the clarifier plays no role. If activated carbon columns are used in the treatment process, their design depends on the diameter of the filter used, and the potency of decontamination is based on the remaining amount of the utilized chemical and their contact time (Kodavasal 2011). A Brief outline of the Design criteria for STP is provided in Table 10.1.

10.8 Role of the Central Pollution Control Board (CPCB)

The CPCB, is a government body, established in September 1974 under the Watter Act, 1974. Later, it was also authorized under the Air Act, 1981. It works under the Ministry of Environment and Forest and climate change to provide technical support in law making. Its main responsibilities account for water prevention and control, cleaning of streams and wells, reduction of water pollution, and improving air quality by reducing and managing air pollution across the nation (MoEFCC 2018).

The CPCB ensures effective governance and regulation. These functions include providing counsel to the Union Government on issues related to water and air pollution prevention and control, along with strategies for improving air quality and designing and executing a nationwide program focused on reducing, controlling, or preventing water and air pollution, which establish the foundation for its operation. Furthermore, the board is responsible for promoting the coordination among State Boards and mediating any disputes that arise between them, along with technical support and advisory services to state boards, undertaking investigations, and

Table 10.1 Design criteria of different STP components (Kodavasal 2011)

STP component	Design criteria
Bar screen	• Flow velocity (maximum) 0.3 m/s • Solids can be captured at 12 mm • A coarse screen is to be placed before the fine screen
Equalization tank	• Detention time (minimum) 4–6 h • Airflow ($1m^3/m^3$) of tank volume to allow mixing and avoid settling and septicity
Raw sewage lift pump	• Capacity to be estimated from 20-h/day working of STP
Aeration tank	• Aeration time: 16 h (minimum) • F/M (Food to microorganism) ratio: 0.12 for 98% BOD removal efficiency • MLSS (Mixed liquor suspended solids): 3500 mg/L • BOD: 250 mg/L • Diffusers: Flux rate 8–12 m^3
Secondary clarifier	• Detention time: 2.5–3.5 h • Solids loading: 2-3 $kg/m^2/h$ • Overflow rate: 12–18 $m^3/m^2/day$
Sand filter	• Loading rate: less than $12m^3/m^2/h$
Activated carbon filter	• Loading rate: less than $10m^3/m^2/h$
Softener	• Hardness removal achieved from 300 to 100–120 mg/L
Excess sludge	• 0.20–0.25 kg excess solids per kg BOD removal

supporting research on water and air pollution, including approaches for their reduction, prevention, and control, to enhance compliance and oversight. The Act also emphasizes capacity-building initiatives by organizing and coordinating training programs for individuals involved in water and air pollution control. It promotes public awareness through mass media campaigns to educate citizens on pollution prevention, control, and mitigation. Additionally, it mandates the collection, organization, and publication of technical and statistical data on pollution levels and management strategies to address national priorities. The Central Board also develops manuals, codes, and guidelines for sewage and industrial effluent treatment, as well as the maintenance of stack gas cleaning devices, stacks, and ducts. Furthermore, it facilitates the dissemination of information on pollution control techniques to support sustainable development. Additionally, in collaboration with the relevant State Governments, establish, update, or revoke standards for water quality in streams or wells, set air quality standards, and conduct any other functions as mandated by the Government of India (MoEFCC 2018) highlighting its commitment in promoting accountability and transparency in various sector.

10.9 Wastewater Treatment Standards and Compliance

The report of 2021 by CPCB highlights that India's water treatment capacity stands at 27.3%, while sewage treatment capacity is 18.6%, with plans for a 5.2% increase. Despite being 20% higher than the global average, this capacity remains insufficient. If sewage treatment is not expanded urgently, significant environmental and public health risks may arise. Government data indicate that 62.5% of wastewater in metropolitan areas remains partially or entirely untreated. Furthermore, a 2019 study concluded that many treatment plants are non-functional under the Ganga and Yamuna Action Plans, and from the 33,000 million liters per day (MLD) of wastewater generated, only 7000 MLD undergo collection and treatment.

10.10 Government Initiatives for Sewage Management in India

10.10.1 National Urban Sanitation Policy, 2008

The Ministry of Urban Development in the year 2008, introduced National Urban Sanitation Policy to improve sanitation in cities. This policy encourages state governments to come up with their initiatives and plans focusing on public health environmental protection and sanitation, which refers to maintaining cleanliness, disposing of human waste safely, along with proper disposal and treatment methods, and practicing hygiene. The policy not only deals with human waste and its impacts on health and the environment instead it also accounts for solid waste management, industrial and hazardous waste, drainage, etc. (MoUD 2008).

According to Census 2001, 27.8% of Indians (286 million people equivalent to 55 million households) lived in urban areas. Projections estimated the urban population would reach 331 million by 2007 and 368 million by 2012. Among urban households, 7.87% (12.04 million) lacked access to latrines, resorting to open defecation, while 8.13% used community latrines and 19.49% relied on shared facilities. Additionally, 18.5% lacked a drainage network, and 39.8% were connected to open drains. Sanitation challenges are particularly severe for the urban poor and slum dwellers. In notified slums, 17% lacked latrines, while this figure rose to 51% in non-notified slums. The availability of septic latrines and underground sewerage was significantly lower in non-notified slums at 35% and 15%, respectively. The unsafe disposal of 37% of urban human excreta poses major public health and environmental risks, especially affecting vulnerable groups like women, children, and the elderly. The economic loss due to sanitation-related diseases for children under 14 was estimated at ₹500 crore (2001 prices). Poor wastewater management has contaminated 75% of surface water in India, highlighting the urgent need for improved urban sanitation systems (MoUD 2008).

The main objective of this policy is to rejuvenate urban areas into cities and towns that are fully sanitized, healthy, and liveable, driven by active community participation. Firstly, Raising Awareness and Encouraging Behavioural This initiative aims to raise public and environmental health awareness about sanitation and promote strategies for promoting healthy sanitation practices among communities and institutions. Secondly, attaining cities free from open defecation requires ensuring universal access to safe and hygienic sanitation facilities while eliminating open defecation. Additionally, institutional reorientation and sanitation integration involve strengthening governance structures to prioritize sanitation through effective planning, implementation, and maintenance, particularly in marginalized and underserved areas. Safe waste disposal aims to ensure the proper management of human excreta and liquid waste by promoting functional sewerage systems, wastewater recycling, and secure sludge disposal. Lastly, the Ministry of Urban Development emphasizes the importance of proper operation and preservation of household, shared, and public cleanliness facilities to sustain long-term sanitation improvements.

10.10.2 Policy Implementation and Challenges

The Government of India will support states in creating and executing innovative strategies to prioritize urban sanitation. States and cities may explore various approaches to meet sanitation objectives by utilizing existing provisions in municipal and other laws to ensure sanitation compliance. It can amend municipal laws, and create byelaws and regulations (such as building codes) to promote sanitation by public and private entities and to prevent untreated sewage discharge into open areas. Further implementation requires establishing systems of incentives and penalties, including charges on polluters where applicable. By adjusting policies to ensure access to improved sanitation facilities for urban poor households and residents of informal settlements. Allocation of land for community and public sanitation facilities is also done for implementation. Collaborators are encouraged with public, private, and non-governmental organizations to improve the delivery, maintenance, and management of sanitation services. Focus is also laid on launching sanitation programs in a mission mode to encourage collaboration between public and non-governmental agencies, such as creating state-level urban sanitation steering committees and city-level task forces and exploring other locally suitable innovations and solutions (MoUD 2008).

10.10.3 State Sewerage and Management Policy, 2016

The Department of Local Self Government (LSG), Rajasthan, introduced the State Sewerage and Wastewater Policy 2016 to align with the objectives of the National Urban Sanitation Policy 2008. It aims to enhance urban public health, particularly for underprivileged communities, by ensuring sustainable sanitation services and environmental protection. It emphasizes safeguarding public health, protecting water resources, and ensuring the efficient operation of sewer systems to prevent dry weather flow in drains and streets. The policy mandates proper sewage and sludge treatment before environmental discharge and facilitates the reuse and recycling of processed wastewater for other purposes. Additionally, it seeks to make sewerage projects both economically viable and environmentally sustainable while fostering inclusive and transparent decision-making. Strengthening institutional capacity and optimizing investments are also key priorities to effectively manage the sector.

The State Sewerage and Wastewater Management Policy, 2016 aims to address water shortages by prioritizing potable water for drinking while promoting the reuse of treated water for other purposes. It seeks to achieve 100% sanitation in urban areas and enhance water supply services with a focus on customer satisfaction, coverage, frequency, and reliability. Given the high cost of potable water and its widespread use for non-drinking purposes, the policy emphasizes the importance of utilizing treated wastewater to supplement limited freshwater sources, particularly during periods of scarcity. Reusing treated sewage alleviates pressure on overburdened water basins and aquifers, contributing to environmental sustainability and preventing resource depletion. Additionally, improper sewage disposal poses significant health risks and contaminates groundwater, making water reuse a crucial strategy for reducing sewage volume and minimizing environmental and health hazards. By promoting resource conservation and reducing pollutant loads, the policy supports ecosystem preservation and sustainable water management.

10.11 Role of Local Authorities and Municipal Governance in Wastewater Management

The Solid Waste Management Plan at the municipal level, provides stepwise guidance about the importance of urban local bodies (ULBs) and other authorities in addressing sludge and sewage management concerns in addition to municipal solid waste. As per the Solid Waste Management Rules 2016, local governments are assigned to prepare comprehensive plans for managing waste streams, including sludge from wastewater treatment facilities and sewage, by environmental and health standards. ULBs are responsible for identifying gaps in current infrastructure, securing financial and technical resources, and implementing plans to reduce the environmental and health consequences of inappropriate sludge management.

Municipalities are critical in implementing safe sludge treatment and disposal technologies such as anaerobic digestion, composting, and co-treatment with municipal solid waste. The guidebook emphasizes the need to guarantee source segregation and establish treatment facilities adapted to each city's needs. It promotes using decentralized solutions where possible, such as small-scale treatment centres in peri-urban settings. To improve the operational and financial effectiveness of sludge management systems, authorities are encouraged to build public–private partnerships (PPPs) with the informal sector. Additionally, municipalities must conduct frequent environmental monitoring of sewage treatment plants (STPs), devise systems for utilizing treated sludge (e.g., as compost or electricity), and verify compliance with statutory environmental criteria. The guidance also advocates for a coordinated approach to sludge management among central, state, and municipal administrations. It describes the functions of state pollution control boards (SPCBs) to observe sludge quality, award treatment facility approvals, and assist ULBs with capacity development. At the local level, officials must actively involve communities, create awareness about proper sludge treatment, and encourage the repurposed use of treated sewage. The handbook anticipates municipalities and ULBs performing a key role toward accomplishing ecological sludge and sewage management, decreasing pollution, and defending public health through a successful strategy, adoption of technology, and stakeholder cooperation. (Central Public Health and Environmental Engineering Organization 2016).

10.12 Global Best Practices in Sewage and Sludge Management

10.12.1 European Union Wastewater Framework Directive

The Water Framework Directive (WFD), introduced in 2000, aims to achieve a good ecological and chemical status for various water bodies by 2027, thereby reducing wastewater generation and pollutant discharge (Official Journal of the European Communities 2000). The Directive 2000/60/EC acknowledges water as a critical resource requiring responsible protection and management. It establishes a comprehensive framework to preserve water quality and prevent environmental degradation, considering the increasing demand for clean water across multiple sectors. This directive builds upon previous commitments, international agreements, and environmental policies that prioritize precautionary measures, pollution prevention, and the "polluter pays" principle. A fundamental component of the directive is the promotion of sustainable water management through integrated planning, encompassing key sectors such as agriculture, energy, and tourism. Protecting aquatic ecosystems, particularly coastal and estuarine areas, is crucial for biodiversity and economic sustainability. To achieve these goals, the directive sets environmental targets, including preventing further deterioration and ensuring good water status for surface and groundwater. Recognizing the challenges of implementation, the directive allows flexibility through phased approaches, extensions, and exemptions under justified conditions. It adopts a river basin management strategy to ensure coordinated efforts, particularly for transboundary waters. This framework reinforces the EU's commitment to environmental protection while aligning with international obligations.

10.12.2 The United States Clean Water Act 1972

The Clean Water Act mandates the proper regulations of pollutant discharge and assures criteria for assessing surface water quality. It aims to "restore and maintain the chemical, physical, and biological integrity of the Nation's waters" (Banerjee and Shirey 2024). This act was enacted in 1948, followed by amendments in 1972 for final implementation. Under this act, EPA has implemented various pollution control strategies and programs to facilitate wastewater standards for different industries and developed regulations defining pollutant limits in surface water. The CWA comprises two major parts, title II which authorizes grants and subsidies for municipal sewage treatment infrastructure, and title VI deals with regulations related to industrial and municipal discharges (Silyok 2001).

10.13 Conclusion and Future Recommendations

Effective sewage treatment and sludge management are critical for sustainable urbanization and environmental protection. While India has a comprehensive legislative and policy framework addressing sewage and sludge management, as detailed in this chapter, significant gaps remain in implementation, enforcement, and monitoring. A clear definition of treated wastewater standards still lack, due to which successful treatment and reuse of wastewater discharge remains inadequate. Although there have been huge increments in wastewater treatment infrastructure in India, starting from 10 to 44% currently, still financial constraints remain, for which the recommendation of forming a wastewater fund should be opted. The monitoring systems and STPs are also fewer in number in comparison to the amount of wastewater generation, due to which more developments need to take place in this interest. The involvement of third parties or private ownerships also promotes better infrastructure for wastewater treatment facilities. Law and policies have remained in existence for a long back, but public involvement and acceptance along with government initiatives and projects is a must for successful implementation and continuous working. Therefore, integration of advanced technologies, empowerment of local bodies, and fostering community participation are required to accomplish the targets of sustainable wastewater management and pollution control in India. The findings in this chapter underscore the importance of bridging the gap between policy formulation and ground-level execution, paving the way for cleaner water bodies and healthier communities.

References

Abhijeet K (2013) Governing water pollution effectively: a comparative study of legal frameworks & their implementation in India & Sweden

Ahmed MB et al (2017) 'Progress in the biological and chemical treatment technologies for emerging contaminant removal from wastewater: a critical review. J Haz Mater 323:274–298. https://doi.org/10.1016/j.jhazmat.2016.04.045

Arzate S et al (2019) Environmental impacts of an advanced oxidation process as tertiary treatment in a wastewater treatment plant. Sci Total Environ 694:133572. https://doi.org/10.1016/j.scitotenv.2019.07.378

Arzu Özkara DA (2018) Environmental pollution and the effects of the pollutants on the ecosystem. Turk J Sci Rev (preprint)

Banerjee P, Shirey PD (2024) State drainage laws muddy the goals of the US clean water act (1972). J Sci Policy Govern 25(1). https://doi.org/10.38126/JSPG250103

Breitenmoser L et al (2022) Perceived drivers and barriers in the governance of wastewater treatment and reuse in India: Insights from a two-round Delphi study. Resour Conserv Recycl 182:106285. https://doi.org/10.1016/j.resconrec.2022.106285

Carson R (2015) The story of silent spring. https://www.nrdc.org/stories/story-silent-spring. Accessed 12 Nov 2024

Central Public Health & Environmental Engineering Organisation (CPHEEO) (2013) Manual on sewerage and sewage treatment systems part C management. Ministry of Housing and Urban Affairs, Government of India. https://cpheeo.gov.in/upload/uploadfiles/files/systems_chapter2.pdf. Accessed 16 Nov 2024

Central Public Health and Environmental Engineering Organization (2016) Municipal solid waste management manual. Ministry of Urban Development, Government of India. https://www.scribd.com/document/356826022/MSW-2-2016-Municipal-Solid-waste-Management-Rules-2016-Vol-2. Accessed 4 Dec 2024

Demirbas A, Edris G, Alalayah WM (2017) Sludge production from municipal wastewater treatment in sewage treatment plant. Energy Sour Part A Recov Utiliz Environ Effects 39(10):999–1006. https://doi.org/10.1080/15567036.2017.1283551

Demirbas A, Taylan O, Kaya D (2016) Biogas production from municipal sewage sludge (MSS). Energy Sour Part A Recov Utiliz Environ Effects 38(20):3027–3033. https://doi.org/10.1080/15567036.2015.1124944

Divan S, Rosencranz A (2001) Environmental law and policy in India: cases, materials and statutes. Oxford University Press, 56 pp (preprint)

Gothwal R, Shashidhar T (2015) Antibiotic pollution in the environment: a review. CLEAN Soil Air Water 43(4):479–489. https://doi.org/10.1002/clen.201300989

India Code-Digital Repository (1853) The shore nuisances (Bombay and Kotaba) ACT. https://www.indiacode.nic.in/bitstream/123456789/19040/1/a1853-11.pdf. Accessed 14 Nov 2024

Jimenez B et al (2004) Sustainable sludge management in developing countries. Water Sci Technol 49(10):251–258. https://doi.org/10.2166/wst.2004.0656

Kacprzak M et al (2017) Sewage sludge disposal strategies for sustainable development. Environ Res 156:39–46. https://doi.org/10.1016/j.envres.2017.03.010

Kodavasal AS (2011) The STP guide, design, operation & maintenance, Karnataka State Pollution Board, Bangalore India. https://www.ielrc.org/content/e1144.pdf. Accessed 15 Nov 2024

Kumar Gupta A (1994) Pollution oriented occupational health problems of limestone crusher workers of Rewa division in India. https://www.researchgate.net/publication/310170610

MoEFCC (1974) The water (prevention and control of pollution) act. India code-digital repository. https://www.indiacode.nic.in/bitstream/123456789/15429/1/the_water_%28prevention_and_control_of_pollution%29_act%2C_1974.pdf. Accessed 14 Nov 2024

MoEFCC (2018) Central pollution control board. Government of India. https://cpcb.nic.in/functions/. Accessed 15 Nov 2024

MoJS (2016) National mission for clean Ganga. Department of Water Resources, River development and Ganga Rejuvination. https://nmcg.nic.in/. Accessed 15 Nov 2024

MoUD (2008) National urban sanitation policy. Government of India. https://www.mohua.gov.in/upload/uploadfiles/files/NUSP_0.pdf. Accessed 15 Nov 2024

Nowak R, Imperowicz A (2016) Liquid waste from septic tanks as a source of microbiological pollution of groundwater. Inżynieria Ekologiczna (47):60–67. https://doi.org/10.12912/23920629/62848

Official Journal of the European Communities (2000) Directive 2000/60/EC 327(43):1–72

Palamuleni LG (2002) Effect of sanitation facilities, domestic solid waste disposal and hygiene practices on water quality in Malawi's urban poor areas: a case study of South Lunzu Township in the city of Blantyre. Phys Chem Earth Parts A/B/C 27(11–22):845–850. https://doi.org/10.1016/S1474-7065(02)00079-7

Paul S (2024) Environmental crisis: cause and effect of Stockholm conference in India. https://thelegalquorum.com/environmental-crisis-cause-and-effect-of-stockholm-conference-in-india/#:~:text=Conference%20on%20Human%20Environment%2C%20held,%2Dlife%20of%20the%20Country.%E2%80%9D. Accessed 12 Nov 2024

Petrie B, Barden R, Kasprzyk-Hordern B (2015) A review on emerging contaminants in wastewaters and the environment: current knowledge, understudied areas and recommendations for future monitoring. Water Res 72:3–27. https://doi.org/10.1016/j.watres.2014.08.053

Rorat A et al (2019) Sanitary and environmental aspects of sewage sludge management. In: Industrial and municipal sludge. Elsevier, pp 155–180. https://doi.org/10.1016/B978-0-12-815907-1.00008-8

Sharma M et al (2022) Wastewater treatment and sludge management strategies for environmental sustainability. In: Circular economy and sustainability. Elsevier, pp 97–112. https://doi.org/10.1016/B978-0-12-821664-4.00027-3

Silyok VA (2001) Environmental Laws summaries of statutes administered by the environmental protection agency. Nova Science Publisher, Inc

Zhang QH et al (2016) Current status of urban wastewater treatment plants in China. Environ Int 92–93:11–22. https://doi.org/10.1016/j.envint.2016.03.024

Chapter 11
Pharmaceutical Contaminants in Sewage and Their Toxicological Considerations

Prashant Upadhyay, Arvind Raghav, Km. Anjali, and Sukirti Upadhyay

Abstract Pharmaceutical contaminants in sewage represent a growing global environmental and public health concern. These bioactive substances originate from domestic discharges, hospital effluents, pharmaceutical manufacturing, and improper disposal of pharmaceuticals and are frequently detected in wastewater. Conventional wastewater treatment plants (WWTPs) are not equipped to effectively remove complex pharmaceutical residues, which results in their persistence in treated effluents and biosolids. Commonly identified contaminants include antibiotics, analgesics, hormones, and psychiatric medications, many of which exhibit high chemical stability and resistance to degradation. Once in the environment, these compounds can disrupt aquatic ecosystems, induce endocrine and behavioral changes in wildlife, and promote the proliferation of antimicrobial resistance (AMR). Moreover, chronic exposure through contaminated water and food sources raises significant human health concerns, including endocrine disruption and immune dysfunction. The chapter explores the environmental fate of pharmaceuticals, their transformation into potentially toxic byproducts, and their bioaccumulation potential. It also highlights advanced analytical techniques such as LC–MS/MS and GC–MS for detection and examines emerging treatment technologies, including membrane filtration, advanced oxidation processes (AOPs), and bioreactors that aim to enhance removal efficiency.

P. Upadhyay (✉) · S. Upadhyay
School of Pharmaceutical Sciences, Faculty of Pharmacy, IFTM University, Lodhipur Rajput, Moradabad, Uttar Pradesh, India
e-mail: prashantupadhyay@iftmuniversity.ac.in

S. Upadhyay
e-mail: sukirtiupadhyay22@gmail.com

A. Raghav (✉)
Department of Pharmaceutics, Teerthanker Mahaveer College of Pharmacy, Teerthanker Mahaveer University, Moradabad, Uttar Pradesh, India
e-mail: arvindraghav.pharmacy@gmail.com

Km. Anjali
Sahu Onkar Saran School of Pharmacy, Faculty of Pharmacy, IFTM University, Moradabad, Uttar Pradesh, India
e-mail: anjali.ved.2303@gmail.com

S. Madhav et al. (eds.), *Municipal Sewage and Sludge: Treatment and Disposal Strategies*, Environmental Science and Engineering,
https://doi.org/10.1007/978-3-032-07597-0_11

However, these technologies often face limitations regarding cost, energy demand, and scalability. The paper concludes by emphasizing the need for integrated strategies, including green chemistry, regulatory reforms, public awareness, and multidisciplinary collaboration, to mitigate pharmaceutical pollution. Addressing this issue requires a global effort to develop sustainable pharmaceutical practices, improve wastewater treatment infrastructure, and strengthen policy frameworks to protect both environmental and human health.

Keywords Drug resistance · Resistance · Environmental toxicology · Pharmaceutical residues · Sewage · Wastewater treatment

11.1 Introduction

Pharmaceutical contaminants of active drug ingredients are being found more frequently than before in various settings, including water treatment plants. These pollutants are derived from many sources: Domestic trash, Hospital sewage, industrial effluents, and the discarding of prescriptions. Antibiotics, analgesics, anti-inflammatory drugs, tranquilizers, and hormones are the most commonly used medicines. Because these substances are biologically active, they cannot be removed from the environment by standard wastewater treatment processes, and as a result, they remain in the environment. The occurrence of pharmaceutical contaminants in sewerage is alarming and raises serious hazards to the sustainability of the ecosystem, and human health (Kookana et al. 2014; Ortúzar et al. 2022a). The strive-to-minimize concentration of pharmaceuticals is designed to have an impact on biological systems, thus even small concentrations in drinking or water are a serious problem. These pollutants, when introduced to the natural environment, interfere with the normal physiological functions of aquatic organisms, resulting in toxic effects and playing a role in the emergence of antimicrobial resistance. In addition, pharmaceutical byproducts can circulate through the food chain, representing a potential risk to humans, through indirect exposure (Fernandes et al. 2021). Pharmaceutical contaminants in sewage became a global environmental hazard due to their persistence in the environment and potential toxicological impact on both ecosystems and human health. These are pollutants such as antibiotics, analgesics, hormones, and psychiatric or psychotropic medicines that are present in wastewater systems from human stools and urine, improper drug disposal, or all types of effluent generated during pharmaceutical production. Traditional wastewater treatment plants cannot fully remove these complex chemicals and they accumulate in aquatic environments. This contamination can harm aquatic ecosystems by affecting the reproductive and behavioral habits of mammals, causing the emergence of antibiotic resistance and modifying the composition of the microbial population. Furthermore, long-term human exposure to hint pharmaceutical residues in consuming water and food resources increases worries approximately continual fitness outcomes such as endocrine disruption and immune device harm. On a worldwide scale, this trouble emphasizes the importance of contemporary wastewater treatment technology, strict pharmaceutical waste control practices, and massive regulatory frameworks for reducing environmental pollutants and defensive public health.

11.1.1 Importance of Understanding Sewage Contamination

Pharmacological and chemical pollutants are often released into sewers, and this phenomenon should be addressed because it is observed quite frequently and may have certain impacts. Conventional wastewater treatment plants (WWTP) inadequately treat and/or remove these hazardous compounds so they are discharged into rivers, lakes, and other water sources (Kumar et al. 2023). It is estimated that as the use of pharmaceuticals across the world increases, so does the discharge of these chemicals into the environment a factor that will further exaggerate the accumulation effects (Okoye et al. 2022). It is essential to investigate the behavior of pharmaceuticals in the context of sewage and wastewater treatment for several reasons. First, it underlines worries to both the environment and the public, like effects on water organisms and the rise of antibiotic-resistant ones (Samal et al. 2022). Second, it contributes to the advancement of information regarding improved wastewater treatment procedures and policies that lower drug contamination levels. Lastly, it gives data to scientists, politicians, as well as corporations to work together toward reducing the tremendous environmental damage created by pharmaceutical waste (Singh et al. 2023). The analytical and toxicological aspects of medicinal pollutants in sewage are examined in detail: from origins, and behavior in sewage systems to the potential toxicologic impact. It looks at several sources of pharmaceutical pollution, including home wastewater, industrial pollution, and contaminants related to healthcare. Using this approach, first, the mechanistic understanding of how these pollutants work in wastewater treatment systems and the factors that determine their capacity to persist in the system are examined including the characteristics of the pollutants and treatment effectiveness (Hejna et al. 2022). The chapter also explores ecotoxicological aspects of the impacts of such pharmaceutical residues on aquatic life including bioaccumulation risk in the food chain and imbalances of hormones in exposed animals. It also looks at human health concerns, including indirect exposure to contaminants in water sources (drinking water or something a treated effluent has touched) (Khan et al. 2023).

11.1.2 Sources of Pharmaceutical Contaminants in Sewage

Active chemical compounds including antibiotics, analgesics, and hormones are identified in wastes from household and industrial sources. The chemical structures of these compounds are supposed to be stable enough to keep their beneficial effects over time. A variety of pharmaceutical compounds, including antibiotics and cytotoxins, are sourced from hospitals and healthcare facilities and can only be partially degraded in wastewater treatment plants (WWTPs), resulting in the release of bioactive residues (Kumar et al. 2022; Patel et al. 2019a). Antibiotics and hormones from animals being treated by veterinary medicine run off onto the ground and into the sewerage system from agricultural runoff. Some of these microbes interact with microbial populations there and some may encourage antibiotic resistance by horizontal gene transfer. Flushing old tablets leads to high concentrations of drugs in the water, enough to thwart the natural decay process that occurs in our waterways (Caneschi et al. 2023). These toxins resist decomposition due to their complex chemical structures which are designed for durability and as a result, they persist in the environment and accumulate in aquatic systems (Manyi-Loh et al. 2018).

11.2 Pharmaceutical Contaminants in Sewage and their Toxicological Considerations

Pharmaceuticals have become ubiquitous environmental pollutants due to their widespread use in human and veterinary medicine. After administration, unmetabolized drug residues are excreted from the body and enter municipal sewage systems via domestic, hospital, and industrial effluents. With the implementation of conventional wastewater treatment technologies, many pharmaceutical compounds persist in treated effluents and sludge due to their complex chemical structures and resistance to biodegradation. Municipal sewage and sludge have emerged as significant reservoirs and vectors of pharmaceutical contaminants, posing toxicological risks to aquatic and terrestrial ecosystems, as well as potential threats to human health through environmental exposure. Pharmaceutical contaminants in municipal sewage primarily originate from a variety of sources. One of the main contributors is domestic discharge, which includes the excretion of unmetabolized drugs from human bodies and the improper disposal of unused or expired medications. Hospitals and other healthcare facilities also significantly contribute to pharmaceutical pollution, as they release concentrated effluents containing a wide range of pharmaceuticals, including antibiotics, analgesics, and diagnostic agents. Moreover, pharmaceutical manufacturing industries discharge wastewater containing active pharmaceutical ingredients (APIs) during production processes (Sharma et al. 2022). Veterinary practices and agricultural activities further exacerbate the issue through runoff from animal farms and aquaculture operations, which often contain antibiotics and hormonal compounds used in livestock and fish farming. These diverse sources lead to the presence of various classes of pharmaceuticals in municipal sewage. Commonly detected compounds include antibiotics such as ciprofloxacin and sulfamethoxazole; analgesics and anti-inflammatory drugs like ibuprofen and diclofenac; anticonvulsants such as carbamazepine; antidepressants like fluoxetine; and hormonal substances such as ethinylestradiol. The widespread and persistent presence of these substances in sewage poses significant challenges for wastewater treatment and environmental protection (Massima Mouele et al. 2021).

11.2.1 Fate and Transport of Pharmaceuticals in Sewage Systems

11.2.1.1 Pharmaceutical Behavior in Wastewater Treatment Plants (WWTPs)

Pharmaceuticals that enter wastewater treatment plants (WWTPs) go through a variety of physical, chemical, and biological processes that affect their behavior and fate. Conventional treatment methods such as activated sludge and biofilm reactors are primarily intended to remove organic debris and nevertheless, their effectiveness in degrading medicines varies. Many drugs particularly those with complex chemical structures such as antibiotics, anti-inflammatory agents, and hormone substances do not undergo complete metabolization. This inadequate elimination is owing to the molecular stability of these chemicals, which have been designed to resist degradation in biological systems (Caneschi et al. 2023; Khalidi-idrissi et al. 2023). Pharmaceuticals in WWTPs may adsorb onto sludge particles that are partially degraded or remain in the aqueous phase. Lipophilic drugs adsorb more readily to sludge whereas hydrophilic chemicals remain in the water. Biodegradation in WWTPs frequently happens through microbial enzymes that target specific functional groups on the pharmaceutical molecules. However, this process is restricted by the drug's structural complexity and resistance to enzymatic attack (Caneschi et al. 2023).

11.2.1.2 Persistence of Pharmaceuticals in Treated Water

Pharmaceuticals that are not eliminated during treatment remain in treated effluent, which is released into aquatic bodies. These chemicals' persistence in the environment is mostly dictated by their physicochemical qualities such as solubility, volatility, and resistance to hydrolysis as well as photolysis. Many drugs including fluoroquinolone antibiotics that possess strong carbon–fluorine bonds are making them very resistant to breakdown mechanisms (Adeleye et al. 2022; Liu and Wong 2013). Furthermore, the aromatic rings and halogen groups in many drug compounds give stability, allowing them to remain in treated water. These persistent chemicals can travel great distances in aquatic settings and potentially bioaccumulate in creatures and enter the food chain (Al-Tohamy et al. 2022).

11.2.1.3 Degradation and Transformation Products

Pharmaceuticals in sewage systems can partially degrade resulting in the creation of transformation products. These transformation products have resulted from chemical reactions such as hydrolysis, oxidation, photolysis, or biodegradation, which modify the parent drug molecule. For example, if pharmaceuticals including ibuprofen are handled under oxidative conditions in prescribed WWTP they may be hydroxylated or carboxylated to yield metabolites that may be more hazardous than the parent drug (Ghazal et al. 2022; Ruziwa et al. 2023). The degradation of drugs often depends on the breakdown of specific structural bonds, such as amide or ester bonds, resulting in the production of more polar and soluble metabolites. Yet, in those particular conditions, they can still possess or even increase their bioactivity and thus become ecotoxiotes for aquatic environments. The prediction of the compound forces and weird transformations resulting from pharmaceutical contaminants is challenging due to the complexity of these degradation processes and the degree range of transformation products (Davison et al. 2014).

11.2.2 Toxicological Considerations

Pharmaceutical contaminants in treated wastewater and sludge pose significant toxicological risks to both aquatic and terrestrial ecosystems. In aquatic environments, chronic exposure to even low concentrations of pharmaceutical compounds can lead to serious ecological consequences. For instance, synthetic estrogens such as ethinylestradiol are known to cause endocrine disruption in fish, leading to feminization and reproductive impairments. Additionally, pharmaceuticals can alter the behavior of aquatic species and contribute to the development of antibiotic resistance among microbial populations. Terrestrial organisms are similarly affected, particularly when contaminated biosolids are applied to agricultural lands. These biosolids can disrupt soil microbial communities, impact plant health, and facilitate the entry of pharmaceuticals into the terrestrial food chain. One of the most pressing concerns related to pharmaceutical contaminants is the promotion of antibiotic resistance. The presence of antibiotics in the environment encourages the selection and proliferation of antibiotic-resistant bacteria (ARB) and the spread of antibiotic-resistant genes (ARGs). Wastewater treatment plants (WWTPs), due to their microbial diversity and the continuous influx of antibiotics, become hotspots for horizontal gene transfer, amplifying public health threats associated with antimicrobial resistance (Ortúzar et al. 2022b; Samal et al. 2022). Another toxicological issue is the potential for bioaccumulation and biomagnification. Lipophilic pharmaceutical compounds can accumulate in the tissues of organisms and magnify the food chain, posing long-term exposure risks to both wildlife and humans. These risks are compounded by the fact that pharmaceuticals are often present as complex mixtures in the environment, which can lead to synergistic or cumulative toxic effects that are not yet fully understood. Although pharmaceutical concentrations in the environment are generally low, ranging from nanograms to micrograms per liter, continuous exposure through multiple pathways raises legitimate concerns for human health. These concerns include endocrine disruption from hormone-based drugs, allergic or hypersensitivity reactions to persistent residues, and the potential toxicity of pharmaceutical mixtures (Patel et al. 2019b). Moreover, contamination of drinking water sources via effluent discharge or the leaching of pharmaceuticals from land-applied biosolids further heightens the risk of human exposure, necessitating urgent attention to monitoring and mitigation strategies.

11.3 Classes of Pharmaceuticals Commonly Found in Sewage

Pharmaceuticals commonly found in sewage systems are divided into several types each of which has a different therapeutic function but comparable environmental impact. Antibiotics and antimicrobials such as tetracycline and ciprofloxacin are often found in sewage systems because they are commonly used to treat infections. These chemicals can cause bacterial populations in sewage to develop antimicrobial resistance (AMR), which can lead to the spread of illnesses that are resistant to drugs. Ibuprofen and diclofenac are two examples of typical analgesics and anti-inflammatory medications because of their stable molecular structure, which enables them to survive in sewage systems (Mutuku et al. 2022; Polianciuc et al. 2020). These medications have the potential to interfere with aquatic creatures' hormone function when they get into lakes and rivers. Antidepressants and psychotropic medications, such as diazepam and fluoxetine, alter GABA receptors or prevent serotonin reuptake, which affects the neurotransmitter system. Aquatic species' behavior and ability to reproduce may be hampered by certain substances found in sewage. Because they imitate or change natural hormone systems, hormonal drugs and endocrine disruptors—especially synthetic estrogens like ethinylestradiol, pose serious ecological risks (Thompson and Vijayan 2022). These chemicals can cause fish and other animals to become feminized, which can lead to problems with reproduction and development. Last but not least, medications that decrease cholesterol and treat heart disease, such as statins (simvastatin) and beta-blockers (atenolol), wind up in the trash. These medications can change the metabolic activity and stress responses of aquatic organisms, which affects their general well-being and ability to survive. The need for better wastewater treatment and management is highlighted by the bioactivity and persistence of several pharmacological groups in sewage (Blonç et al. 2023) (Table 11.1).

Table 11.1 Classes of pharmaceuticals commonly found in sewage and their mechanism of toxic interactions with fauna

Classes of pharmaceuticals	Example	Mechanism	References
Antibiotics and antimicrobials	Tetracyclines, ciprofloxacin, sulfonamides	Bacterial mortality or growth inhibition means that bacteria disrupt cell walls or protein synthesis. However, these methods may enhance antimicrobial resistance (e.g. AMR) in sewage by mediating abnormalities and horizontal gene transfer among bacterial populations	Kümmerer (2009) and Rizzo et al. (2013)
Analgesics and anti-inflammatory drugs	Ibuprofen, diclofenac, naproxen	Prostaglandin production is decreased and therefore we feel uncomfortable, by inhibiting the enzyme cyclooxygenase (COX). Strong structure allows them to survive in sewage and can cause endocrine disruption to aquatic organisms	Santos et al. (2010) and Vieno et al. (2007)
Antidepressants and psychiatric medications	Fluoxetine, sertraline, diazepam	In the management of mood and anxiety, they use selective serotonin reuptake inhibitors (SSRIs) or the gamma-aminobutyric acid (GABA) receptors. They can affect the behavior and reproduction of aquatic species by affecting their neurotransmitter system through these compounds in sewage	Brooks et al. (2003) and Fong and Ford (2014)
Hormonal drugs and endocrine disruptors	Ethinylestradiol, progesterone	Endocrine problems in people and wildlife are caused by synthetic hormones that mimic or interfere with natural estrogen and androgen receptors. Fish and other aquatic creatures may become feminized as a result of certain substances found in sewage	Kidd et al. (2007) and Liney et al. (2005)
Cardiovascular and lipid-lowering drugs	Atenolol, metoprolol, simvastatin	Beta-blockers (atenolol) and statins (such as simvastatin) inhibit HMG-CoA reductase to lower cholesterol, while both block (beta-adrenergic receptors). Since these drugs change the lipid and stress pathways, they reside in sewage and affect aquatic animal metabolism	Carlsson et al. (2006)

11.4 Analytical Methods for Detecting Pharmaceutical Contaminants

The analysis of drug contamination in such samples is so complex that advanced analytical techniques are necessary. Sampling strategies play an important part in the precise, representative sampling of sludge and wastewater. The three common ways of acquiring pharmaceuticals over time and across places are grab sampling, composite sampling, and passive sampling. These pollutants, as such, show a worrying polar, volatile, and solubility composition, which is the key to how these pollutants enter the aqueous and solid phases of wastewater and sludge (Petrie et al. 2015). Watching a sample of the amount of pollutants at a given time is a standard approach to obtaining a rapid sample, but it cannot capture changes over time. By combining many samples over a given period, composite sampling gives a more realistic picture of variable pollutant levels. In particular, this procedure is useful when monitoring drugs that may break down or otherwise fluctuate due to environmental conditions, e.g. pH, and temperature (Cornman et al. 2018). For more common pollutants, passive sampling is the rule; a device sucks up the medication over time, identifying trace amounts in minute amounts (ng/L or less). Drug hydrophobicity or hydrophilicity affects drug molecular interactions with absorptive materials and their partitioning in the sample media. Sewage sludge exceeds the hurdles, because of its highly complex organic composition. Analysis requires solid phase extraction (SPE) because the hydrophobic medicines are immiscible with sludge and bind out to organic components. Molecular mimics can be used by SPE to remove contaminants from the sludge matrix, enhancing the detection of drugs strongly associated with the sludge matrix. Improvements in molecular sampling technologies allow for drug quantification particularly drugs with limited water solubility or which degrade quite rapidly (Godlewska et al. 2020; Šauer et al. 2023). However, instrumental methods of measurement, e.g. Liquid Chromatography with Tandem Mass Spectrometry (LC–MS/MS), Gas Chromatography-Mass Spectrometry (GC–MS), and High-Performance Liquid Chromatography (HPLC) are commonly used to identify and measure drug residues. While GC–MS is usually best suited for volatile or semi-volatile drugs, LC–MS/MS is ideal for detecting trace levels of polar and thermally volatile substances. HPLC, because of its high separation efficiency, is widely used in combined applications with mass spectrometry to attain more exact identification (Patel et al. 2010; Pitt 2009; Stachniuk and Fornal 2016).

Sewage, as well as sludge, is a challenging matrix because the presence of matrices containing organic components, salts, and contaminates can interfere with drug identification from low concentrations (ng/L to µg/L). Matrix effects such as ion suppression or amplification limit the accuracy of mass spectrometry analysis, at the same time requiring rigorous calibration and sample preparation processes. It wished to further develop creative methodologies and tactics to increase sensitivity and accuracy. UHPLC coupled with powerful mass spectrometry and biosensors for real-time monitoring can produce promising results. Moreover, the capability to detect drugs is enhanced by these advancements in microextraction methods and nanotechnology for drug presence in environmental samples (Ohoro et al. 2019; Pérez-Lemus et al. 2019).

11.5 Ecotoxicological Considerations of Pharmaceutical Contaminants

Pharmaceutical pollutants in sewage have severe ecotoxicological effects on aquatic life, including fish, crustaceans, and algae. These pollutants, which include antibiotics, antidepressants, and hormone medications, can impair biological activities even at low amounts. Fish exposed to antidepressants demonstrate changed behavior and reproductive patterns, whereas algae and invertebrates may undergo reduced development and metabolism due to exposure to different pharmaceuticals (Munzhelele et al. 2024).

Pharmaceutical harmful substances have complicated physiological and biochemical effects on aquatic life, including fish, invertebrates, and algae. In fish, antidepressants such as fluoxetine (an SSRI) inhibit serotonin reuptake in the brain resulting in behavioral changes such as altered eating, aggressiveness, and reproductive behavior. These molecular disruptions affect neurotransmitter signaling pathways which influence fish behavior at extremely low concentrations (Fent et al. 2006; Santos et al. 2022). Anti-inflammatory drugs like diclofenac inhibit cyclooxygenase (COX) enzymes which are essential for the production of prostaglandins that regulate inflammation and immune responses in fish, leading to organ damage, particularly in the kidneys and gills. In invertebrates, antibiotics such as tetracyclines inhibit bacterial ribosomal protein synthesis by binding to the 30S subunit of bacterial ribosomes, but they also affect beneficial microbial communities that invertebrates rely on for digestion and nutrient cycling. This disruption leads to weakened immune responses and decreased growth in organisms like crustaceans and mollusks. Moreover, hormonal drugs like ethinylestradiol can interfere with invertebrate endocrine systems by binding to estrogen receptors which disrupt normal reproductive and developmental processes (Gunaydin and Bilge 2018; Khan et al. 2019). For algae, pharmaceuticals such as beta-blockers and lipid-lowering agents like statins interfere with cellular respiration and photosynthesis by inhibiting ATP production and electron transport chains in chloroplasts (Broniarek et al. 2020). Pharmaceuticals in sewage frequently exist in complex combinations, and their synergistic effects can be more hazardous than the individual drugs. At the molecular level, several drugs can interact with aquatic species enhancing their toxicity. Antidepressants, antibiotics, and NSAIDs for example, can all have an effect on different physiological pathways at the same time. Antidepressants such as fluoxetine may limit serotonin absorption, antibiotics might affect the bacterial flora required for digestion in invertebrates and NSAIDs may inhibit cyclooxygenase (COX) enzymes, resulting in immunological suppression (Xiong and Huang 2018). When these drugs are combined, they can have unexpected toxicological effects because the presence of one can increase the absorption, bioavailability, or toxicity of the other. This causes several disturbances and alters systems such as the immunological response, growth, and reproduction. These synergistic effects are typically difficult to anticipate, as drugs interact not only with their primary targets but also with other metabolic processes potentially inflicting more harm to aquatic species than when drugs are present in isolation (Baek et al. 2019).

11.6 Advanced Wastewater Treatment Technology

Advanced wastewater treatment methods are required to remove pharmaceutical pollutants that traditional wastewater treatment techniques cannot effectively eradicate. Pharmaceuticals due to their persistence and complicated chemical structures can provide problems to treatment systems and need specialized approaches to ensure their removal from sewage and wastewater streams. Wastewater treatment involves several stages, typically categorized into preliminary, physical, chemical, and biological methods. Each stage is designed to progressively eliminate impurities, enabling the safe discharge or reuse of treated water. This essay explores the advanced technologies integrated into each of these stages to enhance the efficiency and sustainability of wastewater treatment processes (Fig. 11.1).

11.6.1 Preliminary Treatment Methods

The initial step in the wastewater treatment process is called preliminary treatment, and its goal is to get rid of big particles, grit, and other things that might harm machinery or make later stages of treatment more difficult. Throughout modern screening systems, simple bar screens gave way to automated fine screens and micro

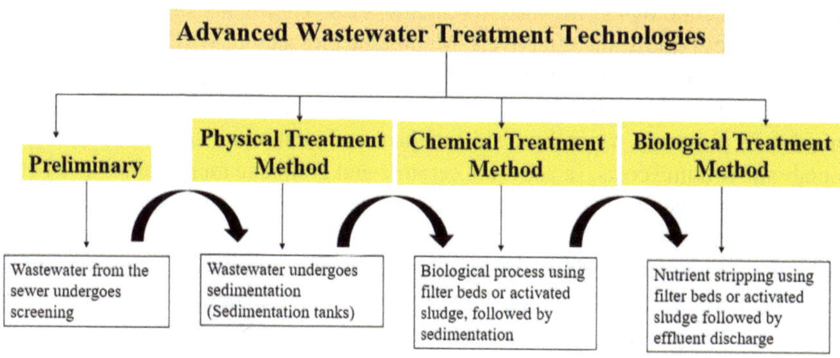

Fig. 11.1 Advanced wastewater treatment technologies

screens capable of hauling in microscopic particles. The reliability is improved through the use of mechanized displays with programmable settings and self-cleaning characteristics which reduces labor costs. Also, in addition to vortex-type grit chambers which separate out the complex grit particles utilizing centrifugal force to remove grit particles from organic waste, additional important devices are used. These methods are commonly used because they protect the downstream equipment and reduce mechanical maintenance requirements. Automated control systems and the use of ultrasonic sensors in flow equalization basins for the increase in performance of the preliminary treatment are additional advances. Flow equalization technology relieves hydraulic fluctuations to maintain constant flow rates, avoiding shock loads that could upset downstream biological processes. In addition, trash and debris compactors have been combined so that less rubbish is thrown into landfills, encouraging ecological practices in the field of waste management. Integration of the newest technologies, such as Internet of Things (IoT) sensors and real-time information analytics, reduces the need for preventive maintenance and allows for operational change that allows for greater overall system efficiency.

11.6.2 Physical Treatment Method

Physical treatment approaches include the use of mechanical methods to remove suspended particles and other particulate matter. Technology has advanced, and traditional procedures such as filtration and sedimentation have significantly enhanced. Lamella clarifiers are highly inventive, using slanted plates to increase surface area, increasing settling efficiency without taking up much space. Dissolved air flotation (DAF) technology is another cutting-edge technology, which is used in cleaning wastewater through the injection of tiny air bubbles suspending particles to the top of wastewater, making the process simple. Because DAF systems are so effective in treating wastewater with a high oil and grease content, DAF systems are ideal for industrial applications. Moreover, physical treatment has been revolutionized by membrane technologies. MF and UF membranes remove bacteria, protozoa, and microparticles rapidly and effectively. Membrane bioreactor (MBR) effluents can be combined with membrane filtration and biological treatment to produce high-quality effluent for reuse. Improved permeability and anti-fouling properties of advanced membrane materials have been able to increase membrane lifespan and reduce membrane lifetime costs. In addition, ceramic and graphene membranes have better durability and chemical resistance and are consequently suited to harsh wastewater conditions. Another major leap forward is the use of advanced filtering technology, including various cloth media filters, disk filters, and other such technologies that allow high solids removal with a low footprint. Heavy materials are separated up to increase the grit and sand removal using the density-based centrifugal separators. However, physical approaches are best paired with automation and real-time monitoring, and the processes become much more efficient, and the use of energy is lessened, for an ecologically friendly wastewater treatment system (Ezugbe and Rathilal 2020; Qalyoubi et al. 2021).

11.6.3 Chemical Treatment Methods

During chemical treatment, chemicals are used to kill both dissolved and suspended contaminants. Classic techniques of flocculation and coagulation have been further enhanced by modern flocculants and coagulants which have improved efficiency and environmental friendliness. Coagulation of small particles and chemical dosages has become more efficient when used in nanomaterials. For example, nano–iron, and nanoalumina are examples of these materials. The larger surface areas for particle aggregation in these materials improve their ability to remove colloidal particles and organic waste. Advanced oxidation techniques offer a significant improvement in chemical treatment technologies. Ozone (O_3), hydrogen peroxide (H_2O_2), and ultraviolet (UV) light interact to create very reactive hydroxyl radicals in AOPs. These radicals can break down pharmaceuticals, endocrine-disrupting chemicals, and persistent organic pollutants that are resistant to standard treatment. UV/H_2O_2, ozone-UV, and photocatalysis using titanium dioxide (TiO_2) have all become widely used for the breakdown of complex organic contaminants. Because these technologies remove micropollutants with such great effectiveness, treated water is safer to reuse or discharge. Electrochemical treatment technologies such as electro-oxidation and electrocoagulation are becoming increasingly popular due to their ability to remove organic pollutants, viruses, and heavy metals. Electrocoagulation generates coagulant species on-site by applying electrical currents to sacrificial metal electrodes (typically iron or aluminum). Conductive electrodes are used in electro-oxidation to generate reactive species that break down pollutants. Because they utilize less energy and produce less chemical sludge, these approaches address disposal difficulties associated with traditional chemical treatments. Ion exchange operations have been improved through the development of selective resins that selectively target specific contaminants, e.g., nitrates, heavy metals, and radionuclides. Real-time dose control systems increased the efficiency of the chemical precipitation methods while reducing the amount of chemicals required and the total chemical consumption. Using these advances in chemical treatment, the usage of chemicals, and their corresponding toxic byproducts, has been minimized which makes wastewater management more sustainable (Hübner et al. 2024).

11.6.4 Biological Treatment Approaches

Biological treatment, which uses microorganisms to break down organic pollutants, is an essential part of wastewater management. Conventional activated sludge technologies have been greatly improved as complex biological systems have evolved.

The Moving Bed Biofilm Reactors (MBBR) are one example of such technology and the biofilm-covered plastic carriers, which float inside the aeration tanks. Treatment capacity can be improved, mechanical and organic stress loads are more resilient and biomass is maintained. Membrane Bioreactors (MBRs) are used to generate outstanding waste products through membrane filtration and use minimal space. A larger concentration of a mixture of liquid solids in suspension (MLSS) and better organic matter removal is achieved due to the membranes' capability of efficient biomass storage. Membrane bioreactors (MBRs) are a particular substitute for wastewater recovery and reuse because of the pathogen-free effluent produced. This advancement represents yet another approach in biological treatment technology: aerobic sludge with particles. Aerobic, anoxic, and even anaerobic activities can be performed in the same granule resulting from this approach; granules are compact and spherical and they settle rapidly. The chemical-resistant design of a multireactor system is multifunctional design eliminating the need for multiple treatment processes and therefore effective withdrawal of nitrogen and phosphorus from only one reactor. Efficient removal of nutrients has been accomplished through the development of advanced biological nutrient removal (BNR) systems. Anammox (anaerobic ammonium oxidation) is a low energy-consuming nitrogen removal technique, independent from organic carbon sources. In oxygen-free environments, Anammox bacteria instantly lower the energy required for aeration and sludge production because they provide anaerobic reduction of ammonia quickly to nitrogen gas. Some bacteria in EBPR accumulate phosphorus that can be recovered and made into fertilizer. Biological treatment options available include algal–bacterial systems and constructed wetlands. Real-time control and monitoring equipment have greatly increased the performance of biological treatment approaches. With modern sensors and automated learning algorithms, dynamic control of sludge recycling, fertilizer dose, and aeration are all possible, improving treatment results and energy efficiency. Also, such analysis of microbial populations with genomes and metagenomics has enabled us to comprehend biologically the mechanisms more thoroughly and manage them for a more resilient and adaptive treatment system (Hassan et al. 2021; Wang et al. 2022).

11.7 Limitations and Efficiency of Current Technologies

The restrictions and efficiency of today's modern wastewater treatment systems vary considerably based on the approach taken, availability of pharmaceuticals, and operational problems.

(a) Microfiltration, ultrafiltration, and reverse osmosis processes have been shown effective in excluding medicinal products by size exclusion with removal rates in the range of 40–90%, depending on the properties of the membrane and the drug. Even though these are not without disadvantages such as high operational costs and membrane fouling that can degrade performance with time and require frequent repairs and membrane service, these are commonly used. Also, dissolved contaminants, because medications with lower molecular weights could make it through some membranes, are less effectively filtered (Ezugbe and Rathilal 2020).

(b) Adsorption, in particular using activated carbon, is a very effective means of removing an array of pharmaceuticals, although the performance of the adsorbent depends on the chemical composition of the pollutant. An example is that hydrophobic chemicals will adsorb more easily; however, the adsorbent materials themselves become saturated and need either replacement or regeneration (Qalyoubi et al. 2021; Yenkie 2019).

(c) Ozonation kills a broad range of medicines by oxidative processes with removal rates in many cases exceeding 80%. However, this technique has major disadvantages, including the manufacture of toxic byproducts such as bromate and the requirement of additional treatment before disposal. Furthermore, the ozonation procedure takes substantial energy and incurs high costs, limiting its practicality for wider application in large treatment facilities (Hübner et al. 2024).

(d) Advanced oxidation procedures (AOPs), which employ hydroxyl (OH) radicals to break down pharmaceuticals, are extremely successful in removing persisting pollutants, often achieving near-complete clearance. However, AOPs were energy-demanding and might result in incomplete mineralization, indicating that some medications may only partly break into the potentially dangerous transitional products (Hübner et al. 2024; Rajasulochana and Preethy 2016).

(e) Bioreactors accelerate the breakdown of microorganisms and can outperform natural systems, particularly when microbe growth parameters are well controlled. However, certain pharmaceuticals continue to degrade poorly, particularly those that resist biodegradation, such as antibiotics or synthetic hormones. Furthermore, the cost of managing a bioreactor system may be rather expensive (Hassan et al. 2021; Wang et al. 2022).

11.8 Future Research Directions

Current pharmaceutical waste treatment research will center on environmental effect identification, monitoring, and reduction. Pharmaceutical waste detection requires more sensitive, fast, and cost-effective analytical approaches. Biosensors, nanomaterial-based detection equipment, and high-throughput screening approaches are being developed to detect medicines in minute quantities in complicated environmental matrices such as sewage and sludge (Gould et al. 2021). Efficient waste management technologies are a primary emphasis. Existing treatment processes typically fail to eliminate medicines, therefore future research should concentrate on environmentally conscious and cost-effective alternatives. Green chemistry ideas, biodegradation technologies, and biotechnological innovations, such as enhanced microbial production or plant-based remediation techniques (phytoremediation), are all viable alternatives for sustainable management. The goal is to remove pharmaceutical residues while minimizing energy use and secondary pollutants (Bienfang et al. 2012). An extensive strategy that incorporates the whole life cycle of the drug is needed in approaches to reduce the environmental impact of medications. Ultimately, this can be accomplished by improving medication formulations that degrade more quickly, improving prescription best practices, and engaging in successful pharmaceutical practice take-back initiatives. The elimination of pharmaceutical contamination, however, requires collaboration between firms, governments, and researchers. Science and business must be part of the dialogue with policymakers to ensure strong enforcement of rigid regulations focused on sustainable pharmaceutical manufacturing and the provision of innovative therapeutic options. The cooperation will promote the implementation of standards of excellence standard and ensure that new research is made into viable and broad appropriate solutions. The scientists working in the area of the environment, chemists, engineers, healthcare personnel, and lawmakers have to work together to combat pharmaceutical pollution. The next critical step should be for future research programs to encourage cross-sector collaboration to produce innovative solutions. It may also support the commercialization of novel therapeutic technologies and environmental drugs through joint endeavors. Initiatives to raise public knowledge about proper medicine disposal and reduce pharmaceutical waste generation are crucial. Future initiatives should employ online platforms and community participation to bring attention to the environmental consequences of improper disposal procedures. Motivating consumers to participate in drug disposal initiatives and practice responsible usage can considerably reduce the number of drugs that enter wastewater systems. Partnerships among government agencies, healthcare specialists, and environmental groups will be important in fostering behavioral change. Climate change may worsen the distribution and permanence of pharmaceutical pollutants by altering precipitation patterns, raising temperatures, and increasing the frequency of extreme weather events. Future studies should investigate the relationship between climate change and pharmaceutical pollution for the purpose of designing adaptive wastewater treatment strategies. It will be vital to design robust wastewater treatment systems that can endure climate change. By including climate threat assessments in infrastructure planning, we can better protect water quality from growing environmental threats. Using green chemistry ideas during the production process can help to reduce environmental contamination at the source. Future medication design may prioritize the production of compostable and environmentally friendly pharmaceuticals that maintain therapeutic efficacy while minimizing environmental impact. Research on prodrugs and personalized drug delivery systems has the potential to minimize the quantity of active pharmaceutical components entering the environment by improving medicine absorption and decreasing unmetabolized chemical excretion.

11.9 Conclusion

Pharmaceutical harmful substances in sewage are becoming a global environmental problem, with far-reaching implications for biodiversity and human health. There are many sources that still continue to push pharmaceuticals into wastewater systems such as household waste, hospital discharges, industrial emissions, and improperly disposed leftover medications. Being incapable of eliminating complex pharmaceutical chemicals, these toxins remain in the environment because of the limitations of conventional wastewater treatment plants (WWTPs). However, these compounds can be chemically altered throughout the treatment process, producing byproducts that could be more dangerous than the original substances to dispose of and the risks to the environment. The most common pharmaceuticals found in sewage are antibiotics, pain relievers, antidepressants, and hormone-related medications. Importantly, each of these groups has its ecotoxicological risks. As an example, antibiotics contribute to the worrisome development of antimicrobial resistance (AMR), which is a serious public health risk. Aquatic creatures can be disrupted by substances that affect their endocrine systems causing growth and reproduction problems. Pharmaceuticals also gear up in aquatic food webs, causing long-term ecological changes and maybe posing a threat to human health through the consumption of contaminated water and the consumption of contaminated aquatic life.

Over the past 40 years, the detection of pharmaceutical pollutants has gone through major development, that is methods like Liquid Chromatography Tandem Mass Spectrometry (LC–MS/MS) and Gas Chromatography Mass Spectrometry (GC–MS) allow for the precise and sensitive detection of trace drug residues in waste and water matrices. While such developments have occurred, the quantification of ultra-low amounts of specific pharmaceutical chemicals remains difficult and there is an ongoing need to innovate in more precise and sensitive analytical procedures. For completing full risk assessments, we need to know the chemical's characteristics, degradation processes, and the environmental lifespan of these pollutants. Ecotoxicological effects of pharmaceutical pollutants are significant. Changes in reproductive processes, and developmental difficulties, have been linked to the use of hormone-altering drugs in fish and amphibians. Widespread distribution of antibiotics in the environment promotes the development of resistant strains of bacteria in such a way that antibiotics are less effective both in medical use and are potentially dangerous for global health. Together, these toxins have cumulative effects that limit ecosystem diversity, degrade ecosystem resilience, and provide indirect human health risks from environmental exposure pathways. The methods by which pharmaceutical waste is currently being regulated vary hugely from region to region without often being thorough enough to solve this all too complicated problem. Consumer disposal of pharmaceutical waste is the subject of many rules, but there are no rules concerning emissions from healthcare institutions and pharmaceutical corporations. Identifying and filling regulatory gaps that will lead to better pharmaceutical contamination solutions is critical. In developing countries with weak or no regulations, standardized and coordinated global approaches are required to ensure pharmaceutical waste is disposed of properly. Promising answers to the contamination from pharmaceuticals are provided by innovative technologies of wastewater treatment. To some degree, ozonation, activated carbon adsorption, membrane filtration and advanced oxidation processes (AOPs) have all been successful in reducing or eliminating pharmaceuticals from wastewater. A growing acceptance of biological therapy approaches (for example usage of specific microbial communities) is because they are able to break down complex pharmaceutical residues. However, because these technologies usually involve high operational costs and technical obstacles, it is essential that the scaleable, economically feasible technological solutions can be widely used. Interdisciplinary study and development to more effectively reduce the harmful effects of pharmaceutical pollutants must be promoted. Given the poor environmental persistence of most existing pesticides, research efforts should be directed at developing environmentally friendly, but innovative candidate pesticides with shorter persistence, strengthening the drug takeback systems to counter the growing problem of incorrect disposal and raising awareness among the public of safe disposal. In addition, pharmaceutical design and manufacturing can adopt green chemistry ideas that may reduce the environmental impact of medications. Technologies to identify pharmaceutical pollution through more sensitive analytical means; and technological innovations aimed at controlling and treating pharmaceutical sewage are essential to minimize the risk.

References

Adeleye AS, Xue J, Zhao Y, Taylor AA, Zenobio JE, Sun Y, Han Z, Salawu OA, Zhu Y (2022) Abundance, fate, and effects of pharmaceuticals and personal care products in aquatic environments. J Hazard Mater 424:127284. https://doi.org/10.1016/J.JHAZMAT.2021.127284

Al-Tohamy R, Ali SS, Li F, Okasha KM, Mahmoud YAG, Elsamahy T, Jiao H, Fu Y, Sun J (2022) A critical review on the treatment of dye-containing wastewater: ecotoxicological and health concerns of textile dyes and possible remediation approaches for environmental safety. Ecotoxicol Environ Saf 231:113160. https://doi.org/10.1016/J.ECOENV.2021.113160

Baek IH, Kim Y, Baik S, Kim J (2019) Investigation of the synergistic toxicity of binary mixtures of pesticides and pharmaceuticals on *Aliivibrio fischeri* in major river basins in South Korea. Int J Environ Res Public Health 16. https://doi.org/10.3390/IJERPH16020208

Bienfang PK, Trapido-Rosenthal H, Laws EA (2012) Bioaccumulation/biomagnifications in food chains. Encyclop Sustain Sci Technol 822–845. https://doi.org/10.1007/978-1-4419-0851-3_50

Blonç M, Lima J, Balasch JC, Tort L, Gravato C, Teles M (2023) Elucidating the effects of the lipids regulators fibrates and statins on the health status of finfish species: a review. Animals 13:792. https://doi.org/10.3390/ANI13050792

Broniarek I, Dominiak K, Galganski L, Jarmuszkiewicz W (2020) The Influence of Statins on the Aerobic Metabolism of Endothelial Cells. Int J Mol Sci 21:1485. https://doi.org/10.3390/IJMS21041485

Brooks BW, Foran CM, Richards SM, Weston J, Turner PK, Stanley JK, Solomon KR, Slattery M, La Point TW (2003) Aquatic ecotoxicology of fluoxetine. Toxicol Lett 142:169–183. https://doi.org/10.1016/S0378-4274(03)00066-3

Caneschi A, Bardhi A, Barbarossa A, Zaghini A (2023) The use of antibiotics and antimicrobial resistance in veterinary medicine, a complex phenomenon: a narrative review. Antibiotics 12:487. https://doi.org/10.3390/ANTIBIOTICS12030487

Carlsson C, Johansson AK, Alvan G, Bergman K, Kühler T (2006) Are pharmaceuticals potent environmental pollutants? Part I: environmental risk assessments of selected active pharmaceutical ingredients. Sci Total Environ 364:67–87. https://doi.org/10.1016/J.SCITOTENV.2005.06.035

Cornman RS, McKenna JE, Fike J, Oyler-McCance SJ, Johnson R (2018) An experimental comparison of composite and grab sampling of stream water for metagenetic analysis of environmental DNA. PeerJ 2018:e5871. https://doi.org/10.7717/PEERJ.5871/SUPP-14

Davison NL, Barrère-de Groot F, Grijpma DW (2014) Degradation of biomaterials. Tissue engineering, 2nd edn, pp 177–215. https://doi.org/10.1016/B978-0-12-420145-3.00006-7

Ezugbe EO, Rathilal S (2020) Membrane technologies in wastewater treatment: a review. Membranes 10:89. https://doi.org/10.3390/MEMBRANES10050089

Fent K, Weston AA, Caminada D (2006) Ecotoxicology of human pharmaceuticals. Aquat Toxicol 76:122–159. https://doi.org/10.1016/J.AQUATOX.2005.09.009

Fernandes JP, Almeida CMR, Salgado MA, Carvalho MF, Mucha AP (2021) Pharmaceutical compounds in aquatic environments—occurrence, fate and bioremediation prospective. Toxics 9:257. https://doi.org/10.3390/TOXICS9100257/S1

Fong PP, Ford AT (2014) The biological effects of antidepressants on the molluscs and crustaceans: a review. Aquat Toxicol 151:4–13. https://doi.org/10.1016/J.AQUATOX.2013.12.003

Ghazal H, Koumaki E, Hoslett J, Malamis S, Katsou E, Barcelo D, Jouhara H (2022) Insights into current physical, chemical and hybrid technologies used for the treatment of wastewater contaminated with pharmaceuticals. J Clean Prod 361:132079. https://doi.org/10.1016/J.JCLEPRO.2022.132079

Godlewska K, Stepnowski P, Paszkiewicz M (2020) Pollutant analysis using passive samplers: principles, sorbents, calibration and applications. A review. Environ Chem Lett 19(1):465–520. https://doi.org/10.1007/S10311-020-01079-6

Gould SL, Winter MJ, Norton WHJ, Tyler CR (2021) The potential for adverse effects in fish exposed to antidepressants in the aquatic environment. Environ Sci Technol 55:16299–16312. https://doi.org/10.1021/ACS.EST.1C04724/ASSET/IMAGES/MEDIUM/ES1C04724_0004.GIF

Gunaydin C, Bilge SS (2018) Effects of nonsteroidal anti-inflammatory drugs at the molecular level. Eurasian J Med 50:116. https://doi.org/10.5152/EURASIANJMED.2018.0010

Hassan I, Chowdhury SR, Prihartato PK, Razzak SA (2021) Wastewater treatment using constructed wetland: current trends and future potential. Processes 9:1917. https://doi.org/10.3390/PR9111917

Hejna M, Kapuścińska D, Aksmann A (2022) Pharmaceuticals in the aquatic environment: a review on eco-toxicology and the remediation potential of algae. Int J Environ Res Public Health 19:7717. https://doi.org/10.3390/IJERPH19137717

Hübner U, Spahr S, Lutze H, Wieland A, Rüting S, Gernjak W, Wenk J (2024) Advanced oxidation processes for water and wastewater treatment—guidance for systematic future research. Heliyon 10:30402. https://doi.org/10.1016/J.HELIYON.2024.E30402

khalidi-idrissi A, Madinzi A, Anouzla A, Pala A, Mouhir L, Kadmi Y, Souabi S (2023) Recent advances in the biological treatment of wastewater rich in emerging pollutants produced by pharmaceutical industrial discharges. Int J Environ Sci Technol 20:11719–11740. https://doi.org/10.1007/S13762-023-04867-Z

Khan AH, Khan NA, Naushad M, Aziz HA (2023) The treatment of pharmaceutical wastewater: innovative technologies and the adaptation of treatment systems. In: The treatment of pharmaceutical wastewater: innovative technologies and the adaptation of treatment systems, pp 1–429. https://doi.org/10.1016/C2021-0-01557-1

Khan S, Andrews KL, Chin-Dusting JPF (2019) Cyclo-oxygenase (COX) inhibitors and cardiovascular risk: are non-steroidal anti-inflammatory drugs really anti-inflammatory? Int J Mol Sci 20:4262. https://doi.org/10.3390/IJMS20174262

Kidd KA, Blanchfield PJ, Mills KH, Palace VP, Evans RE, Lazorchak JM, Flick RW (2007) Collapse of a fish population after exposure to a synthetic estrogen. Proc Natl Acad Sci USA 104:8897–8901. https://doi.org/10.1073/PNAS.0609568104

Kookana RS, Williams M, Boxall ABA, Larsson DGJ, Gaw S, Choi K, Yamamoto H, Thatikonda S, Zhu YG, Carriquiriborde P (2014) Potential ecological footprints of active pharmaceutical ingredients: an examination of risk factors in low-, middle- and high-income countries. Philos Trans R Soc B Biol Sci 369:20130586. https://doi.org/10.1098/RSTB.2013.0586

Kumar V, Bansal V, Madhavan A, Kumar M, Sindhu R, Awasthi MK, Binod P, Saran S (2022) Active pharmaceutical ingredient (API) chemicals: a critical review of current biotechnological approaches. Bioengineered 13:4309. https://doi.org/10.1080/21655979.2022.2031412

Kumar V, Lakkaboyana SK, Sharma N, Chakraborty P, Umesh M, Pasrija R, Thomas J, Kalebar VU, Jayaraj I, Awasthi MK, Das T, Oladipo AA, Barcelo D, Dumee LF (2023) A critical assessment of technical advances in pharmaceutical removal from wastewater—a critical review. Case Stud Chem Environ Eng 8:100363. https://doi.org/10.1016/J.CSCEE.2023.100363

Kümmerer K (2009) The presence of pharmaceuticals in the environment due to human use—present knowledge and future challenges. J Environ Manage 90:2354–2366. https://doi.org/10.1016/J.JENVMAN.2009.01.023

Liney KA, Jobling S, Sheers JA, Simpson P, Tyler CR (2005) Assessing the sensitivity of different life stages for sexual disruption in roach (*Rutilus rutilus*) exposed to effluents from wastewater treatment works. Environ Health Perspect 113:1299–1307. https://doi.org/10.1289/EHP.7921/ASSET/451FF59C-43E7-4B89-A4F5-87FB3A1D1796/ASSETS/GRAPHIC/EHP0113-001299F6.JPG

Liu JL, Wong MH (2013) Pharmaceuticals and personal care products (PPCPs): a review on environmental contamination in China. Environ Int 59:208–224. https://doi.org/10.1016/J.ENVINT.2013.06.012

Manyi-Loh C, Mamphweli S, Meyer E, Okoh A (2018) Antibiotic use in agriculture and its consequential resistance in environmental sources: potential public health implications. Mol J Syn Chem Nat Prod Chem 23:795. https://doi.org/10.3390/MOLECULES23040795

Massima Mouele ES, Tijani JO, Badmus KO, Pereao O, Babajide O, Zhang C, Shao T, Sosnin E, Tarasenko V, Fatoba OO, Laatikainen K, Petrik LF (2021) Removal of pharmaceutical residues from water and wastewater using dielectric barrier discharge methods—a review. Int J Environ Res Public Health 18:1683. https://doi.org/10.3390/IJERPH18041683

Munzhelele EP, Mudzielwana R, Ayinde WB, Gitari WM (2024) Pharmaceutical contaminants in wastewater and receiving water bodies of South Africa: a review of sources, pathways, occurrence, effects, and geographical distribution. Water 16:796. https://doi.org/10.3390/W16060796/S1

Mutuku C, Gazdag Z, Melegh S (2022) Occurrence of antibiotics and bacterial resistance genes in wastewater: resistance mechanisms and antimicrobial resistance control approaches. World J Microbiol Biotechnol 38:152. https://doi.org/10.1007/S11274-022-03334-0

Ohoro CR, Adeniji AO, Okoh AI, Okoh OO (2019) Distribution and chemical analysis of pharmaceuticals and personal care products (PPCPs) in the environmental systems: a review. Int J Environ Res Public Health 16:3026. https://doi.org/10.3390/IJERPH16173026

Okoye CO, Okeke ES, Okoye KC, Echude D, Andong FA, Chukwudozie KI, Okoye HU, Ezeonyejiaku CD (2022) Occurrence and fate of pharmaceuticals, personal care products (PPCPs) and pesticides in African water systems: a need for timely intervention. Heliyon 8:e09143. https://doi.org/10.1016/J.HELIYON.2022.E09143

Ortúzar M, Esterhuizen M, Olicón-Hernández DR, González-López J, Aranda E (2022a) Pharmaceutical pollution in aquatic environments: a concise review of environmental impacts and bioremediation systems. Front Microbiol 13:869332. https://doi.org/10.3389/FMICB.2022.869332/FULL

Patel KN, Patel JK, Patel MP, Rajput GC, Patel HA (2010) Introduction to hyphenated techniques and their applications in pharmacy. Pharm Methods 1:2. https://doi.org/10.4103/2229-4708.72222

Patel M, Kumar R, Kishor K, Mlsna T, Pittman CU, Mohan D (2019a) Pharmaceuticals of emerging concern in aquatic systems: chemistry, occurrence, effects, and removal methods. Chem Rev 119:3510–3673. https://doi.org/10.1021/ACS.CHEMREV.8B00299

Patel M, Kumar R, Kishor K, Mlsna T, Pittman CU, Mohan D (2019b) Pharmaceuticals of emerging concern in aquatic systems: chemistry, occurrence, effects, and removal methods. Chem Rev 119:3510–3673. https://doi.org/10.1021/ACS.CHEMREV.8B00299/ASSET/IMAGES/MEDIUM/CR-2018-00299Y_M010.GIF

Pérez-Lemus N, López-Serna R, Pérez-Elvira SI, Barrado E (2019) Analytical methodologies for the determination of pharmaceuticals and personal care products (PPCPs) in sewage sludge: a critical review. Anal Chim Acta 1083:19–40. https://doi.org/10.1016/J.ACA.2019.06.044

Petrie B, Barden R, Kasprzyk-Hordern B (2015) A review on emerging contaminants in wastewaters and the environment: current knowledge, understudied areas and recommendations for future monitoring. Water Res 72:3–27. https://doi.org/10.1016/J.WATRES.2014.08.053

Pitt JJ (2009) Principles and applications of liquid chromatography-mass spectrometry in clinical biochemistry. Clin Biochem Rev 30:19

Polianciuc SI, Gurzău AE, Kiss B, Georgia Ştefan M, Loghin F (2020) Antibiotics in the environment: causes and consequences. Med Pharm Rep 93:231. https://doi.org/10.15386/MPR-1742

Qalyoubi L, Al-Othman A, Al-Asheh S (2021) Recent progress and challenges on adsorptive membranes for the removal of pollutants from wastewater. Part I: fundamentals and classification of membranes. Case Stud Chem Environ Eng 3:100086. https://doi.org/10.1016/J.CSCEE.2021.100086

Rajasulochana P, Preethy V (2016) Comparison on efficiency of various techniques in treatment of waste and sewage water—a comprehensive review. Resour Effic Technol 2:175–184. https://doi.org/10.1016/J.REFFIT.2016.09.004

Rizzo L, Manaia C, Merlin C, Schwartz T, Dagot C, Ploy MC, Michael I, Fatta-Kassinos D (2013) Urban wastewater treatment plants as hotspots for antibiotic resistant bacteria and genes spread into the environment: a review. Sci Total Environ 447:345–360. https://doi.org/10.1016/J.SCITOTENV.2013.01.032

Ruziwa DT, Oluwalana AE, Mupa M, Meili L, Selvasembian R, Nindi MM, Sillanpaa M, Gwenzi W, Chaukura N (2023) Pharmaceuticals in wastewater and their photocatalytic degradation using nano-enabled photocatalysts. J Water Process Eng 54:103880. https://doi.org/10.1016/J.JWPE.2023.103880

Samal K, Mahapatra S, Hibzur Ali M (2022) Pharmaceutical wastewater as emerging contaminants (EC): treatment technologies, impact on environment and human health. Energy Nexus 6:100076. https://doi.org/10.1016/J.NEXUS.2022.100076

Santos GOS, Goulart LL, Cordeiro-Junior PMP, Sánchez-Montes I, Lanza MVM (2022) Pharmaceutical contaminants: ecotoxicological aspects and recent advances in oxidation technologies for their removal in aqueous matrices. J Environ Chem Eng 10:108932. https://doi.org/10.1016/J.JECE.2022.108932

Santos LHMLM, Araújo AN, Fachini A, Pena A, Delerue-Matos C, Montenegro MCBSM (2010) Ecotoxicological aspects related to the presence of pharmaceuticals in the aquatic environment. J Hazard Mater 175:45–95. https://doi.org/10.1016/J.JHAZMAT.2009.10.100

Šauer P, Vrana B, Escher BI, Grabic R, Toušová Z, Krauss M, von der Ohe PC, König M, Grabicová K, Mikušová P, Prokeš R, Sobotka J, Fialová P, Novák J, Brack W, Hilscherová K (2023) Bioanalytical and chemical characterization of organic micropollutant mixtures in long-term exposed passive samplers from the Joint Danube Survey 4: setting a baseline for water quality monitoring. Environ Int 178:107957. https://doi.org/10.1016/J.ENVINT.2023.107957

Sharma J, Joshi M, Bhatnagar A, Chaurasia AK, Nigam S (2022) Pharmaceutical residues: one of the significant problems in achieving 'clean water for all' and its solution. Environ Res 215:114219. https://doi.org/10.1016/J.ENVRES.2022.114219

Singh BJ, Chakraborty A, Sehgal R (2023) A systematic review of industrial wastewater management: evaluating challenges and enablers. J Environ Manage 348:119230. https://doi.org/10.1016/J.JENVMAN.2023.119230

Stachniuk A, Fornal E (2016) Liquid chromatography-mass spectrometry in the analysis of pesticide residues in food. Food Anal Methods 9:1654–1665. https://doi.org/10.1007/S12161-015-0342-0/FIGURES/5

Thompson WA, Vijayan MM (2022) Antidepressants as endocrine disrupting compounds in fish. Front Endocrinol 13:895064. https://doi.org/10.3389/FENDO.2022.895064

Vieno NM, Härkki H, Tuhkanen T, Kronberg L (2007) Occurrence of pharmaceuticals in river water and their elimination in a pilot-scale drinking water treatment plant. Environ Sci Technol 41:5077–5084. https://doi.org/10.1021/ES062720X

Wang J, Long Y, Yu G, Wang G, Zhou Z, Li P, Zhang Y, Yang K, Wang S (2022) A review on microorganisms in constructed wetlands for typical pollutant removal: species, function, and diversity. Front Microbiol 13:845725. https://doi.org/10.3389/FMICB.2022.845725

Yenkie KM (2019) Integrating the three E's in wastewater treatment: efficient design, economic viability, and environmental sustainability. Curr Opin Chem Eng 26:131–138. https://doi.org/10.1016/J.COCHE.2019.09.002

Chapter 12
Bioremediation and Bioenergy Generation Potential of Facultative Anaerobic Bacterial Isolates for Treatment of Municipal Waste

Ankita Saxena and Varsha Gupta

Abstract The remorseless heightening population growth and their enriching anthropogenic needs is the reason for development of several wastes with mismanagement disposal pattern. Hence the need of the hour is to search for efficient waste processing system. Current research and recent waste management strategies is focusing on the targeted waste minimization, reuse, recycling and remediation methodologies rather than only emphasizing on their disposal management system. Furthermore, need for substitute energy resource should also mitigate the developmental needs with clear candidature of ecologically reliable and economically sustainable one. On this aspect, waste-to-energy protocol is one of the great approaches that utilization of waste by microorganisms to promote the generation of energy. Biological or microbial fuel cell (BFC/MFC) is an efficient and viable technology to achieve more sustainable environment. MFC is used to convert stored energy in form of organic and inorganic compounds by the sequential redox reactions into electrical energy. These microorganisms include bacteria, fungi, green algae, and cyanobacteria. The only method that can contribute to the resolution of the global warming problem is the use of renewable energy sources. This process not only reduce the amount of secondary sludge production but also remove the secondary toxic pollutants along with nitrogen and phosphorus and also reduce down the biological oxygen demand of municipal waste. They also help in microbial mediated electro synthesis of related by products and can also function as active biosensor for easy detection of pollutants from waste water system from different sources that can be practically a sustainable approach for environmental stability.

Keywords Bio electrification · Microbial fuel cells · Microbial consortium · Municipal waste · Waste-to-energy

A. Saxena · V. Gupta (✉)
Department of Microbiology, JECRC University, Jaipur, India
e-mail: varsha.gupta@jecrcu.edu.in

S. Madhav et al. (eds.), *Municipal Sewage and Sludge: Treatment and Disposal Strategies*, Environmental Science and Engineering,
https://doi.org/10.1007/978-3-032-07597-0_12

12.1 Introduction

The prodigious entity of xenobiotic waste developed either from industrial sector, agricultural sector or domestic areas is an uninterrupted phenomenon including both biodegradable and non-biodegradable parts. Basically the elevating evolution in socialisation and industrialization with advancement in pharmaceutical and agriculture sector, to meet the market demand of growing anthropogenic needs along with the diminishing crude oil quantities have engendered several detrimental impacts to the surrounding environment through elevation of profligate waste generation rate and their noxious release of many harmful xenobiotics in an unsustainable manner in the surrounding environment (Langa et al. 2021). This facilitates the accumulation of toxic pollutants and other hazardous elements in the hydrosphere, lithosphere and atmosphere, thereby altering the prime balance within the ecology and development of many serious threats to the concerned living community. This issue has forced to shift focus towards development of step wise synchronised waste management strategies is focusing on the targeted waste minimization, reuse, recycling and remediation methodologies rather than only emphasizing on their disposal management system along with development of renewable and sustainable energy resources (Kumar et al. 2021).

Hence the need of the hour is to search for an effective and efficient processing system for all the types of generated waste at its source of generation and after release. The recent trend in the above situation is way more tilted for production and implementation of green biological study and its applications in proper manner for the entire process of disposal and remediation of generated waste (Boas et al. 2022). In the current scenario, microbial fuel cell (MFC) is considered to be the efficient and effective technical applications for dual purpose including both waste management system and their proper utilisation for generation of bioenergy.

The one-way research on the line of improvements in degradation methodologies applicable for removal of toxic metals and recalcitrant dyes will pave the path for future role, targeted toxic pollutants, contaminated medium with newer developed bioremediation techniques with less time consumption and less sludge formation. Conventional methods for treating wastewater using physicochemical techniques are not recommended since they take too long and use less effective techniques (Jobby et al. 2019). This has paved the path for utilization of potent biological methods for elimination of these hazardous pollutants without any generation of secondary sludge. On this panorama, the bioremediation, individually or in clubbed format are found to be effective (Nigri et al. 2020).

12.1.1 Bioremediation

Bioremediation is referred to as treatment of toxic pollutants using biological systems such as plant, microbes, etc. or biological products such as enzymes, polysaccharide derived from microbes (Marco and Abram 2019a, b). Bioremediation is a sustainable approach for complete or partial eliminate toxic pollutants from the surrounding medium (Nigri et al. 2020). The sustainable use of microbial consortium can effectively remove toxic heavy metals and pollutants without producing associated sludges, despite the need for coordinated and synergistic techniques (Singh et al. 2017).

12.2 Microbial Remediation of Waste Water

The microbial bioremediation method mostly relies on the application of selected microbes, individually or in consortium form with alteration of their metabolic pathways and is dependent on the modern synthetic biology for developing synthetic genetic biosensors or circuits (Marco and Abram 2019a, b). There are several benefits of utilization of microbial bioremediation processes including unaltered target site structure, low-cost treatment, minimal direct or indirect hazardous impact on population and can be achieved in lesser time duration (Mishra and Maiti 2018; Verma et al. 2019; Yin et al. 2019; Madhav et al. 2024).

Such toxic pollutants and the related secondary sludge must be utilised in a proper manner, so as to get rid of it simultaneously along with mitigating another vital issue like managing the energy crisis by availing the possible replenishable energy sources (Caizán-Juanarena et al. 2020; Nigri et al. 2020) by managing the toxic pollutants which are basically discharged to open landfill area or water bodies creating pollution and generation of maximum wastewater system (Shi et al. 2021; Elhenawy et al. 2022). To ensure the preservation of sustainable and balanced environmental elements by limiting pollution accumulation, the problem calls for an environmentally friendly, biologically influenced strategy (Nie et al. 2021). Microbial mitigation techniques are being developed for efficient incorporation of inorganic pollutants are immobilised (Tarekegn et al. 2020). Microbial colony profiling with their identification has vital significance in toxic metal ion detoxification process (Sharma 2021; Varjani et al. 2021). By creating a waste-to-energy protocol that facilitates energy generation and extends the accessible waste biologically, a comprehensive strategy to mitigate potential energy constraint sustainably has led to the utilisation of less hazardous, widely available biological agents (Moradian et al. 2021; Hoang et al. 2022) by utilisation of microbial fuel cells (MFCs) through bio-electrification (Obileke et al. 2021; Elshobary et al. 2021; Uddin et al. 2021).

12.3 Microbial Fuel Cells

Microbial fuel cells generate electrical energy by combining microbial metabolic and electrochemical reactive pathways while reducing the amount of waste that is available (Gadkari et al. 2019). It is a bio-catalytically structured device (Fig. 12.1) for electrical power generation through the decomposition of organic matter by the assistance of either bacteria or by the enzymes under slight reaction conditions (Elhenawy et al. 2022). These are structurally composed of basic compartment system: one anode chambers and other one is the cathode chamber. The material of cathode chamber can be glass, polycarbonate, lead oxide with sulphide or sulphur whereas Anode chamber material can be glass or polycarbonate. Anode is made of different materials i.e. Graphite, Reticulated vitreous carbon, Carbon paper, Stainless steel, Graphite granules bed, Manganese oxide; lead dioxide and cobalt complexes, Nafion. Cathode is composed of Carbon paper, Graphite, Reticulated vitreous carbon.

In the anode compartment the desired microbes are being placed for biocatalysis of the organic waste samples for energy generation by production, transportation and transfer of electrons through soluble electron transfer mediator (Ardakani and Gholikandi 2020).

The cathode compartment is aerobic in nature while anode one is anaerobic. Both the chambers are partitioned with a proton exchange membrane or an ion exchange membrane. The proton exchange membrane also regulates influx of selected substrate alongside the diffusion rate of oxygen (Ardakani and Gholikandi 2020). All the

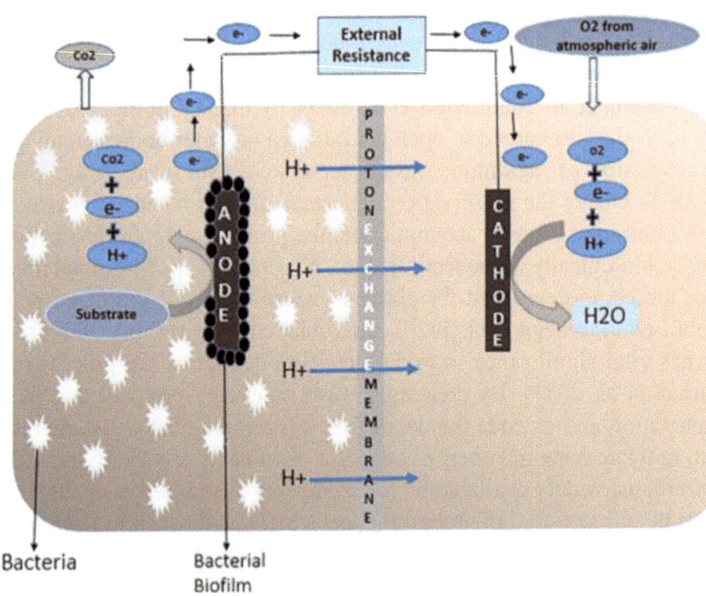

Fig. 12.1 Structure of microbial fuel cell

available sources of organic compounds including the wastewater are being oxidised to release electrons and protons via the activities of selected microorganisms followed by their movement into the cathode chamber for energy production (Moradian et al. 2021). However, the proton exchange membrane also regulates influx of selected substrate alongside the diffusion rate of oxygen (Ardakani and Gholikandi 2020). In the following step, the amount of oxygen is delimited. However, oxygen is one of the most suited electron acceptor for MFCs and besides it some other effective electron acceptor is hydrogen peroxide (Moradian et al. 2021; Hoang et al. 2022). The proton displacement is greatly influenced by the basic concentration and polarization efficacy of proton exchange membrane. Entire biochemical reaction steps occurring within microbial fuel cells are depicted below (Saeed et al. 2015). During the provided steps, the protons get diffused through proton exchange membranes towards cathode followed by combining with the available oxygen and electron available in provided waste water compounds (Dharmalingam et al. 2019). Within anodic compartment, to the given substrate specifically selected exo-electrogens bacteria and anode respiring bacteria are provided, which are facilitating their individual growth followed by oxidation of the substrate provided to remove the electrons leading to be accepted by an electron acceptor.

Chemical Interactions of Microbial Fuel Cells

Step-1: Anodic interaction

$$(CH_2O)_n + nH_2O \rightarrow nCO_2 + 4ne^- + 4nH^+$$

Step-2: Cathodic interaction

$$O_2 + 4ne^- + 4nH^+ \rightarrow 2H_2O$$

12.4 Microorganisms Used in MFCs

The microbes within the microbial fuel cell are a vital point that has major impact on the production of electricity. Various microbes are having the efficiency of electron diffusion from organic waste water bioprocessing within the microbial fuel cell (Elhenawy et al. 2022). Several microbes play a major role in microbial fuel cells including *Escherichia coli, Sachrromyces cerevisiea, Shewanella putrifaciens, Aeromonous hydrofili, Pseudomonous*spp., *Geobacter sulfureducens, Halobacterium salinarum, Gluconobacter oxydans, Klebsiella pneumonia, Micrococcus luteus, Gluconobacter oxydans, Rhodoferax ferrireducens, Streptococcus lactis, Enterococcus faecium.* Through electron generations that diffuse to the cathodic compartment, the bacteria use as a biological stimulators to speed up the degradation of waste water that contains organic molecules (Obileke et al. 2021; Prathiba et al. 2022).

The exo-electrogens facilitate electricity generation through the substrate (organic compound) oxidation. Several microbes are available which are having greater potency in terms of electron diffusion from any suitable organic matter through microbial metabolic functions towards anode chamber. For instance, microbes like *Proteobacteria* spp., *Firmicutes*spp. and *Acidobacteria*spp. are super-efficient fuel cells in generation of bioelectricity (Pant et al. 2010). Moreover, those microbes, which are efficient in wastewater management system along with removal of some highly toxic pollutants, are the best suited ones as these microbes facilitates both utilization of some unwanted organic compounds as substrate from wastewater thereby facilitating the microbial mediated bioremediation of available toxic metals and ions (Obileke et al. 2021; Elhenawy et al. 2022).

12.4.1 Advantages of Microbial Fuel Cells

The microbial fuel cells (MFCs) have several potent benefits over many conventional methodologies for development of a least expensive and high yield throughout the process for energy generation (Ahn and Logan 2010; Prathiba et al. 2022). Some of these advantages are listed below.

They can efficiently produce bioelectricity using organic matters as substrates.
Voltage produced is utilised in electrolysis cells to produce H_2/H_2O_2.
They remain active with the altering surrounding temperature.
They do not need external aeration.
They can work taking different microorganisms.
MFCs facilitate waste water management process.
These are used as potential biosensor for pollutant detection.
MFCs mostly do not produce any secondary sludge.

12.5 Microbe Assisted Waste Water Treatment

On this panorama, the utilisation of waste water can play a very crucial role as its content is very enormous with easy availability. Furthermore, the all total cost expenditure in collecting and utilising the waste water for bioelectricity generation through microbial fuel cell utilisation is also minimal. The importance of any energy-rich compound is gradually elevating due to which the excessive biological load in wastewaters is not considered as waste recently (Ahn et al. 2014). Though several physical and chemical techniques have high potency but they require large cost expenditure and readily produce secondary pollution and sludge, while biological method has the advantages of minimal cost and no production of any kind of toxic pollution (Pant et al. 2010). To exploit these biological substrates degradation, microbes are used as cells for generation of energy. This also has a dual role the prime one related to waste water management and the secondary goal is to generate bioelectricity efficiently along with release of pollutant free water for further utilisation (Prathiba et al. 2022).

In this process an-oxygenic condition is maintained to carry out the entire process without interference of oxygen molecules to diffuse the electrons (Obileke et al. 2021). In the mediator regulated mechanism, the mediator catches and releases the electrons. Conventional wastewater management methods utilising the municipal and industrial waste water in a strict aerobic condition is quite expensive in terms of consumption of excess amount of energy and available resources. However, when we are comparing the conventional methods with the microbial mediated technologies utilising the microbes, it was found to be much more effective. Bacterial fuel cells generate bioelectricity, remove toxic pollutants, and act as efficient biological detectors for identification of contaminants in waste water systems.

12.6 Microbial Aided Electro-Synthesis from Waste Water

The quality assessment of water level and the constraints arising due to deposited and/or absorbed pollutants within it is one of the principal reasons that profoundly striking on human life. Peculiarly, the groundwater quality in a specific area is greatly influenced by several anthropogenic factors like inadequate domestic, agricultural and industrial waste management system and some plebeian factors like certain geological interventions or soil disintegration (Kazi et al. 2009). The waste water released from several industrial and/or domestic waste disposal systems is one of the major concerns now-a-days. On this context bacterial fuel cells act as sustainable, effective and feasible methodology for waste water treatment as compared to several traditional protocols available by far. Microbial aided cells are a potent technology that produce electrical energy by converting the potential energy during substrate oxidation via microorganisms. This allows wastewater to be treated, which addresses water treatment and energy shortages by generating electricity (Ren et al. 2014). MFCs are a popular method for producing clean and renewable energy from wastewater, offering a new way to manage wastewater effectively (Obileke et al. 2021).

The over exploited population growth, unsustainable utilisation of renewable and non-renewable natural products and continuous improvement in mitigating the increasing anthropogenic needs have facilitated production of maximum xenobiotics leading to several harmful toxicant depositions within the water bodies (Peña et al. 2013). This is due to the fact that the water bodies are capable of dissolving and absorbing maximum of the xenobiotics leading towards the generation of waste water system in larger quantity. This issue is even more critical near the industrial sectors and has a bold impact on the ground water system in those areas as toxic accumulation rate within the ground water through soil level is more rapid than any other medium. On this instance the adverse impact of polluted ground water on human health is a major concern these days. To monitor the ground water quality and to reduce the accumulated toxicity rate within it, the application of microbes is found to be more suitable than any other method of toxic degradation. Many studies are going on towards managing these waste water quantities. The waste water management system including physical, chemical or biological treatment of polluted water bodies is being continuing since ages. However, to facilitate a better response during waste water management, the biological composition of waste water should be validated with determination of each and every toxicant available within it (Marco and Abram 2019a, b).

12.7 Microbial Assisted Bioremediation of Waste Water

The increasing population density and their needs have led to continuous development of several agricultural and industrial sectors, resulting in production of associated wastes. The random waste disposal in any landfill areas or water bodies creates many serious threats to the concerned living community and elevating hazardous pollutant amount within the environment (Keshavarzi and Kumar 2018). The use of conventional methods for waste treatment is unfavourable due to financial ineffectiveness and accessory waste generation. This has paved the path for microbial bio remediation methods for treatment of industrial effluents. Bioremediation, a economical and environment-friendly method of treating toxic pollutants through plants, microbes, and microbial products, is a significant environmental solution (Jobby et al. 2019). Wastewater is the major source of generation of pollutants in several water bodies including the ground water system (Gude 2016). The elevated concentration of pollutants and organic matters as in form of domestic, agricultural or any other industrial waste has led into increasing amount of stored energy within the waste water and facilitates the pollutant degradation process a bit tedious. All the available traditional methodologies for waste water treatment and remediation of toxic pollutants are either prolonged ones or require higher cost expenditure.

Microbial consortia and plant species can effectively remove toxic sources from wastewater, utilizing organic compounds for bioremediation and electricity generation through microbial fuel cells. Several studies are available for waste water treatment utilising microbes as the main agent (Marashi and Kariminia 2015; Gude 2016; Sawasdee and Pisutpaisal 2016; Singh et al. 2019; Srivastava et al. 2020; Vélez-Pérez et al. 2020; Yadav et al. 2020) and using microbial fuel cells (Table 12.1). The study utilized benthic microbial fuel cells (BMFCs) to generate replenishable energy and bioprocess the wastewater containing aromatic hydrocarbons (Umar et al. 2021). The study utilized an integrated drip hydroponics-microbial fuel cell system for wastewater treatment (Yadav et al. 2020). A study found that after biofilm removal, the power supply in a sea-water fuel cell with a stainless steel cathode and platinum anode decreases by up to 2.8 mW/m^2 (Shanmuganathan et al. 2018). The study revealed that the formation of biofilm accelerates oxygen reduction and eliminates the unpleasant odour.

Table 12.1 Different types of microbial fuel cells used for wastewater treatment

Sl No.	Type of MFCs used	Wastewater	References
1	50 Stacked triple chambers	Wastewater collected from municipal area	Liang et al. (2018)
2	3 Dual stacked chambers	Municipal wastewater	Mehravanfar et al. (2019)
3	Single chambered MFCs	Wastewater collected from Tannery	Sawasdee and Pisutpaisal (2016)
4		Waste collected from Petroleum refinery	Mohanakrishna et al. (2018)
5		Waste from wood hydrothermal	Kloch et al. (2020)
6	Up-flow constructed wetland-microbial fuel cell	Wastewater collected from Synthetic site	Nguyen et al. (2017)
7	Up-flow single chamber	Dairy wastewater	Marassi et al. (2020)
8	Double chambered MFCs	Wastewater from industrial Area	Rozendal et al. (2006)
9		Wastewater collected from Brewery	Ahn and Logan (2010)
10		Wastewater from industrial acid mine and municipal	Vélez-Pérez et al. (2020)

Liang and his colleagues built a research in 2018 that treated actual municipal wastewater using a 1000 L modularized MFC system. For over a year, this MFC system was run in two municipal wastewater treatment plants (MWTP) with two distinct water flow connections to test its capacity to treat wastewater with both low (average 80 mg L^{-1}) and high initial COD concentrations. This study concludes with energy recovery of 0.033 ± 0.005 kWh per m^3, the COD concentration in the MFC system's effluent stayed below 50 mg L^{-1} at a removal rate of 70–90%, whereas in 2019 Mehravanfar et al., used 3 dual stacked chambers to treat Municipal wastewater.

Single chambered MFCs were used to treat different types of wastewaters collected from various sources like Wastewater collected from Tannery, Petroleum refinery, Wood hydrothermal etc., Sawasdee and Pisutpaisal 2016; Mohanakrishna et al. 2018; Kloch and Toczyłowska-Maminska, 2020 respectively likewise Double chambered MFCs were also used to treat the Wastewater collected from Industrial Area, Brewery waste, Industrial acid mine and municipal by Rozendal et al. 2006; Ahn and Logan, 2010; Vélez-Pérez et al. 2020 respectively.

Due to the high organic load of dairy wastewater, the Up-flow single chamber Microbial Fuel cells was operated in 2020 by Marassi et al., which removes the total chemical and biochemical oxygen demand (TCOD and TBOD) by 94 and 96%, under external resistance of 1.0 kΩ, and 96 and 97%, under external resistance of 0.3 kΩ. In addition, the MFC treatment provided high efficiency for the removal of organic nitrogen (47%), phosphorus (95%), nitrate (100%), and sulfate (75%).

Depending upon their source of generation, all the types of waste water system contain different quantity of carbon content in form of organic matter. While industrial, domestic and agricultural waste water contain maximum organic matter, the brewery wastewater, pharmaceutical wastewater contains a little lesser quantity (Gul et al. 2021). These variable organic carbon sources also facilitate the rate of bioremediation. The utilization of multiple organic sources within bacterial fuel cells significantly enhances the bioremediation rate. Larrosa-Guerrero et al (2010) has employed a double-chambered microbial fuel cell with a mixed waste water pattern that includes both home and brewery wastewater, and the mixed waste water system has produced better results. The general steps involved in waste water treatment process include two major steps namely primary and secondary treatment. The primary treatment comprises the addition of certain coagulants with the goal of eliminating grits, coarse particles, oil, and grease, if any are present. Solids settle (sludge) during preliminary treatment. The second stage transforms organic materials in sewage into stable forms using biological processes, resulting in secondary sedimentation by the employment of numerous bacteria such as *Bifidobacterium* spp., *Clostridium*spp., *Propionibacterium*spp., *and Staphylococcus*spp. (Srivastava et al. 2020).

12.8 Bioenergy Generation

The uniqueness of potent bacterial fuel cell is its ability to directly convert the potential carbon source energy into electrical energy (Fig. 12.2). These cells operate optimally in minimal temperature range, facilitating maximum waste water treatment (Gul et al. 2021). As anaerobic mode of reaction, the bacterial fuel cells do not require any additional carbon dioxide supply or enrichment due to the fact that these are the off-gas mechanisms. Water treatment, toxicant remediation, and energy shortages can all be addressed by using microbial fuel cells to cleanse wastewater and generate electricity. One important factor that controls the energy output rates of microbial fuel cells using numerous types of accessible bacteria is the development design of the microbial fuel cell system and the nature of substrate employed (Parkash 2016).

The bacterial fuel cells are bio-electrochemical, energy producing device that help in conversion of chemical energy from organic carbon source into direct electrical form taking organic wastewater as substrates through the applications of microbes (Idris et al. 2016). Several new approaches towards accomplishing a sustainable treatment and management protocol for waste water system to generate purified water bodies have been developed globally (Zhang et al. 2012; Zhao et al. 2018; Sekar et al. 2019; Jatoi et al. 2020; Gul et al. 2021). As wastewater is intrinsically fertile in bioavailability of energy, these energy sources can be channelized for bioelectricity generation. Out of many available methodologies for energy generation from waste water, microbial fuel cells are found to be the most evident ones with higher practical application due to their efficacy in conversion of potential carbon energy source directly into electrical energy (Sekar et al. 2019; Jatoi et al. 2020). Even the microbial

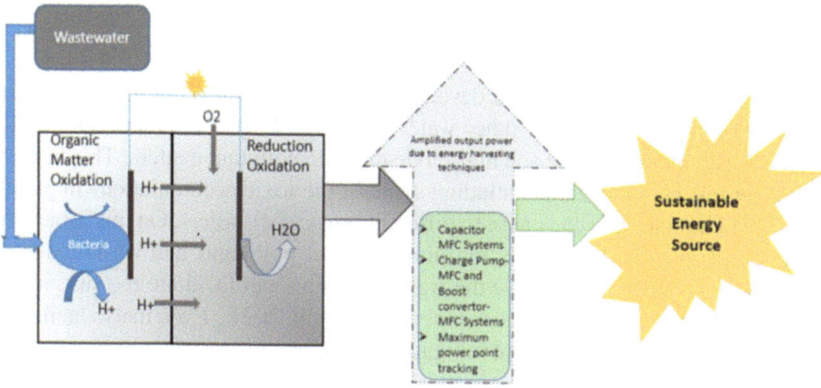

Fig. 12.2 Applications of bacterial fuel cell for energy harvesting systems

fuel cells can act in a consortium format to facilitate the bioremediation process coupled with bioelectricity generation.

The efficacy of energy scavenging systems can be increased by lowering the number of components. Energy scavenging potency can be elevated by obtaining a validated fuel cell design that effectively takes waste water with utilisation of minimum power consumption. Active generation of energy can improve up to two hundred and forty percent power output, up to forty five percent current output, and up to one hundred- and forty-one percent columbic efficiency under various power criteria (Lu et al. 2019). Microbial fuel cell technology is one of the effective and eco-friendly approach to generate bioenergy and also workable to manage the wastewater. Different categories of wastewater in different can be used up as carbon source within the fuel cells (Zhao et al. 2018; Umar et al. 2021). Microbial fuel cells are being emphasized for commercial use, particularly in wastewater treatment and green energy generation, with electrode stability being a critical consideration. A pilot-scale microbial fuel cells system was successfully established in brewery waste, Yatala, Australia (Gajda et al. 2018). Several steps can be taken for generation of higher electricity effectively from the bacterial fuel cells. The design of the cell can help in regulating lower internal resistance capacity. The process of electron transfer can be enhanced through the utilisation of nanoparticles. Even the genetically altered microbes can also be utilised those can produce maximum biofilm conductivity for electricity generation (Gul et al. 2021). This review aims to enhance commercial energy generation by introducing new methods for self-powered energy harvesting models in microbial fuel cells to make them more sustainable.

12.9 Societal Implication

The harsh reality of today's rapid development includes several unmanaged waste disposals to ground water and other water bodies leading to the increasing pollution level with biodegradable and non-biodegradable pollutant matters. This leads to several serious health hazards including some of the toxic accumulations in plants, animals and humans and can cause to different types of disorders. On this aspect the utilisation of microbial fuel cells for elimination of pollutants and waste management in water bodies can manage the water pollution and facilitate the sustainable utilisation of water in a systematic manner (Umar et al. 2021). Even it also facilitates energy production that can help in mitigation of energy crisis in an effective manner (Gul et al. 2021).

The scientific conflict on production of bioenergy through application of microbial fuel cells has basically concentrated on several climate and environmental change (Searchinger et al. 2008); along with societal and associated economic factors as an integral holistic approach for sustainability (Fritsche et al. 2018). As battening unanimity on the potential socio-economic impacts can facilitate the sustainability criteria fulfilment; hence the validation of overall factors associated with production of electricity through implications of microbial cells can facilitate the sustainability.

Different kinds of conventional wastewater treatment processes are basically associated in processing and managing the waste and has little or no significance to the development of other related societal issues and other environmental impacts. Conventional methods involved in waste water treatment includes formation of secondary sludge, formation of other toxic pollutants, lengthy post treatment disposal management, excessive time duration and many more (Verma et al. 2019; Yin et al. 2019). However, no attention is provided generally to other allied aspects that can harm the socio-cultural aspects thereby elevating the associated concerns (Raman and Kanmani 2016; Sharma 2021).

In most cases the lack in considering socio-cultural factors during enabling the steps of wastewater treatment can create serious havoc and minimize the feasibility of the applied technology. This can be due to the pressure generated by industrial stakeholders, politicians, influencers or even some bunch of people considering individual growth more than the social development. On this aspect, the utilisation of bacterial isolates for remediation of waste water is one of the pivotal role that consider not only clearing the waste water (Yin et al. 2019) but also facilitates the major global crisis like energy limitations. The utilisation of bacterial isolates through use of microbial fuel cells can facilitate the eased out methods of waste water treatment without any secondary sludge or pollutant generation and also facilitate the production of economically feasible electricity to mitigate the global crisis at the same time (Sharma 2021).

12.10 Future Prospects of Microbial Fuel Cells

With potential uses in fields like biofuel production, environmental remediation, medicine, and food production, microbial cell research appears to have a bright future. This is largely due to developments in genetic engineering, synthetic biology, and sophisticated analytical techniques, which enable researchers to better understand and manipulate microbial communities in order to address complex issues like climate change and emerging diseases.

Microbial cell research seems to have a bright future because of its potential applications in industries like food production, medicine, environmental cleanup, and biofuel production. This is mostly because to advancements in synthetic biology, genetic engineering, and advanced analytical methods that allow scientists to better understand and work with microbial communities to address complicated problems like new diseases and climate change.

Creating microorganisms with specialized roles, such as probiotics for human health or enzymes for industrial processes, using CRISPR technology and other genetic engineering methods.

Investigating and improving the relationships between various microbial species in communities in order to improve intended results, such as nutrient cycling or bioremediation.

Creating effective MFCs to use microbes to produce power from waste biomass, so supporting sustainable energy production.

12.11 Conclusion

The increased organic matter content of waste water resources, their ease of access, and the greater efficacy of microbes in metabolically degrading them have prepared the road for two-way benefits from a single targeted source. At one point, waste water can be effectively treated by degrading accumulated hazardous contaminants, which can also be used to generate some bioelectricity. Traditional wastewater treatment procedures are thought to be costly due to excessive usage of time duration, energy and available resources. This produces a considerable amount of undesired sludge. However, microbial mediated methods that use bacteria have been shown to be far more effective, cheaper, and time efficient.

Bacterial fuel cells not only aid in bioelectricity generation, but also in the elimination of a variety of harmful pollutants from waste water, as well as associated toxicant elimination. These bacterial cells not only aid in microbial mediated energy production, but also in the formation of pollutant-free water for reuse, so contributing to the conservation of natural resources. The futuristic technique might be tailored such that improvements in commercial applications of microbial fuel cells become more viable, as well as sustainable applications that safeguard natural resources.

References

Ahn Y, Logan BE (2010) Effectiveness of domestic wastewater treatment using microbial fuel cells at ambient and mesophilic temperatures. Bioresour Technol 101:469–475

Ardakani MN, Gholikandi GB (2020) Microbial fuel cells (MFCs) in integration with anaerobic treatment processes (AnTPs) and membrane bioreactors (MBRs) for simultaneous efficient wastewater/sludge treatment and energy recovery—a state-of-the-art review. Biomass Bioenergy 141:105726

Boas JV, Oliveira VB, Simões M, Pinto MFRA (2022) Review on microbial fuel cells applications, developments and costs. J Environ Manag 307:114525. ISSN 0301-4797

Caizán-Juanarena L, Borsje C, Sleutels T, Yntema D, Santoro C, Ieropoulos I, Soavi F, TerHeijne A (2020) Combination of bio-electrochemical systems and electrochemical capacitors: principles, analysis and opportunities. Biotechnol Adv 39:107456

Dharmalingam S, Kugarajan V, Sugumar M (2019) Membranes for microbial fuel cells. https://doi.org/10.1016/B978-0-444-64052-9.00007-8. https://www.researchgate.net/publication/330049875

Elhenawy S, Khraisheh M, Al Momani F, Al-Ghouti M, Hassan MK (2022) From waste to waters: updates on key applications of microbial fuel cells in wastewater treatment and energy production. Sustainability 14:955

Elshobary ME, Zabed HM, Yun J, Zhang G, Qi X (2021) Recent insights into microalgae-assisted microbial fuel cells for generating sustainable bioelectricity. Int J Hydrog Energy 46:3135–3159

Fritsche UR, Eppler U, Fehrenbach H, Giegrich J (2018) Linkages between the sustainable development goals (SDGs) and the GBEP sustainability indicators for bioenergy (GSI). Int Inst Sustain Anal Strateg. Berlin, Germany

Gadkari S, Shemfe M, Sadhukhan J (2019) Microbial fuel cells: A fast converging dynamic model for assessing system performance based on bioanode kinetics. Int J Hydrog Energy 44:15377–15386

Gajda I, Greenman J, Ieropoulos IA (2018) Recent advancements in real-world microbial fuel cell applications. Curr Opin Electrochem 11:78–83

Gude VG (2016) Wastewater treatment in microbial fuel cells—an overview. J Clean Prod 122:287–307

Gul H, Raza W, Lee J, Azam M, Ashraf M, Ki-Hyun K (2021) Progress in microbial fuel cell technology for wastewater treatment and energy harvesting. Chemosphere 281:130828

Hoang AT, Nižetić S, Ng KH, Papadopoulos AM, Le AT, Kumar S, Hadiyanto H, Pham VV (2022) Microbial fuel cells for bioelectricity production from waste as sustainable prospect of future energy sector. Chemosphere 287:132285

Idris SA, Esat FN, Rahim AA, Rizzqi WA, Ruzlee W, Razali MZ (2016) Electricity generation from the mud by using microbial fuel cell. MATEC Web Conf 69:02001

Jatoi AS, Akhter F, Mazari SA, Sabzoi N, Aziz S, Soomro SA, Mubarak NM, Baloch H, Memon AQ, Ahmed S (2020) Advanced microbial fuel cell for waste water treatment—a review. Environ Sci Pollut Control Ser 1–15

Jobby R, Jha P, Gupta A, Gupte A, Desai N (2019) Biotransformation of chromium by root nodule bacteria *Sinorhizobium* sp. SAR1. PLoS ONE 14(7):e0219387

Kazi TG, Arain MB, Jamali MK, Jalbani N, Afridi HI, Sarfraz RA, Baig JA, Shah-Abdul Q (2009) Assessment of water quality of polluted lake using multivariate statistical techniques: a case study. Ecotoxicol Environ Saf 72:301–309

Keshavarzi A, Kumar V (2018) Ecological risk assessment and source apportionment of heavy metal contamination in agricultural soils of North eastern. Iran Int J Environ Health Res 1–17

Kloch M, Toczyłowska-Maminska R (2020) Toward optimization of wood industry wastewater treatment in microbial fuel cells—mixed wastewaters approach. Energies 13:263

Kumar M, Seth A, Singh AK, Rajput MS, Sikandar M (2021) Remediation strategies for heavy metals contaminated ecosystem: a review. Environ Sustain Indicat 12:100–155

Langa C, Hara J, Wang J, Nakamura K, Watanabe N, Komai T (2021) Dynamic evaluation method for planning sustainable landfills using GIS and multi-criteria in areas of urban sprawl with land-use conflicts. PLoS ONE 16(8):e0254441

Larrosa-Guerrero AK, Scott IM, Head F, Mateo A, Ginesta C, Godínez B (2010) Effect of temperature on the performance of microbial fuel cells. Fuel 89:3985–3994

Liang P, Duan R, Jiang Y, Zhang X, Qiu Y, Huang X (2018) One-year operation of 1000-L modularized microbial fuel cell for municipal wastewater treatment. Water Res 141:1–8

Lu L, Lobo FL, Xing D, Ren ZJ (2019) Active harvesting enhances energy recovery and function of electroactive microbiomes in microbial fuel cells. Appl Energy 247:492–502

Madhav S, Mishra R, Kumari A, Srivastav AL, Ahamad A, Singh P et al (2024) A review on sources identification of heavy metals in soil and remediation measures by phytoremediation-induced methods. Int J Environ Sci Technol 21(1):1099–1120

Marashi SKF, Kariminia HR (2015) Performance of a single chamber microbial fuel cell at different organic loads and pH values using purified terephthalic acid wastewater. J Environ Health Sci Eng 13:27

Marassi RJ, Queiroz LG, Silva DCVR, da Silva FT, Silva GC, de Paiva TCB (2020) Performance and toxicity assessment of an up-flow tubular microbial fuel cell during long-term operation with high-strength dairy wastewater. J Clean Prod 259:120882

Marco DE, Abram F (2019a) Using genomics, metagenomics and other "omics" to assess valuable microbial ecosystem services and novel biotechnological applications. Front Microbiol 10:151

Marco DE, Abram F (2019b) Using genomics, metagenomics and other "omics" to assess valuable microbial ecosystem services and novel biotechnological applications. Front Microbiol 10:151

Mehravanfar H, Mahdavi MA, Gheshlaghi R (2019) Economic optimization of stacked microbial fuel cells to maximize power generation and treatment of wastewater with minimal operating costs. Int J Hydrog Energy 44:20355–20367

Mishra S, Maiti A (2018) The efficacy of bacterial species to decolourise reactive azo, anthroquinone and triphenylmethane dyes from wastewater: a review. Environ Sci Pollut Res 5:8286–8314

Mohanakrishna G, Abu-Reesh IM, Kondaveeti S, Al-Raoush RI, He Z (2018) Enhanced treatment of petroleum refinery wastewater by short-term applied voltage in single chamber microbial fuel cell. Bioresour Technol 253:16–21

Moradian JM, Fang Z, Yang-Chun Y (2021) Recent advances on biomass-fueled microbial fuel cell. Bioresour Bioprocess 8:14

Nguyen HTH, Kakarla R, Min B (2017) Algae cathode microbial fuel cells for electricity generation and nutrient removal from landfill leachate wastewater. Int J Hydrog Energy 42:29433–29442

Nie A, Kung SS, Li H, Zhang L, He X, Kung CC (2021) An environmental and economic assessment from bioenergy production and biochar application. J Saudi Chem Soc 25:101173

Nigri EM, Santos ALA, Rocha SDF (2020) Removal of organic compounds, calcium and strontium from petroleum industry effluent by simultaneous electrocoagulation and adsorption. J. Water Process Eng. 37:101442

Obileke K, Onyeaka H, Meyer EL, Nwokolo N (2021) Microbial fuel cells, a renewable energy technology for bio-electricity generation: a mini-review. Electrochem Commun 125:107003

Pant D, Bogaert GV, Diels L, Vanbroekhoven K (2010) A review of the substrates used in microbial fuel cells (MFCs) for sustainable energy production. Bioresour Technol 101:1533–1543

Parkash A (2016) Microbial fuel cells: a source of bioenergy. J Microb Biochem Technol 8(3):247–255

Peña M, Ducci J, Zamora V (2013) Tratamiento de aguas residuales en México.—Banco Interamericano de Desarrollo, Divisi6n de Agua y Saneamiento, IDB-TN, p 524

Prathiba S, Kumar PS, Dai-Viet NV (2022) Recent advancements in microbial fuel cells: a review on its electron transfer mechanisms, microbial community, types of substrates and design for bio-electrochemical treatment. Chemosphere 286:131856

Ren L, Ahn Y, Logan BE (2014) A two-stage microbial fuel cell and anaerobic fluidized bedmembrane bioreactor (MFC-AFMBR) system for effective domestic wastewater treatment. Environ Sci Technol 48:4199–4206

Rozendal RA, Hamelers HVM, Buisman CJN (2006) Effects of membrane cation transport on pH and microbial fuel cell performance. Environ Sci Technol

Saeed HM, Husseini GA, Yousef S, Saif J, Al-Asheh S, Abu Fara A, Azzam S, Khawaga R, Aidan A (2015) Microbial desalination cell technology: a review and a case study. Desalination 359:1–13

Sawasdee V, Pisutpaisal N (2016) Simultaneous pollution treatment and electricity generation of tannery wastewater in air-cathode single chamber MFC. Int J Hydrog Energy 41:15632–15637

Searchinger T, Heimlich R, Houghton RA, Dong F, Elobeid A, Fabiosa J, Tokgoz S, Hayes D, Yu T H (2008) Use of U.S. croplands for biofuels increases greenhouse gases through emissions from land-use change. Science 319(5867):1238–1240. https://doi.org/10.1126/science.1151861

Sekar AD, Jayabalan T, Muthukumar H, Chandrasekaran NI, Mohamed SN, Matheswaran M (2019) Enhancing power generation and treatment of dairy waste water in microbial fuel cell using Cu-doped iron oxide nanoparticles decorated anode. Energy 172:173–180

Shanmuganathan P, Rajasulochana P, Ramachandra A (2018) Factors affecting the performance of microbial fuel cell. Int J Mech Eng Technol 9(9):137–148

Sharma P (2021) Efficiency of bacteria and bacterial assisted phytoremediation of heavy metals: an update. Bioresour Technol 124835

Shi J, Huang W, Han H, Xu C (2021) Pollution control of wastewater from the coal chemical industry in China: environmental management policy and technical standards. Renew Sustain Energy Rev 143:110883

Singh P, Jain R, Srivastava N, Borthakur A, Pal DB, Singh R et al (2017) Current and emerging trends in bioremediation of petrochemical waste: a review. Crit Rev Environ Sci Technol 47(3):155–201

Singh HM, Pathak AK, Chopra K, Tyagi VV, Anand S, Kothari R (2019) Microbial fuel cells: a sustainable solution for bioelectricity generation and wastewater treatment'. Biofuels 10:11–31

Srivastava P, Abbassi R, Yadav AK, Garaniya V, Khan F (2020) Microbial fuel cell–integrated wastewater treatment systems. In: Integrated microbial fuel cells for wastewater treatment. Elsevier, Amsterdam, pp 29–46

Tarekegn MM, Zewdu SF, Ishetu AI (2020) Microbes used as a tool for bioremediation of heavy metal from the environment. Cogent Food Agric 6(1):1783174

Uddin MJ, Jeong YK, Lee W (2021) Microbial fuel cells for bioelectricity generation through reduction of hexavalent chromium in wastewater: a review. Int J Hydrog Energy 46:11458–11481

Umar MF, Rafatullah M, Abbas SZ, Mohamad Ibrahim MN, Ismail N (2021) Advancement in benthic microbial fuel cells toward sustainable bioremediation and renewable energy production. Int J Environ Res Public Health 18:3811

Varjani S, Pandey A, Upasani VN (2021) Petroleum sludge polluted soil remediation: integrated approach involving novel bacterial consortium and nutrient application. Sci Total Environ 750:142934

Vélez-Pérez LS, Ramirez-Nava J, Hernández-Flores G, Talavera-Mendoza O, Escamilla-Alvarado C, Poggi-Varaldo HM, Solorza-Feria O, López-Díaz JA (2020) Industrial acid mine drainage and municipal wastewater co-treatment by dual-chamber microbial fuel cells. Int J Hydrog Energy 45:13757–13766

Verma K, Saha G, Kundu LM, Dubey VK (2019) Biochemical characterization of a stable azo reductase enzyme from *Chromobacterium violaceum*: application in industrial effluent dye degradation. Int J Biol Macromol 121:1011–1018

Yadav RK, Chiranjeevi P, Sukarampal-Patil SA (2020) Integrated drip hydroponics-microbial fuel cell system for wastewater treatment and resource recovery. Bioresour Technol Rep. 9:100392

Yin K, Wang Q, Lv M, Chen L (2019) Microorganism remediation strategies towards heavy metals. Chem Eng J 360:1553–1563

Zhang D, Yang F, Shimotori T, Wang KC, Huang Y (2012) Performance evaluation of power management systems in microbial fuel cell-based energy harvesting applications for driving small electronic devices. J Power Sources 217:65–71

Zhao N, Jiang Y, Alvarado-Morales M, Treu L, Angelidaki I, Zhang Y (2018) Electricity generation and microbial communities in microbial fuel cell powered by macroalgal biomass. Bioelectrochemistry 123:145–149

Chapter 13
Artificial Intelligence (AI) as a Cutting-Edge Technology in Municipal Solid Waste Management

Naveen Kumar, Devesh Singh, Piyush Sharma, and Pankaj Kumar Chaurasia

Abstract Urbanization and population growth have resulted in a global solid waste crisis, where municipal solid waste (MSW) is expected to become 3.40 billion tons per year by 2050. Inadequate waste management and traditional treatment methods endanger the environment, causing pollution, ecosystem destruction, human health problems, and resource depletion, especially when it is stored in densely populated areas or near water or sewage systems. Complex nonlinear parameters make traditional methods difficult to model and optimize. Hence there is a need for reliable technologies for efficient waste monitoring, planning and management. In this context, artificial intelligence (AI) as well as machine learning (ML) technologies has become a powerful tool as alternative computational approaches for managing solid waste. Nowadays it is an upcoming technology and is entering the waste treatment sector due to its efficiency, speed and autonomy from human operation. Various AI and hybrid technologies have been fruitfully working in various MSW management areas, like generation, collection, transportation, treatment and final disposal of waste. Artificial neural networks, decision trees, support vector machines, and genetic algorithms are the predominantly used AI models in the field of solid waste management systems. This chapter focuses on the potential of AI in forecasting of waste generation, classification, sorting, smart bins detection, vehicle routing, and resource recovery, MSW

N. Kumar (✉)
Department of Chemistry, Sanatan Dharm College Muzaffarnagar, Muzaffarnagar, Uttar Pradesh, India
e-mail: naveenkumariitr@gmail.com

D. Singh
Department of Chemistry, DAV College Muzaffarnagar, Muzaffarnagar, Uttar Pradesh, India

P. Sharma
Department of Zoology, Sanatan Dharm College Muzaffarnagar, Muzaffarnagar, Uttar Pradesh, India

P. K. Chaurasia
P.G. Department of Chemistry, L.S. College, B.R. Ambedkar Bihar University, Muzaffarpur, India

S. Madhav et al. (eds.), *Municipal Sewage and Sludge: Treatment and Disposal Strategies*, Environmental Science and Engineering,
https://doi.org/10.1007/978-3-032-07597-0_13

management planning and final disposal with cost-savings and process-efficiency benefits.

Keywords Municipal solid waste · Artificial intelligence · Machine learning · Waste monitoring · Vehicle routing · Smart bins detection

13.1 Introduction

Municipal solid waste (MSW) is the solid and semisolid garbage generated in urban areas. Global population growth, industrialization as well as rapid urbanization have resulted in huge rise in MSW generation. This waste consisting of garbage, refuse and litter, is generated in a variety of domestic and industrial sectors and is primarily composed of organic matter, plastic, glass and metal. The yearly worldwide solid waste production ranges from 7 to 9 billion tonnes, of which MSW accounted for approximately 2 billion tonnes in 2016 (Ihsanullah et al. 2022; Qureshi et al. 2023). World Bank suggests a significant enhancement in global MSW production, from 2.01 billion tonnes in 2018 to 3.40 billion tonnes by 2050 (Abdallah et al. 2020). Unfortunately, globally more than 43% of solid waste generated is mismanaged and disposed of in illegal dumpsites, incineration and open burning (Samadhiya et al. 2025). These improper and inefficient waste discarding create several environmental and health problems such as surface and groundwater pollution, ecosystem destruction, depletion of natural resources, risk of cancer, congenital disabilities and birth anomalies (Triassi et al. 2015; Fang et al. 2023). Further, these improper waste practices are also related to climate alteration problems (e.g. greenhouse gas emissions etc.) (Askr et al. 2024; Toha and Rahman 2023). Thus, effective and efficient MSW management is the most serious and difficult challenges in the world today.

The most important focus of the MSW management system is to diminish adverse effects of MSW on the eco-health while conserving resources. Poor MSW management is caused by lack of knowledge, inadequate waste management planning, absence of sustainable waste management strategies, lack of strict regulation and insufficient funding (Andeobu et al. 2022). Therefore, there is a need of effective and more sustainable MSW management systems. In this context, Artificial Intelligence (AI) technology is recently been integrated into MSW management for sustainability and profitability (Abdallah et al. 2020; Ihsanullah et al. 2022; Xia et al. 2022; Mounadel et al. 2023; Olawade et al. 2024; Samadhiya et al. 2025). Many developed countries used AI techniques in the waste generation, sorting, collection, processing and final disposal with optimization of resource utilization (Andeobu et al. 2022; Fang et al. 2023). The importance of this chapter is that it offers comprehensive analysis of the impact of AI in revolutionizing MSW management at global level and provides valuable insights for policy makers, researchers, industry professionals by incorporating current research and development. Thus this study may provide a platform for further innovations in AI-driven waste management by identifying technology gap and promising areas for innovation. The present chapter is focussed to discuss the application of AI tools in predictive waste production models, waste-sorting robots, intelligent bin systems, sensor-based waste monitoring, and MSW treatment practices. The chapter also discusses the advantages, limitations and challenges in implementing AI methods and presents future recommendations.

13.2 Overview of Artificial Intelligence and AI Models

Artificial intelligence (AI), also known as machine intelligence, that allow machine to learn with human-level intelligence, observe, understand and act (Andeobu et al. 2022; Abdallah et al. 2020). AI approaches have been used in mathematics, statistics, data science, computer science, medical science, agriculture, geology, water treatment, information engineering, environmental engineering, space exploration, and many more (Malviya and Jaspal 2021; Ihsanullah et al. 2022). Likewise, AI technologies have been used to improve the MSW management processes (Fig. 13.1). Moreover, these are also used in a variety of MSW disposal processes like composting, landfilling and incineration to aid in energy recovery, which can contribute to sustainable progress by reducing waste and utilizing resources (Ihsanullah et al. 2022; Abdallah et al. 2020; Xia et al. 2022; Mounadel et al. 2023; Olawade et al. 2024; Samadhiya et al. 2025).

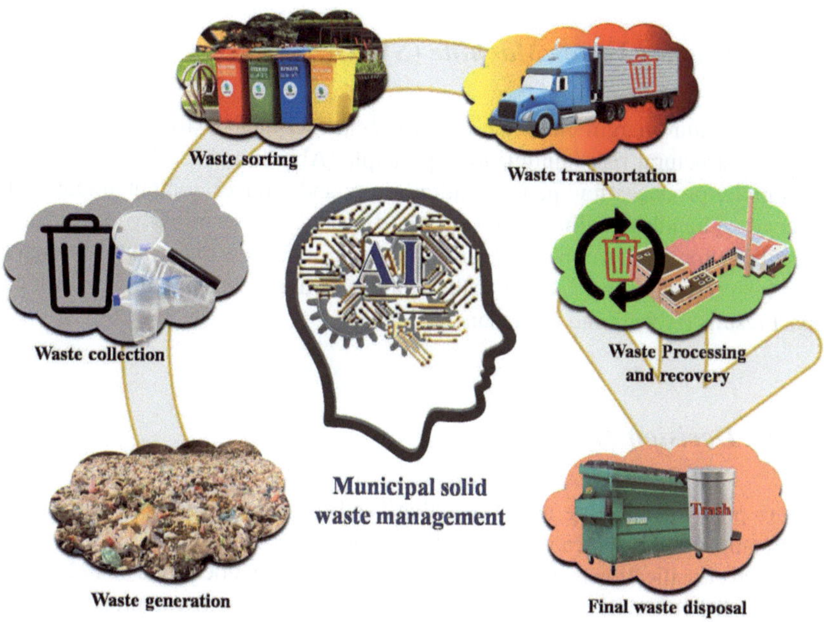

Fig. 13.1 Applicationof AI technologies in MSW management

Various AI models are being utilized to forecast, classify and treatment of MSW and each model has its own advantages and limitations (Abdallah et al. 2020; Ihsanullah et al. 2022). Table 13.1 summarizes the advantages and limitations of frequently used AI models in MSW management.

13.2.1 Artificial Neural Networks (ANNs)

It is a popular mathematical tool for modelling and controlling environmental processes and has been widely used in different stages of the MSW management processes. ANNs can mimic the brain's neural connections and recognize patterns in data and be trained from them. These networks contain three layers (input, hidden and output layers). Input layers represent sensory receivers, hidden layers mimic the biological neural networks, and output layers represent decision outputs (Oliveira et al. 2019). Feed-forward multilayer ANNs consist of multiple layers of neurons, each of which processes information before passing it on to the next (Oliveira et al. 2019). ANN has many advantages over other traditional methods, such as the ability to understand dynamics through the use of grouped variables. However, ANNs have inherent limitations, such as the requirement for a specific network structure and a specific network parameter (Mounadel et al. 2023).

13.2.2 Support Vector Machine (SVM)

It is a maximum margin classifier which uses kernel method for data separation. It follows structural risk minimization principle (Abbasi and El Hanandeh 2016). It is simple, produces few generalization errors and reduces computational costs. Moreover, it handles nonlinear data effectively, reduces susceptibility to overfitting and suitable for small sample problems. However, it has few disadvantages such as needs long training time, feature scaling, and sensitive to missing data, which can impact results (Abbasi and El Hanandeh 2016).

13.2.3 Decision Trees (DT)

DTs are popular modeling techniques that handle non-linear data, simplify result interpretation, and produce results similar to those of a human professional. It is a supervised classification method for extracting rules from unidentified data (Kannangara et al. 2018). It reduces the computational costs and easy to interpret and understand the results. However, DTs often overfit data and require more time for training (Kannangara et al. 2018; Abdallah et al. 2020).

Table 13.1 Advantages and disadvantages of frequently used AI models in MSW management (Ihsanullah et al. 2022; Abdallah et al. 2020; Xia et al. 2022; Mounadel et al. 2023; Samadhiya et al. 2025)

AI model	Advantages	Disadvantages
Artificial neural network (ANN)	1. Energetic to noise in training data 2. Fault tolerance ensures functionality even with faults in cells 3. Distributed memory system 4. Suitable for modeling intricate multivariate systems 5. Strong robustness and reliability 6. Effective for nonlinear relationships 7. Requires fewer calibration parameters than deterministic models	1. Requires high processing power and multiple processors 2. Unexplained and undefined network behavior and structure 3. Weak in handling logical and arithmetic problems requiring precision 4. Necessitates data pre-processing and struggles with non-stationary datasets 5. Prone to overfitting and correlation challenges among variables 6. Results are difficult to interpret, with a lack of clarity and interpretability 7. High parameter requirements
Support vector machine (SVM)	1. Effective in maximizing class separation while minimizing classification errors 2. Performs well in high-dimensional spaces with minimal computational cost 3. Handles nonlinear data effectively and avoids local minima 4. Exhibits low generalization error and reduces susceptibility to overfitting 5. Memory efficient and suitable for small sample problems 6. Simultaneously minimizes error estimates and model dimensions	1. Not suitable for large or noisy datasets 2. Highly sensitive to kernel selection and tuning variables 3. Needs feature scaling for effective performance 4. Long training times are often required 5. Sensitive to missing data, which can impact results

(continued)

Table 13.1 (continued)

AI model	Advantages	Disadvantages
Decision trees (DT)	1. Requires minimal effort for data preprocessing, normalization, and scaling 2. Results are easy to interpret and understand 3. Handles missing values and irrelevant features effectively 4. Low computational costs and high efficiency	1. Requires more time for training and has high computational complexity 2. Unsuitable for prediction with continuous values 3. Not ideal for large datasets 4. Prone to overfitting 5. Ignores feature correlation
Genetic algorithms (GA)	1. Easy to understand and program 2. Performs well with both discrete and continuous data 3. Effectively handles local minima and maxima 4. High precision and accuracy 5. Supports multi-purpose optimization	1. Computationally expensive and time-consuming 2. Cannot utilize gradients 3. Requires meticulous design, as incorrect operator selection can adversely affect results
Linear regression (LR)	1. Easy interpretation of results 2. Low computational costs	1. Not suitable for modeling nonlinear data
Convoluted neural network (CNN)	1. High image recognition accuracy 2. Extracts features from images automatically	1. Require more data 2. Unable to encoding object orientation 3. Needs parameter tuning 4. Relies solely on information extracted from the image
Recurrent neural network (RNN)	1. Uses a parameter-sharing approach 2. Good for time series prediction 3. Involves recurrent connections 4. Effective for sequence data and memory maintenance	1. Hard to train RNN 2. Incapable with long sequences of data 3. Prone to gradient exploding and vanishing problems 4. Requires significant computational resources
K nearest neighbor (kNN)	1. Easy to understand and implement 2. Fast convergence 3. Few parameters needed	1. Not suitable for large data sets 2. Unable with high-dimensional data 3. Requires feature scaling 4. Sensitive to noise and outliers
Adaptive neuro-fuzzy inference system (ANFIS)	1. Merges benefits of neural networks and fuzzy logic	1. Not ideal for high-dimensional features 2. Computationally intensive and time-consuming

(continued)

Table 13.1 (continued)

AI model	Advantages	Disadvantages
Multi-layer perceptron (MLP)	1. Easy to set up and train 2. Suitable for large datasets 3. Supports adaptive learning 4. Good for nonlinear problems	1. Computationally more complex 2. Prone to overfitting 3. Can cause redundancy due to many parameters
Particle swarm optimization (PSO)	1. Capable of performing parallel computation 2. Easy to use 3. Vigorous control above parameters	1. Complexity in analysis and evaluation 2. Weak in local searchability for optimal solutions

13.2.4 Genetic Algorithms (GA)

It is a type of metaheuristic search algorithm that mimics natural evolution by using historical data (Abdallah et al. 2020). Crossover, mutation, and selection of the fittest are the three key components of GA. It start with a fixed population size representing potential solutions to a problem, and then combine two parents of individuals to create new offspring. The crossover operator integrate two individuals' parents, whereas the mutation operator randomly changes certain genes on a chromosome until the optimal solution is found (Oliveira et al. 2019; Mounadel et al. 2023). It is easy to programming, understanding, with high precision. However, it is computationally expensive, time-consuming and requires meticulous design (Abdallah et al. 2020).

13.2.5 Linear Regression (LR)

It can be used for single linear regression or multiple variables and multiple linear regressions are more suitable for MSW management. This technique models a target value using independent predictors (Chu et al. 2022). It is computationally inexpensive and easy to interpret and understand the results. However, it is unsuitable for modeling nonlinear data (Andeobu et al. 2022).

13.2.6 Hybrid Techniques

These techniques integrated multiple AI techniques and outperformed the individual AI technique (Ihsanullah et al. 2022). For example, ANN was integrated with GA, MLP, PSO, Sine cosine algorithm (SCA), Archimedes optimisation algorithm (AOA) etc. to enhance the forecast accuracy of MSW generation (Liang et al. 2021; Samadhiya et al. 2025). Likewise, SVM was integrated with Fuzzy logic (FL), Least square

(LS), and Wavelet theory (WT) to predict MSW generation (Samadhiya et al. 2025; Ihsanullah et al. 2022). ANFIS, which combines fuzzy logic and ANN, is a popular hybrid technique used in MSW management. Hybrid systems are able to analyze huge amount of nonlinear data from dynamic systems (Abbasi and El Hanandeh 2016).

13.3 AI Applications Approaches in MSW Management

Prediction of waste generation, collection, sorting as well as recycling of waste is the major step in the MSW management cycle. AI technologies have been used in all stages of MSW management including recycling and recovery of energy from waste, which is essential for sustainable development (Ihsanullah et al. 2022; Abdallah et al. 2020; Xia et al. 2022; Mounadel et al. 2023; Olawade et al. 2024; Samadhiya et al. 2025). Figure 13.2 depicts the application of the AI technology in all stages of MSW management.

13.3.1 Prediction of MSW Generation

Accurate projections of MSW quantities are essential for planning to efficient waste scheduling, storage, collection, vehicle routing, waste recycling and treatment. It can help to implement waste management policies, and suggest recycling and recovery

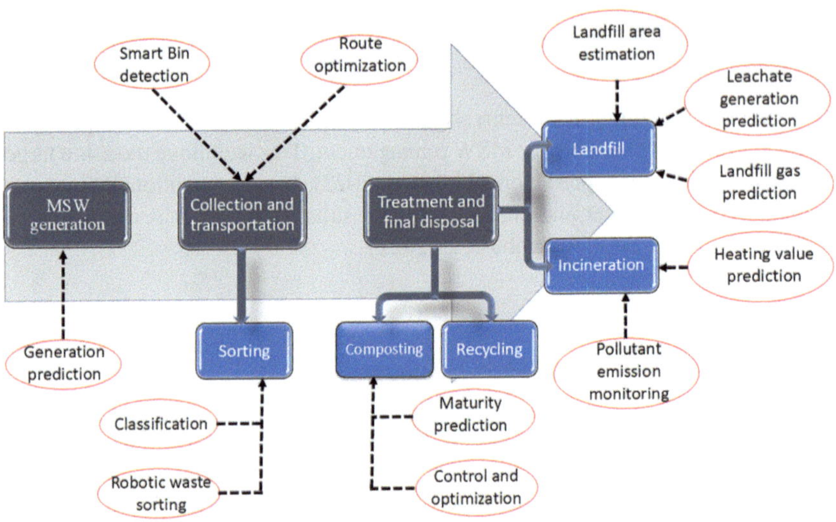

Fig. 13.2 Applications of the AI technique in each stage of the MSW management processes

opportunities (Kannangara et al. 2018; Soni et al. 2019; Abbasi and El Hanandeh 2016). The generation of MSW influenced by various factors including total population, electricity consumption, administrative, socio-economical, technical, political, legal, cultural, tourists' influx, environmental as well as climate related factors such as CO_2 emissions, rainfall, minimum or maximum temperature etc. (Jassim et al. 2022; Liang et al. 2021; Lin et al. 2021). To predict MSW in mainland China, Wu et al. (2020) developed ANN model. They used disposable income, GDP per capita, and common public budget expenditure as inputs parameters and it was observed that regional differences significantly impact MSW prediction accuracy with more than 0.94 in R^2 (determination coefficient) and less than 20 in RMSE (root means square error) (Wu et al. 2020). Ayeleru et al. conducted a long-term prediction (up to 2050) study comparing the performance of 11 ML (6 SVMs and 5 ANNs) algorithms to forecast annual MSW generation in Johannesburg (South Africa). The results showed a linear increase in MSW generation over time which is expected to reach 1.95×10^6 tonnes by 2050 with an R^2 value up to 99.9% (Ayeleru et al. 2021). Likewise, Chhay et al. (2018) showing the forecast of 24,666.65 tons of MSW generation in China by 2030 using ANN model with $R^2 = 0.931$, RMSE = 450.84, MAPE (Mean absolute percentage error) = 0.0143%, and MAE (Mean absolute error) = 228.53. They used FL, GM (1, 1), and LR models also but ANN model outperformed the others (Chhay et al. 2018). Kannangara et al. (2018) applied ANN and DT models to guess the MSW production in residents of 220 municipalities in Ontario, Canada and reported that ANN model showed better performance than DT with R^2 value of 0.72 (Kannangara et al. 2018). In a study in Logan City, Australia, SVM, ANN, ANFIS, and kNN models showed well performance (R^2 value of > 0.8) to guess monthly waste generation. However, the ANFIS model outperformed the other models with $R^2 = 0.98$, MAE = 52.16, RMSE = 175.88, and MAPE = 0.008 (Abbasi and El Hanandeh 2016). To predict the amount of MSW in Shanghai Lin et al. (2021) proposed a new hybrid neuronal network structure of 1D-CNN and Long Short-Term Memory (LSTM) denoted by 1D-CLA, which outperformed any single model with 0.95 of R^2 value (Lin et al. 2021). Liang et al. (2021) integrate inclusive multiple model (IIM) with various hybrid models to forecast monthly solid waste production in Iran. The proposed approach showed better prediction performance with R^2 value of 0.93, as it integrated the advantages of all models (Liang et al. 2021). Soni et al. (2019) used ANN and some hybrid models to predict the garbage generation in New Delhi (India). They calculated index of agreement (IA) (0.0982 to 0.864), RMSE (95.7 to 393.8), and R^2 (0.36 to 0.87), values for each model, and reported that the GA-ANN hybrid model was found to be the most accurate with the highest IA (0.864), lowest RMSE (95.7), and highest R^2 (0.87) values (Soni et al. 2019). To guess the yearly domestic packaging waste from 42 municipalities in Portugal, Oliveira et al. (2019) used ANN model with GA using various variables such as population education, age, size, economic power, poverty, urbanization, and waste collection service. The ANN model outperforms the best regression model by 34%, with a R^2 value of 0.98, indicating superior explanatory power compared to traditional regression techniques ($R^2 = 0.73$)(Oliveira et al. 2019). To forecast the amount of hospital waste in Karaj metropolis (Iran) Golbaz et al. (2019) used various

Neuron- and Kernel-based ML methods which achieved R^2 value of 0.82–0.86 and MSE value of 0.003–0.008 (Golbaz et al. 2019). To predict the MSW amount in Bahrain MLP-ANN and RBF-SVR (Radial basis function- Support Vector Regression) models was used by Jassim et al. (2022). The RBF-SVR model outperformed the MLP-ANN model with R^2 and RMSE values of 0.97 and 0.164 respectively. The MLP-ANN model was also effective with an R^2 and RMSE values of 0.94 and 0.151(Jassim et al. 2022). Ghanbari et al. (2021) used ANN, RF (Random Forest), and MARS (Multivariate adaptive regression splines) methods to estimate the solid waste generated in Tehran, Iran. The MARS model outperformed other standalone models. Further to increase the model's accuracy the solid waste generation data was analysed with crow search algorithm (CSA). Finally the performance of optimized hybrid model (MARS-CSA) was convincing, with $R^2 = 0.82$, RMSE $= 10.908$ and Nash–Sutcliffe efficiency (E_{NS}) $= 0.56$ (Ghanbari et al. 2021). Nguyen et al. compared the performance of SVM, RF, LM, Cubist, KNN, and DNN (Deep Neural Network) for MSW amount in housing areas of Vietnam and reported that the RF and KNN algorithms had superior projecting ability with an R^2 value > 0.96 (Nguyen et al. 2021). Niu et al. used ARIMA (Autoregressive integrated moving average), LSTM, and ANN models for MSW production forecasting in Suzhou, China and the LSTM outperforming the ARIMA and traditional ANN models with 0.92 and 935.08 R^2 and RMSE values, respectively (Niu et al. 2021).

13.3.2 Collection and Transportation of Waste

13.3.2.1 Smart Bin Systems

Maintenance of cleanliness around bins, prevention of illegal waste dumping, and prevention of overflowing is essential. Overflow bins can cause nuisances like smell, flies, and scavenging animals. Overflow bins can also lead to residents setting fires, which pollute the environment with toxic smoke (Hannan et al. 2013). To address these problems advanced information technologies i.e. wireless sensor networks (WSN), Internet of Things (IoT), global system for mobile (GSM), Geographical information system (GIS), global positioning systems (GPSs) etc. provide immediate action by monitoring trucks as well as bins in real-time (Hannan et al. 2013; Praveen et al. 2020; Ramson et al. 2021; Rahman et al. 2022; Vishnu et al. 2022; Arthur et al. 2024). These modules of the smart bin send a message to authorities for garbage collection, if communication is urgent. This mechanism helps determine the status of the bins, dangerous objects, or gas leaks by facilitates device location or tracking (Vishnu et al. 2022; Arthur et al. 2024). Smart bin systems are integrating advanced sensors such as Ultrasonic, Infrared, Electronic, Metal, Plastic, Gas, Weight, Rain, Temperature and Humidity sensors to ensure accurate data on fill levels. The IoT module connects to sensors of the garbage bin through a microcontroller for collecting and processing information to send to cloud storage, gateways, web applications, and garbage collectors (Rahman et al. 2022). This information is transmitted via wireless

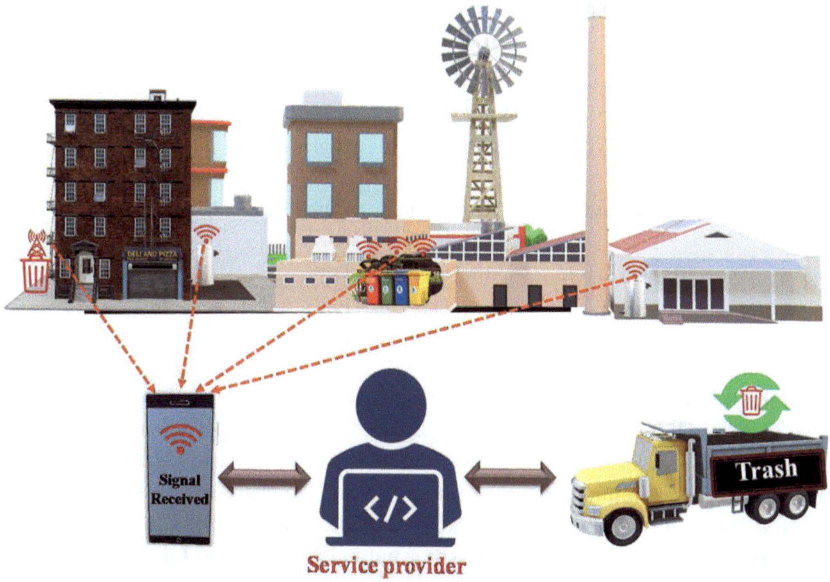

Fig. 13.3 Applications of IoT based technologies in MSW management

communication methods such as Bluetooth, Wi-Fi, Zigbee, Long Range (LoRa) Wide Area Network (LoRaWan), etc. (Ramson et al. 2021; Vishnu et al. 2022; Qureshi et al. 2023). Figure 13.3 illustrates the general flow process of the technological-based IoT model for MSW management.

Bluetooth is wireless and short-range technique intended to device-to-device connections. It is the fastest way through which smart bins can communicate with other bins or movable users for data to be transmitted consecutively to a mobile app or cloud storage (Laha et al. 2023). Wi-Fi offers high-speed wireless internet connectivity in specific locations. It simplifies the process of setting up communications by transferring data to a specific app or cloud server using the available Wi-Fi networks (Laha et al. 2023). ZigBee communication is a short-range, low-power, and low-data-rate protocol designed for IoT applications. It uses a mesh topology in which several devices can work as routers (Laha et al. 2023). LoRa (Long Range) is the preferred mechanism of communication of IoT-enabled smart bins because it offers long-range data transmission with low energy consumption and high-quality service (Laha et al. 2023; Vishnu et al. 2022). LoRaWAN, is based on Long Range (LoRa) Wide Area Network (LoRaWan) protocol, offers a nice stability among long-range communication, low power utilization, and scalability (Table 13.2). Even image processing solutions have also been developed for the solid waste collection. However, in the capture of bin images, it is difficult to position the camera so that there is a bin area's centered image obtained (Laha et al. 2023). In this context, Islam et al. (2014) proposed an effective image processing algorithm for solid waste collection using a MLP as classifier, Gabor wavelet (GW) and Dynamic time warping (DTW) as texture

Table 13.2 An overview of comparison of multiple network system (Laha et al. 2023)

Network name	Range (km)	Energy consumption	Transmission consumption	Minimum coupling loss
Bluetooth	Up to 0.1	Low	Low	– 90 to – 100
Wi-Fi	Up to 0.1	Moderate to high	High	40 to – 80
Cellular (4G/5G)	Several	Moderate to high	High	– 20 to – 60
ZigBee	Up to 0.15	Low	Low	– 100 to – 120
LoRaWAN	Several	Low	Low	– 140 to 148

extraction and bin area recognition. The approach was successful in predicting the status at the bin level and produced very accurate results with 98.5% accuracy (Islam et al. 2014). Hannan et al. proposed a system for bin level detection using RFID, GPS, GPRS, and GIS communication technologies with Gabor wavelet (GW) filter which achieved 94.29% classification accuracy (Hannan et al. 2013). Likewise, Hannan et al. (2016) used kNN and MLP classifier with GW filters and GLAM (Gray level aura matrix) as feature extraction for the bin level classification and reported that the KNN classifier outperformed the MLP with accuracy of 97 and 95%, respectively (Hannan et al. 2016). Rehman et al. (2022) developed a system which employed an architectural model using Raspberry Pi and a camera module for waste classification. They used an IoT-based smart trash box and CNN containing Bluetooth connectivity to monitor waste in real time. The accuracy of the system was 95.3125%, with an 86% System Usability Scale score (Rahman et al. 2022).

13.3.2.2 Route Planning

Advanced techniques with deep learning predict traffic patterns and adjust the operations according to dynamic conditions, helping to minimize costs and travel time and provide efficient waste collection services (Vu et al. 2019). For example, Amal et al. developed a model called SGA, which integrates a Spatial Geographic Information System (GIS)-with GA to routes optimization for MSW collection. The SGA integrates a modified Dijkstra's algorithm within the GIS to obtain solutions that minimised vehicle travel time and distance. Subsequently the optimal route is presented using a map interface by ArcGIS software. This approach was verified with real data from Sfax City, Tunisia, and demonstrated improved total travel time compared with existing practical routes (Amal et al. 2018). Vu et al. (2019) studied the impact of waste materials composition on air emissions, truck route time and distance using data from Austin, Texas (USA). They integrated an ANN model with a GIS-based route optimization tool that showed good performance with MAP values of 10.92–16.51%. Modified and non-modified scenarios of waste composition were created, and through GIS analysis, optimal routes with travel distance changes up to 19.9% (Vu et al. 2019). Krol et al. (2016) integrated GA with FL to measure household satisfaction with take-back services by optimization the routes of household e-waste collection to reduce costs by minimizing route distances, number of vehicles, and workforce. The performance of proposed model was convincing which may be helpful for waste collecting companies (Król et al. 2016).

13.3.3 Sorting and Classification of Waste

Waste classification or sorting is very important for effective treatment, disposal, and recycling. By categorizing waste, useful materials can be recycled, improving waste quality and minimizing health and environmental risks. Traditional manual sorting is inefficient and slow, leading to high workloads and errors. Advances in AI and ML have introduced intelligent waste recognition techniques that automate sorting, addressing these challenges (Xia et al. 2022; Samadhiya et al. 2025).

13.3.3.1 Automated Sorting Technologies

Several studies used ANNs for waste classification in automated systems, for example, Tehrani and Karbasi (2017) applied ANNs with hyperspectral imaging to sort different types of plastics in e-waste and achieved a high classification accuracy of 99% (Tehrani and Karbasi 2017). Costa et al. used VGG-16, kNN, SVM and RF models to capture and classify garbage images into paper, plastic glass and metal. The results of VGG-16 model was convincing with 93% accuracy (Costa et al. 2018). Chu et al. (2018) developed an automatic hybrid model by using CNN to extract features from images and using MLP for classification of wastes into recyclables and non-recyclables materials. This strategy achieved classification rates up to 98.2% when testing with 50 different waste items (Chu et al. 2018). Likewise a hybrid Vision Transformer-based Multilayer CNN model outperformed the VGG 19 and ResNET-50 models with respect to recyclables and non-recyclables classification and showed accuracy of 95.8% with the tests on the TrashNet dataset (Alrayes et al. 2023). Kuritcyn et al. (2015) used RF, Nu-SVM and C-LibSVM models to classify construction and demolition waste (CDW). These models showed high total recognition rates of 94.7–97.2% based on the DIN benchmark (Kuritcyn et al. 2015). Vo et al. (2019) developed the DNN-TC model for trash classification. Further the performance of the model enhanced by incorporating modifications to the REsNext architecture. The results demonstrated that the model achieved 94% accuracy on Trashnet and 98% accuracy on VN-trash dataset (Vo et al. 2019). Togaçar et al. designed a model by integrated AutoEncoder structures with CNNs as image processing, Ridge Regression (RR) as feature selection and SVMs as classifier. This model classified waste as organic and recycled types with an excellent accuracy of 99.95% on a public dataset (Toğaçar et al. 2020). Liang et al. designed a Multi-task learning architecture (MTLA) model for multi-label waste recognition and categorization. It achieved an accuracy of 0.9716, precision of 0.9628, recall of 0.9672, and F1 score of 0.9550. Additionally, its average precision for waste localization (@IoU = 0.5) was over 81.50%, highlighting its effectiveness (Liang and Gu 2021). Adedeji et al. used a pre-trained ResNet-50 CNN as a feature extractor and SVM for classifying waste into paper, plastic, glass and metal. The performance of model was convincing with an accuracy of 87% (Adedeji and Wang 2019). Mao et al. (2021) used DenseNet121 model for classification of wastes on TrashNet dataset and the model achieved 99.6% accuracy in optimized conditions. Moreover, Data augmentation techniques, like horizontal and vertical flipping and random 25° rotation, also enhanced the classification efficiency of the model (Mao et al. 2021). Mookkaiah et al. (2022) developed ResNet V2 model which classifies biodegradable wastages with high accuracy. This model showed 94.44% accuracy which was 19.08% higher than other existing models (Mookkaiah et al. 2022). Ma et al. applied Back propagation neural network (BPNN) model to examine the physical composition of MSW in China and demonstrated that the waste has 53.7% organic fraction, 16.9% paper, 13.6% plastic and rubber as major materials with R^2 values of 0.26–0.29 (Ma et al. 2020).

13.3.3.2 Robotic Sorting Systems

Robotic technology can greatly improve waste sorting, especially for waste streams with hazardous substances, by enabling fine sorting without human involvement. Robotic sorting systems use robotic arms with sensors, cameras, and AI for waste sorting automation (Gal et al. 2021; Koskinopoulou et al. 2021). Xiao et al. developed a robotic system by using NIR hyperspectral camera, a 3D camera and a truss-type robot for construction and demolition waste (CDW) classification. This robotic system achieved nearly 100% classification accuracy with a sorting efficiency of 2028 peaks per hour (Xiao et al. 2020). Wang et al. used Faster R-CNN and Mask R-CNN integrated with Simultaneous Localization and Mapping (SLAM) to recognize pipes and cables in CDW management. The SLAM computational framework with RGB-D camera offers robot navigation and 3D reconstruction for effective construction site management (Wang et al. 2020). Similarly, Chen et al. (2022a, b) developed an automatic patrolling robotic system with RGB-D camera and 3D LiDAR for real time navigation using SLAM process for CDW recycling. Further Mask R-CNN recognition model with a 3D grasping strategy enhanced the collection and sorting precision of proposed system (Chen et al. 2022b). Mao et al. used and trained Yolo-v3 model using TrashNet dataset and compared its specialty with the TRWD (Taiwan Recycled Waste) trained database. TRWD-trained Yolo-v3 model performance was convincing with a mean average precision (mAP@0.5) of 92.12% for detecting of household waste (Mao et al. 2022). Leveziel et al. (2022) developed an energy efficient robot by integrated a gripper into the 4-DoF parallel robotic structure. The proposed model minimizes platform inertia and mass (Leveziel et al. 2022). Liu et al. trained a GarbageNet system using grasp localization method with integration on a mobile manipulator. This system showed real time garbage classification, localization and recycling through a whole-body control framework (Liu et al. 2021). Kshirsagar et al. (2022) created an autonomous industrial robot by integration of motion control, gripping, and material classification using an optimized LeNet model for sorting reusable materials such as glass and plastic. It efficiently classified garbage and improved performance with multimodal end effectors, reducing errors (Kshirsagar et al. 2022).

13.3.4 Waste Treatment and Final Disposal

Municipal waste is generally managed through recycling, incineration, composting, as well as landfills. Waste recycling means is collecting, treating, and reusing the waste materials. Composting is the aerobic decomposition of organic matter using microorganisms, producing substrates and fertilizers. Incineration reduces waste volume and eliminates hazardous substances through high-temperature pyrolysis and gasification, converting carbon-based materials into hydrogen and carbon monoxide. Landfilling involves depositing waste into pits with treatments like anti-seepage and drainage to minimize environmental impact (Fang et al. 2023). The choice of MSW treatment technology depends on techno-economic viability, environmental safety and sustainability.

13.3.4.1 Recycling

Recycling helps reduce dependency on raw materials, decrease waste, and minimize landfills. After recycling, the metals may be reutilized. The plastics diverted from landfills, and biodegradable waste used as fertilizers. Diverting plastics from landfills could save energy comparable to 60 million barrels of oil and decrease landfill volume by about 20% (Gupta et al. 2019). AI-driven systems efficiently distinguish recyclables and non-recyclables waste, automate the waste sorting and ensure the effective recycling and treatment of waste. By using advanced sensors, image recognition, robotics, and machine learning, AI enhances sorting accuracy, speed, and adaptability, reducing contamination and increasing recycling efficiency (Olawade et al. 2024). Yan et al. developed laser-induced breakdown spectroscopy (LIBS) based hybrid model to distinguish and characterize inorganic and organic compounds in residual wastes (RW). Under optimal conditions, it achieved prediction accuracies of 96.70% for 'C' content, 92.21% for 'H' content, 87.11% for 'O' content, and 94.28% for the lower heating value (Yan et al. 2021). Ziouzios et al. (2020) trained and used MobileNet model to identify and separate five types of waste and achieved excellent classification accuracy of 96.57% (Ziouzios et al. 2020). Duhayyim et al. used Deep reinforcement learning (DRL) and Deep learning (DL) based hybrid model which was based on Mask RCCN for object detection and DQLN for classification. A DenseNet model serves as the baseline for Mask RCCN along with a hyperparametrs optimization using DFA (Dragonfly algorithm). This approach efficiently detected the different waste objects with different scales and orientations (Duhayyim et al. 2022). In 2011, Zen Robotics, a Finnish company, introduced robotic waste classification using AI, computer vision, and machine learning. Their system filters and retrieves recyclable materials from mobile conveyor belts using metal sensors, 3D laser cameras, and display cameras. They integrated non-reusable channels into the neural network. Other companies in the North American robotic business, like AMP, BHS, and Machinex, also use similar technology. In Spain, Sadako Technologies works on AI-based waste disposal at a lower rate. Clark, a company that launched in 2016, uses AI to locate and separate boxes for recycling at high speeds, improving recycling efficiency and reducing human involvement (Ahmed and Asadullah 2020). Plastic recycling robots, such as the samurAI developed by Machinex, use AI to detect recyclable items on moving belts, including boxes, plastic bottles, and containers. The robot uses suction to remove detected items and place them in appropriate containers, operating at a higher speed (up to 70 items per minute) compared to human workers (35 items per minute). This significantly reduces operating costs and increases productivity. SamurAI has been successfully implemented at recycling facilities like Lockshore Recycling Systems in Illinois. Additionally, laser technology developed by the University of Osaka helps robots accurately distinguish different plastics for recycling (Ahmed and Asadullah 2020). These AI based technologies also enables more accurate extraction of precious materials from waste, leading to innovative recycling methods and higher-quality recycled products (Olawade et al. 2024).

13.3.4.2 Composting

Composting is a cost-efficient and effective method for treating organic MSW, transforming it into hygienic, odourless humus that supports harmlessness, waste reduction, and recycling. It degrades organic components through aerobic processes, producing organic fertilizer for soil improvement (Guo et al. 2021). Efficient composting of sewage sludge is essential, with early maturity identification. Compost maturity determines its safe use in agriculture and suitability for space-saving storage in tall heaps. For compost maturity key evaluation indexes include physical factors like temperature, texture, colour, odour, moisture content etc.). Chemical factors are carbon-to-nitrogen ratio, total nitrogen or carbon content, humus content cellulose and lignin content etc. And biological factors such as microbial biomass microbial activity, enzymatic activity, germination index, and pathogenic bacteria etc. (Gao et al. 2007; Kujawa et al. 2014). Kujawa et al. applied CNN to classify the early maturity of compost, achieving classification errors from 0.51 to 17.77% (Kujawa et al. 2020). Sharma et al. (2021) used ANN model to optimize the maturity parameters of vermicomposting using floral waste and cow manure and the optimum mixture was found to be 1325 g floral waste and 500 g cattle dung. This vermicompost had higher germination index, low CO_2 evolution, low electrical conductivity (3.39 mS/

cm) with a slightly basic pH (7.10), and sufficient nutrients such as potassium, phosphorus, and calcium, making it suitable for agricultural use. In this study the ANN model achieved R^2 value approximately equal to 1 (Sharma et al. 2021). Likewise, Gao et al. (2007) used Wavelet Neural Network (WNN) modeling for compost maturity calculation and the effectiveness of model was convincing with 87.5% accuracy (Gao et al. 2007). Xue et al. developed a compost dataset by collecting about 30,000 images from different composting stages and used CNN to estimate the compost maturity. The findings showed the model achieved > 99.5% accuracy rates (Xue et al. 2019). During the process of composting harmful gases like CH_4, N_2O, CO_2 and NH_3 are released, which is affected by temperature, pH, air, C/N ratio etc. and lead to global warming, soil acidification, eutrophication and environmental pollution (Díaz et al. 2012; Li et al. 2022). To deal with these problems, AI approaches have been used to predict toxic gas emissions, pollutants and optimize the composting process. For example Díaz et al. used ANFIS for optimal composting conditions and reported that the low aeration (0.2 L air $(\text{min kg})^{-1}$), medium moisture (55%), and medium-to-low particle size (1–3 cm) produced the best pH, temperature, CO_2 evolution conditions (Díaz et al. 2012). To predict NH_3 emissions during composting of a mixture of sewage sludge, cereal straw, sawdust and tree bark Boniecki et al. (2012) applied ANN modeling which achieved R^2 values of 0.972–0.981 (Boniecki et al. 2012). Likewise, to forecast the CO_2 emissions from green waste composting RF model achieved 88% accuracy (Li et al. 2022). In a study during vermicomposting of MSW the recovery of total nitrogen and total phosphorus was studied by Hosseinzadeh et al. using ANN and MLP models. The ANN model achieved higher R^2 values of 0.9983 and 0.9991 for total nitrogen and total phosphorus recovery, respectively (Hosseinzadeh et al. 2020).

13.3.4.3 Incineration

This process provides effective energy recovery by burning waste at high temperatures of 800–1000 °C with minimal land requirement in short treatment time (Kumar et al. 2024). It reduces waste volume more than 85% and mass by more than 75%, making it a key method of MSW disposal (Chen et al. 2022a). Additionally, it supports energy generation through waste-to-energy (WTE) conversion, which addresses waste management and provides valuable energy resources (Bagheri et al. 2019). However, it emits harmful pollutants, including toxic heavy metals, with greenhouse gases (Kumar et al. 2024). The thermal efficiency of MSW treatment by incineration method depends heavily on its heating value (HV) (Bagheri et al. 2019). Therefore, accurate, reliable, and timely heating value data is essential for optimal designing and operating WTE systems. In this context, Olatunji et al. (2019) developed an MLP-ANN model using Levenberg Marquardt (LM) backpropagation and Resilience backpropagation (RP) algorithm for high heating value (HHV) prediction of MSW. The LM-based model outperformed the RP-based model with R^2 value of 0.986 and RMSE value of 3.095 (Olatunji et al. 2019). Likewise in a study in Taiwan, the low heating value (LHV) of MSW was predicted by MLP-ANN hybrid model which demonstrated good performance 98% accuracy with 0.93 of R^2 value (Shu et al. 2006). Tao et al. (2020) integrated IR spectroscopy with SVM to forecast LHV of waste and achieved 92.49% accuracy. The proposed method utilized a poly kernel for SVM model and also predicted the elemental composition of waste with 95.54, 92.40% and 85.53% accuracies for carbon, hydrogen and oxygen content, respectively (Tao et al. 2020). Bagheri et al. (2019) used FFNN (Feed-forward neural network), SVM, and GEP (Gene expression programming) modeling to predict the HHV of MSW and achieved high precision with R^2 values of 0.978, 0.973, and 0.966 respectively, and RMSE values of 1.25, 1.44, and 1.57respectively (Bagheri et al. 2019). To predict the LHV of MSW Birgen et al. (2021) applied Gaussian Processes Regression modeling which showed high accuracy in both unoptimized (MAE = 0.592 MJ/kg, MAPE = 5.23%) and optimized (MAE = 0.0.688 MJ/kg, MAPE = 6.05%) condition (Birgen et al. 2021). In a study the ANN model showed good HHVs prediction of MSW with a relative deviation of only 9.76% (Zhu and Yang 2022). Taki et al. studied the performance of RBF-ANN, MLP-ANN, SVM, and ANFIS for prediction of HHV of MSW and reported the MAPE values of 0.45%, 7.3%, 11.77% and 23.76% respectively. The performance of RBF-ANN was convincing with its best topology (6-17-1) and spread factor of 0.8 (Taki and Rohani 2022). Chen et al. designed a hybrid model based on feature selection and CNN-BiLSTM which offered more accurate and efficient results for the incineration processes in WTE plants (Chen et al. 2022a). You et al. examined the HVs forecasting performance of MLP, SVM, ANFIS and RF model and reported that ANFIS model outperformed other models with 94% accuracy; but, it had the longer training process. However, all other models also demonstrated quick and precise HV predictions (You et al. 2017). In addition, Cho et al. (2021) used computational fluid dynamics simulations with ANN to minimize the NOx emissions in a fluidized bed reactor designed for TNT explosive waste disposal. The applied model achieved 99% accuracy and analysis showed that the gas velocity and bed temperature as the most impactful on NOx emissions (Cho et al. 2021).

13.3.4.4 Landfills

Landfills are the predominant waste disposal method in many countries due to their ease of operation. Globally, about 60% of solid waste ends up in landfills, causing significant health and environmental issues. In 2021, India generated 160,038.9 metric tons of solid waste daily, with 50% treated, 18.4% disposed of in landfills, and 31.7% unaccounted for (Singh et al. 2024). Landfills produce leachate and landfill gases during waste degradation, with leachate containing hazardous chemicals, heavy metals, and other pollutants (Abunama et al. 2019). This can lead to soil, groundwater, and surface water contamination, posing risks to local resources and ecosystems (Bagheri et al. 2017). To address these problems AI methods offers reliable and effective solutions. For example, MLP-ANN and RBF-ANN modeling with prediction accuracies of 97% and 95%, respectively highlighted the pollution risk of heavy metals associated with penetration of leachate into groundwater

(Bagheri et al. 2017). In one study MLP-ANN model predicted the leachate generation rates at a landfill site in Selangor (Malaysia) with highest R^2 value of 0.956 and lowest RMSE of 398.5 (Abunama et al. 2019). Likewise, at a sanitary landfill site the leachate generation prediction performance of ANFIS model was studied by Abunama et al. (2018). The model showed good performance with a correlation coefficient value of 0.952 (Abunama et al. 2018).

Landfills biodegradation produce significant amounts of CH_4 and CO_2 through anaerobic degradation of waste. This exothermic process generates heat that raises landfill temperatures and creates hazardous conditions. CH_4 poses a risk of underground fires if its ratio with O_2 reaches flammability limits, however, CH_4 can also serve as an energy source. Meanwhile, CO_2 emissions contribute to the greenhouse effect. Therefore, accurate computational models to predict landfill temperatures and gas emissions (CH_4, CO_2, and O_2) are essential for safe landfill management (Li et al. 2011; Fallah et al. 2020). Recently, various AI models are widely used to study greenhouse gas concentrations and improve landfill gas management. For example, Alcibahy et al. (2025) applied an advanced XGBoost model to create high resolution maps of CO_2 and CH_4 concentrations over the Arabian Peninsula. The proposed model achieved high accuracies for XCO_2 ($R^2 = 0.98$, RMSE = 0.58 ppm) and XCH_4 ($R^2 = 0.63$, RMSE = 13.26 ppb) (Alcibahy et al. 2025). Ozkaya et al. (2007) used ANN to forecast the CH_4 fractions in landfill gas using waste temperature, alkalinity, pH, sulphate, chloride, COD and conductivity as input variables. The model operated with (C1) and without (C2) leachate recirculation and achieved R^2 values of 0.92 (C1) and 0.90 (C2) with RMSE of 0.050 and 0.051, respectively (Ozkaya et al. 2007). Qdais et al. (2010) applied ANN and GA model to optimize the biogas production at the Russaifah biogas plant in Jordan. The ANN model accurately forecasted the CH_4 generation with R2 value of 0.87. Furthermore, integrating ANN into the GA model increased the CH_4 yield to 6.9%. The yield of CH_4 reached 77% under optimized conditions (36 0C temperature, 6.6% total solids, 52.8% volatile solids and 6.4 pH) (Qdais et al. 2010). Fallah et al. (2020) designed a NARX-ANN model to examine the CH_4 emission rate from a landfill in Regina (Canada). The proposed model showed excellent performance even with limited data with R^2 value of 0.85 and MAPE % of 3.08 (Fallah et al. 2020). Li et al. developed an ANN model using back-propagation (BP) algorithm to make short-term predictions for temperature, CH_4, CO_2, and O_2 concentration profiles in a southern California landfill. They reported that the proposed model provided accurate predictions and could assist operators with short-term planning, identifying enhanced oxygen zones, detecting cover damage, and locating potential underground fire areas (Li et al. 2011). Toha and Rahman (2023) used ZODM, FODM, Land-GEM-V-3.02 models to develop CH_4 emission inventories for Matuail (ML) and Amin-Bazar (ABL) landfills in Dhaka. In 2021, The ZODM estimated CH_4 emissions at 92.75 and 83.3 Gg/year for ML and ABL, while in-situ methods reported much lower values of 19.02 and 18.89 Gg/year. The authors predict significant methane generation from these landfills until 2030 and recommends implementing technology to capture and utilize this gas as green energy (Toha and Rahman 2023). Recently, Askr et al. (2024) proposed a hybrid Inception-ResNet model to predict CH_4 emissions from landfills and estimating sustainable electricity from captured CH_4. They tested 23,628 images across four waste categories (organic, paper/textiles, wood, and food) and achieved a classification accuracy of 93%. The proposed method estimated 39,695.03 kg of CH_4 emissions and 474,449.7 kW/month of electricity production in a landfill case study (Askr et al. 2024). Thus the AI based approaches may helps to identify high-emission areas and optimize gas collection systems, supporting green energy production and reducing green house gas emissions to align with global sustainability goals.

Furthermore, with increasing MSW generation, suitable landfill site selection is also crucial for maintaining a healthy environment, especially in rapidly growing cities of developing countries (Younes et al. 2016; Hoque and Rahman 2020). In this context, to optimize solid waste disposal and reduce land requirements for Malaysia's waste disposal strategy ANFIS model was applied by Younes et al. (2016). The authors reported that implementation of the model reduced land requirements for waste disposal up to 43% and achieved high prediction accuracy with fewer input variables. Validation results demonstrated the excellent performance of model, with an R^2 value of 0.99 and RMSE values of 0.2678 (Younes et al. 2016). Likewise, Hoque and Rahman (2020) used ANN model to forecast the future landfill area requirements for the Matuail landfill site in Bangladesh by using 120 months of solid waste and vehicle trip data. The performance of model was convincing with R^2 of 0.85 and 0.86 for training and testing data, respectively. Furthermore, it was reported that using the proposed method the landfill land area was reduced by 28.6% from its original size of 90 hectares (Hoque and Rahman 2020). Tables 13.3 summarize application of different AI technique in MSW managements.

13.4 Challenges and Limitations

Although AI offers significant advantages in municipal solid waste (MSW) management, numerous challenges still limit its full potential. The key issues in applying AI to MSW management are outlined below, reflecting the current trends and challenges in this field. Figure 13.4 illustrates the challenges as well as future recommendations for application of AI for MSW management.

13.4.1 Data Availability and Quality

Insufficient data is one of the momentous challenges for implementing AI approaches in waste management. The lack of reliable as well as complete data, especially in developing countries, limits model training and validation. This often leads to overfitting and reduced efficiency, as outdated or inadequate data cannot address diverse problems (Olawade et al. 2024).

Table 13.3 Summary of application of various AI models used in MSW management

Application	AI Model	Study area	Model performance indicators	References
MSW prediction	ANN	Mainland China	ANN: $R^2 = 0.94$, RMSE = 20	Wu et al. (2020)
MSW generation	ANN, SVM	Johannesburg, South Africa	ANN:$R^2 = 99.9\%$	Ayeleru et al. (2021)
MSW generation	ANN, FL, LR	China	ANN: $R^2 = 0.931$, RMSE = 450.84, MAE = 228.53, MAPE = 0.0143%,	Chhay et al. (2018)
MSW production	ANN, DT	Ontario, Canada	ANN: $R^2 = 0.72$	Kannangara et al. (2018)
MSW production	SVM, ANN, ANFIS, kNN	Logan City, Queensland, Australia	ANFIS: $R^2 =$ 0.98, MAE = 52.16, RMSE = 175.88, and MAPE = 0.008	Abbasi and El Hanandeh (2016)
MSW prediction	1D-CLA	Shanghai	1D-CLA: $R^2 =$ 0.95	Lin et al. (2021)
MSW prediction	IMM	Iran	IMM: $R^2 = 0.93$	Liang et al. (2021)
MSW generation	GA-ANN	New Delhi (India)	GA-ANN: IA = 0.864, $R^2 = 0.87$, RMSE = 95.7	Soni et al. (2019)
Prediction of household packaging waste	ANN	Portugal	$R^2 = 0.98$	Oliveira et al. (2019)
Hospital waste generation	MLR	Karaj metropolis, Iran	$R^2 = 0.82$–0.86, RMSE = 0.003–0.008	Golbaz et al. (2019)
Prediction of MSW	MLP-ANN, RBF-SVR	Bahrain	MLP-ANN: RMSE = 0.151, $R^2 = 0.94$ RBF-SVR: RMSE = 0.164, $R^2 =$ 0.97	Jassim et al. (2022)
Solid waste prediction	RF, ANN, MARS, MARS-CSA	Tehran, Iran	MARS-CSA: R^2 = 0.82, RMSE = 10.908	Ghanbari et al. (2021)
Prediction of MSW	SVM, RF, LM, Cubist, KNN, DNN	Vietnam	RF and KNN: R^2 > 0.96, MAE = 121.5–125.0	Nguyen et al. (2021)
MSW generation	LSTM, ARIMA, ANN	Suzhou, China	LSTM: $R^2 = 0.92$, RMSE = 935.08	Niu et al. (2021)

(continued)

Table 13.3 (continued)

Application	AI Model	Study area	Model performance indicators	References
Bin level detection	MLP, DTW, GW	–	Accuracy = 98.5%	Islam et al. (2014)
Bin-level detection	GW	–	Accuracy = 94.29%	Hannan et al. (2013)
Bin level detection	KNN, MLP		KNN: accuracy = 97% MLP: accuracy = 95%	Hannan et al. (2016)
Bins' empty levels	CNN		Accuracy = 95.3125%,	Rahman et al. (2022)
Route optimization	ANN	Austin, Texas, USA	MAE = 10.92–16.51%	Vo et al. (2019)
Waste classification	VGG-16, AlexNet, SVM, KNN, and RF		VGG-16: accuracy = 93%,	Costa et al. (2018)
Waste classification	ANN		Accuracy = 99%,	Tehrani and Karbasi (2017)
Waste classification	CNN, MLP		CNN: accuracy = 98.2% MLP: accuracy = 91.6%	Chu et al. (2018)
Construction and demolition waste classification	RF, Nu-SVM, C-LibSVM		Accuracy > 90%	Kuritcyn et al. (2015)
Trash classification	DNN-TC		On Trashnet: accuracy = 94% On VN-trash: accuracy = 98%	Vo et al. (2019)
Classification of waste	CNNs SVMs, RR		Accuracy = 99.95%	Toğaçar et al. (2020)
Waste classification	MTLA, CNN		Accuracy = 0.9716	Liang and Gu (2021)
Waste classification	ResNet-50 CNN, SVM		Accuracy = 87%	Adedeji and Wang (2019)
Garbage classification	VT-MLH-CNN		Accuracy = 95.8%	Alrayes et al. (2023)
Waste classification	GA		Accuracy = 99.6%	Mao et al. (2021)
Waste identification and classification	CNN, ResNet V2		CNN: accuracy = 87.99% ResNet V2: accuracy = 94.44%	Mookkaiah et al. (2022)

(continued)

Table 13.3 (continued)

Application	AI Model	Study area	Model performance indicators	References
Prediction of physical composition of MSW	BPNN	China	$R^2 = 0.26–0.29$	Ma et al. (2020)
Waste-sorting robots	Yolo-v3	Taiwan	Accuracy = 92.12%	Mao et al. (2022)
Composition of residual wastes	LIBS-ANN		Accuracy = 96.70%	Yan et al. (2021)
Waste identification and classification	CNN	–	Accuracy = 96.57%	Ziouzios et al. (2020)
Compost maturity	ANN		$R^2 \approx 1$	Sharma et al. (2021)
Compost maturity	WNN		Accuracy = 87.5%	Gao et al. (2007)
Compost maturity	CNN		Accuracy = 99.4–99.7%	Xue et al. (2019)
Optimal composting conditions	ANFIS	–	–	Díaz et al. (2012)
NH_3 emission during the composting	ANN		$R^2 = 0.972$ to 0.981	Boniecki et al. (2012)
Predicting CO_2 emissions from green waste composting	RF		RF: accuracy = 88%	Li et al. (2022)
Recovery of total phosphorus and total nitrogen in vermicomposting of MSW	ANN, MLR		ANN: $R^2 =$ 0.9983–0.9991, RMSE = 0.0130–4.3000 MLR: $R^2 =$ 0.834–0.729, RMSE = 0.0928 – 71.4900	Hosseinzadeh et al. (2020)
Predict the LHV of MSW	MLP-ANN	Taiwan	$R^2 = 0.93$, Accuracy = 98%	Shu et al. (2006)

(continued)

Table 13.3 (continued)

Application	AI Model	Study area	Model performance indicators	References
Composition and LHV prediction of waste	SVM	–	Carbon content: accuracy = 95.54% Hydrogen content: accuracy = 85.53% Oxygen content: accuracy = 92.40% LHV: accuracy = 92.49%	Tao et al. (2020)
HHV prediction of solid wastes	GEP, SVM, FFNN	–	GEP: $R^2 = 0.966$, RMSE = 1.57 SVM: $R^2 = 0.973$, RMSE = 1.44 FFNN: $R^2 = 0.978$, RMSE = 1.25	Bagheri et al. (2019)
LHV prediction of MSW	GPR		Unoptimized GPR: MAE = 0.592, MAPE = 5.23% Optimized GPR: MAE = 0.688, MAPE = 6.05%	Birgen et al. (2021)
HHV prediction of MSW	MLP-ANN	–	$R^2 = 0.986$, RMSE = 3.095	Olatunji et al. (2019)
HHV prediction of MSW	ANN, PSO		–	Zhu and Yang (2022)
HHV prediction of MSW	RBF-ANN, MLP-ANN, SVM, ANFIS	–	RBF-ANN: MAPE = 0.45% MLP-ANN: MAPE = 7.3% SVM: MAPE = 11.77% ANFIS: MAPE = 23.76%	Taki and Rohani (2022)
HVs prediction of MSW	MLP, SVM, ANFIS, RF	–	ANFIS: accuracy = 94%	You et al. (2017)
Minimizing NOx emissions for TNT explosive waste disposal	ANN	–	Accuracy = 99%	Cho et al. (2021)

(continued)

Table 13.3 (continued)

Application	AI Model	Study area	Model performance indicators	References
Leachate generation rates	MLP-ANN, RBF-ANN	–	MLP-ANN: accuracy = 97% RBF-ANN: accuracy = 95%	Bagheri et al. (2017)
Leachate generation rates	MLP-ANN, SVM		MLP-ANN: R^2 = 0.956, RMSE = 398.5	Abunama et al. (2019)
Leachate generation rates	ANFIS		ANFIS: R^2 = 0.952	Abunama et al. (2018)
Biogas production optimization	ANN, GA	–	Correlation coefficient = 0.87	Younes et al. (2016)
Predict the CH_4 fractions in landfill gas	ANN		ANN: R^2 = 0.90–0.92, RMSE = 0.051 – 0.050	Ozkaya et al. (2007)
Estimation of methane generation rates	NARX-ANN		R^2 = 0.85 MAPE = 3.08%	Fallah et al. (2020)
To create high-resolution maps of CO_2 and CH_4	XGBoost	Arabian Peninsula	XCO_2: R^2 = 0.98, RMSE = 0.58 ppm XCH_4: R^2 = 0.63, RMSE = 13.26 ppb	Alcibahy et al. (2025)
Predictions for temperature, CH_4, CO_2, and O_2 concentration	ANN	Southern California landfill	–	Li et al. (2011)
Methane emission prediction	ZODM, FODM, MTM, Land-GEM-V-3.02	Matuail and Amin-Bazar landfill site in Dhaka, Bangladesh	–	Toha and Rahman (2023)
CH_4 emissions prediction	Hybrid Inception-ResNet	–	Accuracy = 93%	Askr et al. (2024)
Optimization of solid waste disposal and selection of suitable landfill site	MANFIS	Malaysia	R^2 = 0.99, RMSE = 0.2678–3.9860	Younes et al. (2016)
Prediction of waste collection and estimate future landfill area requirements	ANN	Matuail landfill site, Dhaka, Bangladesh	ANN: R^2 = 0.85–0.86	Hoque and Rahman (2020)

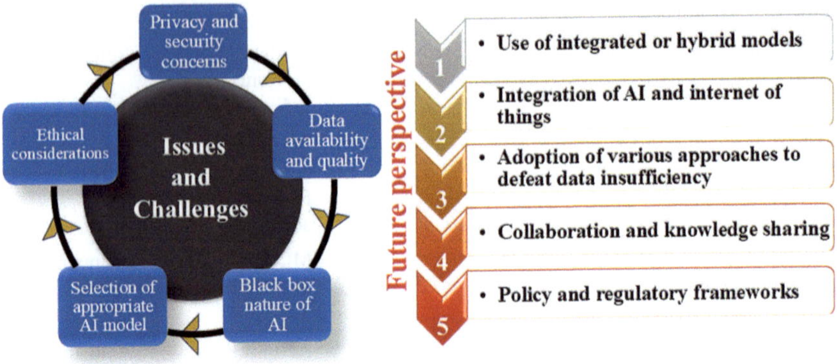

Fig. 13.4 Challenges as well as future recommendations in application of AI for MSW management

13.4.2 Data Privacy and Security Concerns

The AI systems often collected personal and sensitive information like patterns of waste generation, user behaviour, household activities and geographic locations to optimize and ensure efficient waste management. This data can expose sensitive information, raising concerns about privacy and security such as unauthorized access and misuses of data. Additionally, the use of sensor networks and IoT devices further increases the risk of cyber attacks. To tackle these problems it is crucial to execute the strong data security methods such as data encryption, regular software updates and transparent policies (Olawade et al. 2024).

13.4.3 Ethical Considerations

Data bias in AI-powered tools may lead to poor MSW management in deprived areas and exacerbate existing inequalities that may represent unfair facility distribution or injustices. Therefore, during the implementation of AI methods in MSW management it is essential to ensure the ethical concerns like accountability, fairness and transparency. Ethical frameworks should guide the use of AI tools to ensure equity and responsible use of data to avoid the biases that may build the public trust regarding waste management using AI technologies (Olawade et al. 2024).

13.4.4 Selection of Appropriate AI Model

To select the suitable AI model for data processing is a crucial issue. Several strategies showed excellent results; however, it is difficult to offer proper direction for future practical applications or research due to lack of comparative research among diverse models (Guo et al. 2021).

13.4.5 Black Box Nature of AI

'Black box' nature of AI techniques refers to the lack of transparency in how AI models make decisions. Many of these models process large amounts of data with complex internal mechanism that is hard to understand and leading difficulty in apply AI in MSW management (Guo et al. 2021; Ihsanullah et al. 2022).

13.5 Future Perspective

13.5.1 Use of Integrated or Hybrid Models

A hybrid model uses multiple models in an integration series for improved accuracy and reliability. Hybrid models offer excellent prediction accuracy, reduced overfitting and making them more efficient for solving complex MSW treatment issues (Ihsanullah et al. 2022; Guo et al. 2021).

13.5.2 Integration of AI and IoTs

The combination of AI with IoT plans like smart bin and sensors generates real time data which can autonomously observe, examine and optimize MSW collection, classification, and treatment (Ihsanullah et al. 2022; Olawade et al. 2024).

13.5.3 Adoption of Various Approaches to Defeat Data Insufficiency

MSW systems often operate with insufficient or missing data, even though large datasets are crucial for training and validating AI models. To tackle these issues, future research could focus on appropriate sensors and devices to improve data quality and quantity for more accurate predictions with limited data. Potential solutions to overcome data scarcity include extracting data from published studies, using orthogonal and response surface methods to enhance machine learning efficiency, employing feature selection techniques, augmenting image data through transformations, and creating algorithms optimized for small datasets (Guo et al. 2021; Samadhiya et al. 2025).

13.5.4 Policy and Regulatory Frameworks

Future should focus on transparent regulatory guidelines to address ethical security, data privacy and transparency. To ensure privacy and data security, encryption techniques, access controls, and user authentication should be adopted. Governments and researchers should work with technology providers and industry stakeholders to develop policies that balance innovation with ethical values, leading environmental sustainability and social equity (Olawade et al. 2024).

13.6 Conclusion

MSW management is a serious issue in international forums and artificial intelligence (AI) offers promising solutions to manage the MSW with higher efficiency and environmental sustainability. This chapter highlights the potential of various AI and ML-based models in the MSW waste generation, collection, classification and final treatment. The AI models practically been used in smart bin detection, waste shorting robots, waste monitoring and processing and showed excellent performance. ANN, ANFIS models showed excellent prediction of MSW generation with determination coefficient (R^2) value up to 99.9%. Overall AI models particularly, ANN, CNN, ANFIS have been used in all stages of MSW management with more than 99% accuracy. The integration of IoT systems with AI further enhanced the efficiency of MSW management process. However, inadequate data availability, poor quality of data, privacy and security concerns currently hinder their full potential in MSW management. Yet, as research improves, these challenges are likely to be resolved, leading smarter MSW management strategies. Cutting-edge technologies like blockchain, and digital twins can sustain the development of sustainable MWM systems for a better future.

References

Abbasi M, El Hanandeh A (2016) Forecasting municipal solid waste generation using artificial intelligence modelling approaches. Waste Manage 56:13–22

Abdallah M, Talib MA, Feroz S, Nasir Q, Abdalla H, Mahfood B (2020) Artificial intelligence applications in solid waste management: a systematic research review. Waste Manage 109:231–246

Abunama T, Othman F, Younes MK (2018) Predicting sanitary landfill leachate generation in humid regions using ANFIS modeling. Environ Monit Assess 190(10):597

Abunama T, Othman F, Ansari M, El-Shafie A (2019) Leachate generation rate modeling using artificial intelligence algorithms aided by input optimization method for an MSW landfill. Environ Sci Pollut Res 26:3368–3381

Adedeji O, Wang Z (2019) Intelligent waste classification system using deep learning convolutional neural network. Proc Manuf 35:607–612

Ahmed AAA, Asadullah A (2020) Artificial intelligence and machine learning in waste management and recycling. Eng Int 8(1):43–52

Alcibahy M, Gafoor FA, Mustafa F, El Fadel M, Al Hashemi H, Al Hammadi A et al (2025) Improved estimation of carbon dioxide and methane using machine learning with satellite observations over the Arabian Peninsula. Sci Rep 15(1):766

Alrayes FS, Asiri MM, Maashi MS, Nour MK, Rizwanullah M, Osman AE et al (2023) Waste classification using vision transformer based on multilayer hybrid convolution neural network. Urban Clim 49:101483

Amal L, Son LH, Chabchoub H (2018) SGA: spatial GIS-based genetic algorithm for route optimization of municipal solid waste collection. Environ Sci Pollut Res 25:27569–27582

Andeobu L, Wibowo S, Grandhi S (2022) Artificial intelligence applications for sustainable solid waste management practices in Australia: a systematic review. Sci Total Environ 834:155389

Arthur MP, Shoba S, Pandey A (2024) A survey of smart dustbin systems using the IoT and deep learning. Artif Intell Rev 57(3):56

Askr H, Gomaa MM, Rizk-Allah RM, Snasel V, Hassanien AE (2024) Prediction of methane emission and electricity generation from landfills: deep learning approach. Energy Rep 12:5462–5472

Ayeleru O, Fajimi L, Oboirien B, Olubambi P (2021) Forecasting municipal solid waste quantity using artificial neural network and supported vector machine techniques: a case study of Johannesburg, South Africa. J Clean Prod 289:125671

Bagheri M, Bazvand A, Ehteshami M (2017) Application of artificial intelligence for the management of landfill leachate penetration into groundwater, and assessment of its environmental impacts. J Clean Prod 149:784–796

Bagheri M, Esfilar R, Golchi MS, Kennedy CA (2019) A comparative data mining approach for the prediction of energy recovery potential from various municipal solid waste. Renew Sustain Energy Rev 116:109423

Birgen C, Magnanelli E, Carlsson P, Skreiberg Ø, Mosby J, Becidan M (2021) Machine learning based modelling for lower heating value prediction of municipal solid waste. Fuel 283:118906

Boniecki P, Dach J, Pilarski K, Piekarska-Boniecka H (2012) Artificial neural networks for modeling ammonia emissions released from sewage sludge composting. Atmos Environ 57:49–54

Chen L, Wang C, Zhong R, Wang J, Zhao Z (2022a) Intelligent modeling of the incineration process in waste incineration power plant based on deep learning. Energies 15(12):4285

Chen X, Huang H, Liu Y, Li J, Liu M (2022) Robot for automatic waste sorting on construction sites. Autom Constr 141:104387

Chhay L, Reyad MAH, Suy R, Islam MR, Mian MM (2018) Municipal solid waste generation in China: influencing factor analysis and multi-model forecasting. J Mater Cycles Waste Manage 20:1761–1770

Cho S, Kim M, Lyu B, Moon I (2021) Optimization of an explosive waste incinerator via an artificial neural network surrogate model. Chem Eng J 407:126659

Chu Y, Huang C, Xie X, Tan B, Kamal S, Xiong X (2018) Multilayer hybrid deep-learning method for waste classification and recycling. Comput Intell Neurosci 2018(1):5060857

Chu C, Boré A, Liu X, Cui J, Wang P, Liu X et al (2022) Modeling the impact of some independent parameters on the syngas characteristics during plasma gasification of municipal solid waste using artificial neural network and stepwise linear regression methods. Renew Sustain Energy Rev 157:112052

Cortes C (1995) Support-vector networks. Mach Learn

Costa BS, Bernardes AC, Pereira JV, Zampa VH, Pereira VA, Matos GF et al (2018) Artificial intelligence in automated sorting in trash recycling. In: Encontro Nacional de Inteligência Artificial e Computacional (ENIAC): SBC), pp 198–205
Díaz MJ, Eugenio M, López F, García JC, Yañez R (2012) Neural models for optimizing lignocellulosic residues composting process. Waste Biomass Valoriz 3:319–331
Duhayyim MA, Elfadil Eisa TA, Al-Wesabi FN, Abdelmaboud A, Hamza MA, Zamani AS et al (2022) Deep reinforcement learning enabled smart city recycling waste object classification. Comput Mater Contin 71(3)
Gao M, Tian J, Jiang W, Li K (2007) Research of sludge compost maturity degree modeling method based on wavelet neural network for sewage treatment. In: Bio-inspired computational intelligence and applications: international conference on life system modeling and simulation, LSMS 2007, Shanghai, China, September 14–17, 2007. Proceedings. Springer, pp 608–618
Fallah B, Ng KTW, Vu HL, Torabi F (2020) Application of a multi-stage neural network approach for time-series landfill gas modeling with missing data imputation. Waste Manage 116:66–78
Fang B, Yu J, Chen Z, Osman AI, Farghali M, Ihara I et al (2023) Artificial intelligence for waste management in smart cities: a review. Environ Chem Lett 21(4):1959–1989
Gal I-A, Ciocîrlan A-C, Mărgăritescu M (2021) State machine-based hybrid position/force control architecture for a waste management mobile robot with 5DOF manipulator. Appl Sci 11(9):4222
Ghanbari F, Kamalan H, Sarraf A (2021) An evolutionary machine learning approach for municipal solid waste generation estimation utilizing socioeconomic components. Arab J Geosci 14(2):92
Golbaz S, Nabizadeh R, Sajadi HS (2019) Comparative study of predicting hospital solid waste generation using multiple linear regression and artificial intelligence. J Environ Health Sci Eng 17:41–51
Guo H-N, Wu S-B, Tian Y-J, Zhang J, Liu H-T (2021) Application of machine learning methods for the prediction of organic solid waste treatment and recycling processes: a review. Biores Technol 319:124114
Gupta PK, Shree V, Hiremath L, Rajendran S (2019) The use of modern technology in smart waste management and recycling: artificial intelligence and machine learning. In: Recent advances in computational intelligence, pp 173–188
Hannan M, Arebey M, Begum RA, Mustafa A, Basri H (2013) An automated solid waste bin level detection system using Gabor wavelet filters and multi-layer perception. Resour Conserv Recycl 72:33–42
Hannan M, Arebey M, Begum R, Basri H, Al Mamun MA (2016) Content-based image retrieval system for solid waste bin level detection and performance evaluation. Waste Manage 50:10–19
Hoque MM, Rahman MTU (2020) Landfill area estimation based on solid waste collection prediction using ANN model and final waste disposal options. J Clean Prod 256:120387
Hosseinzadeh A, Baziar M, Alidadi H, Zhou JL, Altaee A, Najafpoor AA et al (2020) Application of artificial neural network and multiple linear regression in modeling nutrient recovery in vermicompost under different conditions. Biores Technol 303:122926
Ihsanullah I, Alam G, Jamal A, Shaik F (2022) Recent advances in applications of artificial intelligence in solid waste management: a review. Chemosphere 309:136631
Islam MS, Hannan M, Basri H, Hussain A, Arebey M (2014) Solid waste bin detection and classification using dynamic time warping and MLP classifier. Waste Manage 34(2):281–290
Jassim MS, Coskuner G, Zontul M (2022) Comparative performance analysis of support vector regression and artificial neural network for prediction of municipal solid waste generation. Waste Manage Res 40(2):195–204
Kannangara M, Dua R, Ahmadi L, Bensebaa F (2018) Modeling and prediction of regional municipal solid waste generation and diversion in Canada using machine learning approaches. Waste Manage 74:3–15
Koskinopoulou M, Raptopoulos F, Papadopoulos G, Mavrakis N, Maniadakis M (2021) Robotic waste sorting technology: toward a vision-based categorization system for the industrial robotic separation of recyclable waste. IEEE Robot Autom Mag 28(2):50–60

Król A, Nowakowski P, Mrówczyńska B (2016) How to improve WEEE management? Novel approach in mobile collection with application of artificial intelligence. Waste Manage 50:222–233

Kshirsagar PR, Kumar N, Almulihi AH, Alassery F, Khan AI, Islam S et al (2022) Artificial intelligence-based robotic technique for reusable waste materials. Comput Intell Neurosci 2022(1):2073482

Kujawa S, Nowakowski K, Tomczak RJ, Dach J, Boniecki P, Weres J et al (2014) Neural image analysis for maturity classification of sewage sludge composted with maize straw. Comput Electron Agric 109:302–310

Kujawa S, Mazurkiewicz J, Czekała W (2020) Using convolutional neural networks to classify the maturity of compost based on sewage sludge and rapeseed straw. J Clean Prod 258:120814

Kumar N, Sharma P, Kumar P (2024) Impact of COVID 19 pandemic on the environmental management processes-3R's (reduce, reuse and recycling) of plastic wastes. In: Plastic and the COVID-19 pandemic: innovative solutions to mitigate plastic pollution. Springer, pp 111–139

Kuritcyn P, Anding K, Linß E, Latyev S (2015) Increasing the safety in recycling of construction and demolition waste by using supervised machine learning. J Phys Conf Ser 012035 (IOP Publishing)

Laha S, Pattanayak B, Pattnaik S, Kumar S (2023) A smart waste management system framework using IoT and LoRa for Green City Project. Int J Recent Innov Trends Comput Commun 11(9):342–357

Leveziel M, Laurent GJ, Haouas W, Gauthier M, Dahmouche R (2022) A 4-DoF parallel robot with a built-in gripper for waste sorting. IEEE Rob Autom Lett 7(4):9834–9841

Li Y, Li S, Sun X, Hao D (2022) Prediction of carbon dioxide production from green waste composting and identification of critical factors using machine learning algorithms. Biores Technol 360:127587

Li H, Sanchez R, Qin SJ, Kavak HI, Webster IA, Tsotsis TT et al (2011) Computer simulation of gas generation and transport in landfills. V: Use of artificial neural network and the genetic algorithm for short-and long-term forecasting and planning. Chem Eng Sci 66(12):2646–2659

Liang S, Gu Y (2021) A deep convolutional neural network to simultaneously localize and recognize waste types in images. Waste Manage 126:247–257

Liang G, Panahi F, Ahmed AN, Ehteram M, Band SS, Elshafie A (2021) Predicting municipal solid waste using a coupled artificial neural network with archimedes optimisation algorithm and socioeconomic components. J Clean Prod 315:128039

Lin K, Zhao Y, Tian L, Zhao C, Zhang M, Zhou T (2021) Estimation of municipal solid waste amount based on one-dimension convolutional neural network and long short-term memory with attention mechanism model: a case study of Shanghai. Sci Total Environ 791:148088

Liu J, Balatti P, Ellis K, Hadjivelichkov D, Stoyanov D, Ajoudani A et al (2021) Garbage collection and sorting with a mobile manipulator using deep learning and whole-body control. In: 2020 IEEE-RAS 20th international conference on humanoid robots (humanoids). IEEE, pp 408–414

Ma S, Zhou C, Chi C, Liu Y, Yang G (2020) Estimating physical composition of municipal solid waste in China by applying artificial neural network method. Environ Sci Technol 54(15):9609–9617

Malviya A, Jaspal D (2021) Artificial intelligence as an upcoming technology in wastewater treatment: a comprehensive review. Environ Technol Rev 10(1):177–187

Mao W-L, Chen W-C, Wang C-T, Lin Y-H (2021) Recycling waste classification using optimized convolutional neural network. Resour Conserv Recycl 164:105132

Mao W-L, Chen W-C, Fathurrahman HIK, Lin Y-H (2022) Deep learning networks for real-time regional domestic waste detection. J Clean Prod 344:131096

Mookkaiah SS, Thangavelu G, Hebbar R, Haldar N, Singh H (2022) Design and development of smart internet of things–based solid waste management system using computer vision. Environ Sci Pollut Res 29(43):64871–64885

Mounadel A, Ech-Cheikh H, Lissane Elhaq S, Rachid A, Sadik M, Abdellaoui B (2023) Application of artificial intelligence techniques in municipal solid waste management: a systematic literature review. Environ Technol Rev 12(1):316–336

Nguyen XC, Nguyen TTH, La DD, Kumar G, Rene ER, Nguyen DD et al (2021) Development of machine learning-based models to forecast solid waste generation in residential areas: a case study from Vietnam. Resour Conserv Recycl 167:105381

Niu D, Wu F, Dai S, He S, Wu B (2021) Detection of long-term effect in forecasting municipal solid waste using a long short-term memory neural network. J Clean Prod 290:125187

Olatunji OO, Akinlabi S, Madushele N, Adedeji PA, Felix I (2019) Multilayer perceptron artificial neural network for the prediction of heating value of municipal solid waste. AIMS Energy 7(6):944–956

Olawade DB, Fapohunda O, Wada OZ, Usman SO, Ige AO, Ajisafe O et al (2024) Smart waste management: a paradigm shift enabled by artificial intelligence. Waste Manag Bull

Oliveira V, Sousa V, Dias-Ferreira C (2019) Artificial neural network modelling of the amount of separately-collected household packaging waste. J Clean Prod 210:401–409

Ozkaya B, Demir A, Bilgili MS (2007) Neural network prediction model for the methane fraction in biogas from field-scale landfill bioreactors. Environ Model Softw 22(6):815–822

Praveen A, Radhika R, Rammohan M, Sidharth D, Ambat S, Anjali T (2020) IoT based smart bin: a swachh-bharat initiative. In: 2020 international conference on electronics and sustainable communication systems (ICESC). IEEE, pp 783–786

Qdais HA, Hani KB, Shatnawi N (2010) Modeling and optimization of biogas production from a waste digester using artificial neural network and genetic algorithm. Resour Conserv Recycl 54(6):359–363

Qureshi KN, Khan A, Jamil SUU, Sharma B, Jeon G (2023) Internet of things enables smart solid waste bin management system for a sustainable environment. Environ Sci Pollut Res 30(60):125188–125196

Rahman MW, Islam R, Hasan A, Bithi NI, Hasan MM, Rahman MM (2022) Intelligent waste management system using deep learning with IoT. J King Saud Univ Comput Inform Sci 34(5):2072–2087

Ramson SJ, Moni DJ, Vishnu S, Anagnostopoulos T, Kirubaraj AA, Fan X (2021) An IoT-based bin level monitoring system for solid waste management. J Mater Cycles Waste Manage 23:516–525

Samadhiya A, Kumar A, Mulat-weldemeskel E, Luthra S, Kumar R (2025) Turning trash into treasure: exploring the potential of AI in municipal waste management—an in-depth review and future prospects. J Environ Manage 373:123658

Sharma D, Pandey AK, Yadav KD, Kumar S (2021) Response surface methodology and artificial neural network modelling for enhancing maturity parameters during vermicomposting of floral waste. Biores Technol 324:124672

Shu H-Y, Lu H-C, Fan H-J, Chang M-C, Chen J-C (2006) Prediction for energy content of Taiwan municipal solid waste using multilayer perceptron neural networks. J Air Waste Manag Assoc 56(6):852–858

Singh M, Singh M, Singh SK (2024) Tackling municipal solid waste crisis in India: insights into cutting-edge technologies and risk assessment. Sci Total Environ 170453

Soni U, Roy A, Verma A, Jain V (2019) Forecasting municipal solid waste generation using artificial intelligence models—a case study in India. SN Appl Sci 1(2):162

Taki M, Rohani A (2022) Machine learning models for prediction the higher heating value (HHV) of municipal solid waste (MSW) for waste-to-energy evaluation. Case Stud Thermal Eng 31:101823

Tao J, Liang R, Li J, Yan B, Chen G, Cheng Z et al (2020) Fast characterization of biomass and waste by infrared spectra and machine learning models. J Hazard Mater 387:121723

Tehrani A, Karbasi H (2017) A novel integration of hyper-spectral imaging and neural networks to process waste electrical and electronic plastics. In: 2017 IEEE conference on technologies for sustainability (SusTech). IEEE, pp 1–5

Toğaçar M, Ergen B, Cömert Z (2020) Waste classification using AutoEncoder network with integrated feature selection method in convolutional neural network models. Measurement 153:107459

Toha M, Rahman MM (2023) Estimation and prediction of methane gas generation from landfill sites in Dhaka city, Bangladesh. Case Stud Chem Environ Eng 7:100302

Triassi M, Alfano R, Illario M, Nardone A, Caporale O, Montuori P (2015) Environmental pollution from illegal waste disposal and health effects: a review on the "Triangle of Death." Int J Environ Res Public Health 12(2):1216–1236

Vishnu S, Ramson SJ, Rukmini MSS, Abu-Mahfouz AM (2022) Sensor-based solid waste handling systems: a survey. Sensors 22(6):2340

Vo AH, Vo MT, Le T (2019) A novel framework for trash classification using deep transfer learning. IEEE Access 7:178631–178639

Vu HL, Bolingbroke D, Ng KTW, Fallah B (2019) Assessment of waste characteristics and their impact on GIS vehicle collection route optimization using ANN waste forecasts. Waste Manage 88:118–130

Wang Z, Li H, Yang X (2020) Vision-based robotic system for on-site construction and demolition waste sorting and recycling. J Build Eng 32:101769

Wu F, Niu D, Dai S, Wu B (2020) New insights into regional differences of the predictions of municipal solid waste generation rates using artificial neural networks. Waste Manage 107:182–190

Xia W, Jiang Y, Chen X, Zhao R (2022) Application of machine learning algorithms in municipal solid waste management: a mini review. Waste Manage Res 40(6):609–624

Xiao W, Yang J, Fang H, Zhuang J, Ku Y, Zhang X (2020) Development of an automatic sorting robot for construction and demolition waste. Clean Technol Environ Policy 22:1829–1841

Xue W, Hu X, Wei Z, Mei X, Chen X, Xu Y (2019) A fast and easy method for predicting agricultural waste compost maturity by image-based deep learning. Biores Technol 290:121761

Yan B, Liang R, Li B, Tao J, Chen G, Cheng Z et al (2021) Fast identification and characterization of residual wastes via laser-induced breakdown spectroscopy and machine learning. Resour Conserv Recycl 174:105851

You H, Ma Z, Tang Y, Wang Y, Yan J, Ni M et al (2017) Comparison of ANN (MLP), ANFIS, SVM, and RF models for the online classification of heating value of burning municipal solid waste in circulating fluidized bed incinerators. Waste Manage 68:186–197

Younes MK, Nopiah Z, Basri NA, Basri H, Abushammala MF, Younes MY (2016) Landfill area estimation based on integrated waste disposal options and solid waste forecasting using modified ANFIS model. Waste Manage 55:3–11

Zhu X, Yang G (2022) Study on HHV prediction of municipal solid wastes: a machine learning approach. Int J Energy Res 46(3):3663–3673

Ziouzios D, Tsiktsiris D, Baras N, Dasygenis M (2020) A distributed architecture for smart recycling using machine learning. Future Internet 12(9):141

Chapter 14
Insight on Treatment Mechanism: Constructed Wetland for Sewage Treatment

Sirajuddin Ahmed, Sumaya Irshad, Mohd. Aamir Mazhar, Suriya Akhtar, and Hina Gull

Abstract Rapid urbanisation has led to a surge in organic wastewater generation, significantly contributing to the contamination of freshwater resources and posing serious threats to environmental and human health. Conventional physicochemical treatment methods, while effective, are often constrained by high operational costs, energy consumption, and the production of hazardous byproducts. In response to increasing environmental concerns and regulatory pressures, biological treatment methods have gained attention for their sustainability and efficacy. Among these, constructed wetlands have been vital natural cleaning systems. Constructed wetlands (CWs) are inexpensive eco-technical solutions that have long been utilized to clean municipal and urban wastewater. However, over the last two decades, their application has expanded for wastewater treatment due to technological advancements and extensive research in the field. The CW is a promising approach for treatment strategy because it provides economic and environmental advantages in addition to effective treatment. These ecosystems offer an environmentally beneficial alternative to traditional treatment techniques by combining human ingenuity with natural processes.

Keywords Constructed wetlands · Wastewater treatment · Phytoremediation · Sustainable remediation · Urban wastewater

---

S. Ahmed (✉) · S. Irshad
Department of Environmental Science, Jamia Millia Islamia, New Delhi, India
e-mail: suahmed@jmi.ac.in

Mohd. A. Mazhar
Department of Civil Engineering, Jamia Millia Islamia, New Delhi, India

S. Akhtar
Government Degree College for Women, Nawakadal, Srinagar, Nawakadal, J&K, India

H. Gull
Division of Environmental Science and Engineering, Pohang University of Science and Technology, Pohang-si, South Korea

S. Madhav et al. (eds.), *Municipal Sewage and Sludge: Treatment and Disposal Strategies*, Environmental Science and Engineering,
https://doi.org/10.1007/978-3-032-07597-0_14

14.1 Introduction

Rapid urbanisation produces massive amounts of organic wastewater, which causes environmental pollutants to accumulate and water quality to deteriorate. In actuality, untreated wastewater discharge has contaminated a significant amount of freshwater bodies, posing threats to human and environmental health as well as reducing food produced, obstructing industrial expansion, and generating freshwater shortages. There are several ways to treat urban wastewater, including biological and physico-chemical techniques. Screening, sedimentation, equalization, aeration,, coagulation flotation, filtration, ion adsorption, chlorination, ozonation, are examples of physico-chemical techniques (Malaviya and Singh 2012). However, these methods have some disadvantages like high costs, the formation of sludge, high power consumption, requirement of trained labour, the formation and disposal of harmful end products (Sharma et al. 2021). Also, with growing environmental concerns and stringent regulations regarding wastewater discharge, industries are under pressure to adopt more sustainable and effective treatment solutions. Microbial fuel cells, vermicomposting (using worms to breakdown of pollutants), mycoremediation (using fungi to remove pollutants), bacterial bioremediation (using oxidation ponds, aerobic/anaerobic lagoons and bioreactors, biological filters, etc.), and phytoremediation (using plants and related microorganisms to remove pollutants, such as rain gardens, constructed wetlands, floating treatment wetlands) (Sharma et al. 2021). Because of their well-documented lower construction, operation, and maintenance costs, lower energy requirements, high nutrient absorption potential, and high process stability during wastewater treatment, constructed wetlands (CWs) are attracting more attention than other phytoremediation techniques for managing contaminated waters (Malaviya and Singh 2012). Because of its ecological advantages, affordability, and capacity to naturally remove a variety of pollutants, CWs have become a viable option for wastewater remediation.

The design, functionality, and suitability of CWs for wastewater treatment are examined in this chapter. This chapter investigates the different kinds of CWs, their performance and their methods of treatment, as well as their drawbacks and prospects for the future.

14.2 Constructed Wetland Overview

Engineered structures called CWs are made to duplicate the natural processes seen in wetlands. CWs are made up of well-designed basin with microorganisms, water, plants, and substrate. The growth of plants and microorganisms, which jointly filter, breakdown, and transform contaminants found in wastewater, is greatly aided by substrates like sand, soil, or gravel (Hussein and Scholz 2017). Macrophytes that grow in CWs are essential for absorbing, breaking, and eliminating contaminants from wastewater. Numerous aquatic and semi-aquatic plant species with phytoremediation potential are found in numerous genera, including Lemna, Spirodella, Wolfia, Phragmites, Typha, Eichhornia, Isoetes, Lobelia, Egeria, and Elodea (Pant and Adholeya 2009). Contaminants are taken up by macrophytes from the matrix around their roots, stored in their tissues and then transported to shoots for harvesting and eventual elimination (Chandanshive et al. 2017). Microorganisms can grow, thrive and build biofilms on the surface of roots, which helps rhizospheric cleanup processes like rhizofiltration and rhizodegradation. Growing roots that penetrate the substrate aid in the breakdown of organic contaminants, the establishment of ideal conditions for the settling suspended materials, and the prevention of blockages by creating channels for water to flow through the CW bed (Malaviya and Singh 2012). A wide variety of microorganisms that coexist with the roots and substrate of the wetland bed consume and breakdown the contaminants when wastewater passes through it. For optimal pollutant removal from various wastewater sources, CWs designs and structures can be tested. Performance can be improved by adjusting operation settings and factors influencing remediation processes (Minakshi et al. 2022; John et al. 2020; Sharma et al. 2018; Popov et al. 2008). Pollutants can be removed from wastewater using both the entire plant and adsorbents made from macrophytes.

14.3 Structure of a Constructed Wetland

The structure of a CWs typically includes:

i. **Inlet Zone**: Where wastewater enters the wetland, often equipped with sedimentation basins or screens to remove large particles and debris.
ii. **Treatment Zone**: The main section of the wetland where the bulk of pollutant removal occurs, involving physical, chemical, and biological processes.
iii. **Outlet Zone**: The final stage where treated water is collected before being discharged into the environment or reused.

Figure 14.1 show the typical cross-section of a CWs.

14.4 Types of Constructed Wetlands

CWs can be divided into surface flow and subsurface flow based on the movement of water. Depending on the direction of water movement, subsurface flow CWs are further divided into horizontal, vertical, and hybrid flow-CWs (Vymazal 2014). Each has unique design features and advantages for wastewater treatment.

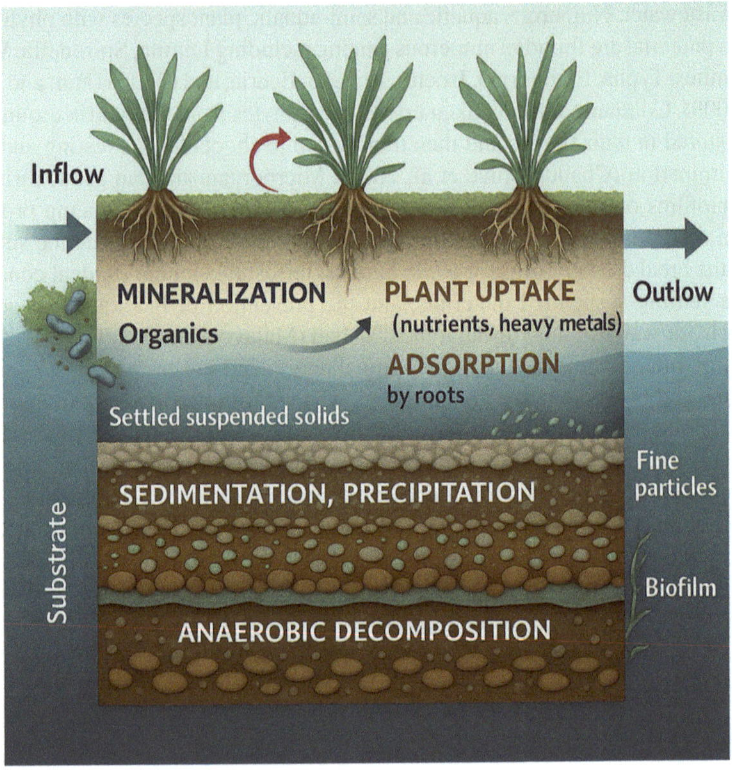

Fig. 14.1 Typical cross-section of a constructed wetland

14.4.1 Surface Flow Constructed Wetlands (SF-CWs)

SF-CWs, also known as free water surface wetlands, allow water to flow above the soil surface. These wetlands resemble natural marshes, with water flowing slowly through a densely vegetated area where pollutants are removed through sedimentation, filtration, plant uptake, and microbial degradation (Lavrnić et al. 2020). A shallow basin filled with aquatic plants such as *Phragmites australis* (common reed) or Typha (cattails). The water level is maintained above the substrate, allowing for the development of aerobic conditions at the water's surface and anaerobic zones beneath. Heavy metals, organic matter, and suspended solids can all be effectively removed by these systems.

14.4.2 *Subsurface Flow Constructed Wetlands (SSF-CWs)*

The influent flows beneath the soil or gravel substrate's surface, which is a characteristic of subsurface flow systems. When the contaminant come into contact with the water-saturated and oxygen-limited substrate surfaces and plant roots, the purification process takes place. Under these circumstances, the covering vegetation and litter layer serve as a thermal insulator, preventing the wetland's performance from drastically declining throughout the winter (Stefanakis et al. 2016).

14.4.2.1 Horizontal Subsurface Flow Constructed Wetlands (HSSF-CWs)

In HSSF-CWs, water moves horizontally through a porous medium (usually gravel or sand) planted with wetland vegetation. Until it reaches the outlet, where it is collected, it moves slowly through the substrate beneath the bed surface, via the pores of the porous substrate, and through roots of the plants. The outlet is situated at the bed's base (Vymazal 2014). The water remains below the surface, somewhere between 5 and 15 cm from the surface of the substrate, which minimises odours and mosquito breeding, thus preventing the spread of diseases, making this system more suitable for urban or industrial areas. Wastewater flows horizontally through the gravel bed, creating a combination of aerobic, anoxic, and anaerobic zones. Plants play a crucial role in oxygen transfer, helping to maintain microbial activity within the bed. The depth of the layer can vary from 30 to 80 cm, and the substrate is frequently composed of gravel or a mixture of gravel and sand. This medium is advantageous for plant growth (Vymazal 2011). The depth of the plant roots determines the depth of substrate, and to stop water from seeping through the base, an impermeable geomembrane is placed over the bottom. The purpose of the bottom slope, which varies from 1 to 3%, is to allow wastewater to flow naturally. Heavy metals, suspended particles and organic matter can all be effectively removed using HSSF-CWs (Khan et al. 2020).

14.4.2.2 Vertical Subsurface Flow Constructed Wetlands (VSSF-CWs)

VSSF-CWs direct water vertically through a filter bed, typically comprising numerous layers of substrate (gravel, sand) planted with macrophytes (Vymazal and Kröpfelová 2011). An underdrain at the substrate surface's bed collects wastewater that is intermittently applied to the surface, allowing it to percolate downwards through the substrate. As a result, wastewater will flow in a direction that is orthogonal to the wetland's spreads. VSSF-CWs can be further classified according to flow direction, saturation level, and saturation length. These are further categorized to include saturated vertical (down-flow, up-flow), recirculating VFCWs, tidal flow, intermittent loading (down-flow), and integrated VFCWs. Combining VFCWs with tidal loading is standard procedure, especially in European nations. This style of operation's main advantage is the formation of a temporary wastewater pond that is 3–5 cm deep on the media surface, this condition is known as an anaerobic mode (Stefanakis et al. 2016). Contact time or retention time are terms used to describe the duration of the aerobic mode. The entrapped air moves much more slowly than usual because of the pressure that the wastewater exerts. Air from the surrounding atmosphere replaces the wastewater as it flows vertically through the porous material due to gravity. The resting period is the period in which it is present. In this type of operation, bed aeration promotes and speeds up the growth of microorganisms, which enhances the oxidation of organic material's nitrification process to avoid blockage (Du et al. 2018). The amount of effluent that is collected depends on the slope of the bottom (which can range from 1 to 2%), as well as the depth of the substrate (which can range from 45 to 120 cm) (Vymazal et al. 2006). Water is introduced at the top of the wetland in batches, ensuring alternating periods of wetness and dryness. This system promotes greater oxygen diffusion, creating aerobic conditions that enhance microbial degradation of pollutants. VSSF-CWs are particularly effective in treating high-strength wastewater with high organic loads. They are known for their efficiency in removing BOD, COD, and nutrients like nitrogen and phosphorus.

14.4.2.3 Hybrid Constructed Wetlands

Hybrid CWs combines various types of wetlands (e.g., horizontal and vertical flow) to enhance treatment performance. The foundation for the hybrid CWs, which were revived at the close of the twentieth century, was established by Seidel, who was the first to try to integrate the different CWs in Germany in the 1960s (Vymazal 2022). These systems are made to solve particular issues and capitalize on the advantages of each type of wetland. Multiple stages are used to treat wastewater, the most popular hybrid systems are VFCWs filter stages followed sequentially by HFCWs filter stage and HFCWs filter stage followed by VFCWs filter stages. The first method is starting with a vertical flow wetland, followed by a horizontal flow system. The purpose of VFCWs is to reduce the quantity of suspended particles and organic materials in the water. VFCW units are followed by HFCW units, which offers a different approach to reducing suspended solids and organic matter and establishing a denitrification friendly environment (Lu et al. 2019). The HFCW units in the second combination are in charge of lowering contaminants and creating ideal conditions for denitrification. Next in like are VFCWs units, which are in charge of nitrification as well as more reduction of suspended particles and organic matter. Depending on the findings of the outflow study, the outflow must be supplied back as an inflow in HFCWs or to another HFCWs unit if the oxidate ammonia nitrogen causes the wastewater effluent after VFCWs to have a higher nitrate content. This is required if there is an excess of nitrate in the effluent. Hybrid systems offer enhanced pollutant removal efficiency and have been shown to significantly improve, especially when combined with pre-treatment processes such as coagulation-flocculation (Soti et al. 2024).

14.5 Advantages of Constructed Wetlands

The CWs have the following advantages:

i. For the regulation of industrial wastewater, CWs are inexpensive eco-technical treatments.
ii. They remove pollutants using biological, chemical and physical processes without the need of external energy.
iii. Compared to other phytoremediation methods for managing contaminated waters, they provide great process stability during wastewater treatment.
iv. They have high nutrient absorption potential.
v. They do not lead to the production of sludge as compared to the conventional method of using physiochemical and biological methods.

14.6 Efficiency of a Constructed Wetland

The CWs efficiency can depend on various factors, such as the initial influent level, the plant species, and the detention time. For example, the removal percentage for chemical oxygen demand (COD) for low concentrations is usually 50% for the control, 59% for the short retention time, and 67% for the long retention time. Another example is CW-MFC, which, with a 3-day HRT, can achieve a maximum removal percentage of 91% for colour and 86% for COD (Hussein 2023). CW-MFCs use a redox gradient which leads to the flow of electrons into the desired route, generating electricity. The use of a combination of plants, bacteria, and electrochemical processes results in higher pollutant removal percentage rates. Gravel beds are more efficient than other substrate beds at eliminating suspended particles because of their unique physical and chemical structure.

14.7 Treatment Mechanisms of Constructed Wetland

The commonly held belief is that the most important aspect of a constructed wetland is the ability of some aquatic plants to assimilate nutrients, and hence directly aid the purification of wastewater. However, this is not exactly true. In fact, intricate physical, chemical and biological processes eliminate contaminants. Compared to traditional biological treatment systems, SSCW have a considerably wider range of biological characteristics (Machineni 2019). In this chapter, an attempt has been made to understand the complex mechanism involved in the functioning of CWs. It discusses the mechanism of treatment for suspended solids, organic matter, nitrogen, phosphorus, metals, and pathogens. This section also discusses the impact of various intrinsic, operational and meteorological factors on the treatment performance of CWs, including filter media, vegetation, organic impurities, oxygen transport through halophytes, microbial degradation, hydraulic retention time, temperature and evapotranspiration.

14.7.1 Theory of Treatment

Organic matter, nutrients, heavy metals, organic and inorganic contaminants, and pathogens can all be effectively removed from wastewater by CWs. Numerous physical, chemical, and biological processes are used in treatment systems to remove a wide range of contaminants from various wastewater types, because of the interactions between macrophytes, microorganisms, substrate, and pollutants, the rhizosphere more specifically, the mycorhizosphere in the active zone of CWs (Sharma et al. 2018). Macrophytes play a vital role in SSCW. They offer surfaces and an appropriate environment for microbial growth and filtration in addition to several operational advantages. There are around 2000 different types of aerobic, anaerobic and facultative microorganisms found in the rhizosphere of CW, whereas none of the other natural systems can boast of so many different types of microorganisms. Because of such high diversity of microorganisms, different types of organic wastes, which are difficult to treat in conventional and other treatment systems, can be effectively biodegraded in CWs (Shahid et al. 2020). With proper acclimatisation, organic industrial effluents can also be treated in a CW. Another significant characteristics of CWs that helps the operation of pollutant removal systems is the oxygen transfer that certain aquatic plants do into the rhizosphere. For the design and operation of any CW and to maximize the removal mechanism needed for the treatment of a specific wastewater, a thorough understanding of the intricate mechanics involved is consequently crucial (Tomenko et al. 2007).

14.7.2 Suspended Solids

Settlable and suspended solids in wastewater are primarily removed through physical processes such as filtration and sedimentation (Sharma and Malaviya 2022). As wastewater flows slowly across the substrate of constructed wetlands (CWs) in a uniform sheet-like manner, it encounters media typically composed of sand, soil, and gravel—materials characterized by substantial void space that facilitates filtration (Galletti et al. 2010). Due to the high possibility of clogging of beds, SSCW is not suitable for effluents bearing high suspended solids (Li et al. 2021). The settlement of suspended particles takes place in the quiescent areas of CW. More than 90% of suspended solids can be effectively retained. Colloidal particles are partially removed by bacterial growth, collisions, and sorption to other solids. Another crucial process step in voids filled with water is filtering. Particles are stretched by flow constrictions and settle into stationary micro-pockets. Emergent macrophytes also aid these processes. The mechanism of enhanced settlement in SSCW is similar to that of conventional settlement vessels, with the addition of an array of inclined plates, also known as tube settlers or lamella separators (Reyes et al. 2022). Suspended particles are intercepted by plant roots and trapped within the pores of the substrate bed, promoting their removal (Nyberg et al. 2014). Sediments are further settled due to gravity and differences in the densities of particulates. Over time, plant roots penetrate the sediment layers and, in conjunction with microbial communities, aid in the decomposition of organic matter (Rehman et al. 2017). This integrated filtration mechanism allows for the efficient removal of over 90% of suspended solids from wastewaters (Ahmed et al. 2008).

14.7.3 Organic Matter (Biochemical Oxygen Demand)

Settleable organic matter is removed by deposition and filtration processes in wetland systems. Dissolved organic compounds in wastewater are effectively degraded through both aerobic and anaerobic microbial processes in constructed wetlands. Anaerobic degradation is important for some kinds of waste and can predominate in organically overloaded systems. The oxygen required for aerobic breakdown comes directly from the atmosphere by diffusion or oxygen leakage from the macrophyte roots in the rhizosphere. When compared to biological decomposition, the macrophytes uptake of organic materials in an SSCW is minimal (Reitsema et al. 2018). The biofilm bacteria play a more significant role in the removal of organic matter as compared to suspended bacteria (Romaní et al. 2013). If lowering the organic content (carbonaceous BOD) is the main goal of treatment, heterotrophic organisms are also significant due to their higher metabolic rate and need for organic material as a carbon source. Two groups of microorganisms control the aerobic breakdown of soluble organic compounds: the chemoautotrophs, which convert ammoniacal nitrogen to nitrite and nitrate (nitrification) and the aerobic chemoheterotrophs, which oxidise organic materials and release ammonia.

$$\begin{aligned} \text{Organic matter} + O_2 &\xrightarrow{\text{Bacteria}} NH_3 \\ &+ CO_2 + H_2O + \text{new cells} \end{aligned}$$

Although both groups degrades organic materials, the heterotrophs higher metabolic rate makes them mostly responsible for the systems decreased BOD. Any SSCWs ability to execute aerobic biological oxidation will be significantly diminished if this group receives an inadequate oxygen supply (Tomenko et al. 2006). However, the amount of active organic material that the organisms have access to will control the aerobic breakdown if there is enough supply of oxygen (Nguyen et al. 2023). In the absence of dissolved oxygen, anaerobic breakdown by anaerobic heterotrophic bacteria takes place in SSCW in two stages. First, organic matter is transformed into organic acids, alcohols and new cells by the acid-forming bacteria. The oxidation is then continued by a second set of methane-forming bacteria, which use some of the organic matter to create new cells while turning the remainder to carbon dioxide, methane and other compounds.

$$\text{Organic matter} \xrightarrow{\text{Bacteria}} \text{alcohols, acids and new cells}$$
$$\xrightarrow{\text{Bacteria}} CH_4 + H_2S + NH_3 + CO_2 + H_2O + \text{new cells}$$

Compared to aerobic degradation, anaerobic degradation happens far more slowly. However, when oxygen supply is limited at high organic loadings or deep within submerged bed matrices or biomass, the anaerobic degradation will predominate (Valipour and Ahn 2016). Evidently the wetlands offers a variety of distinct pathways for organic compound breakdown. Bacteria affixed to plant roots and the filter matrix's surfaces play a major role in these activities (Saeed and Sun 2017). Microorganisms are essential for breaking down of complex organic components and lowering BOD (Lv et al. 2024). By offering large surface areas for microbial colonization, taking up nutrients from the surrounding soil solution, and releasing oxygen into the rhizosphere, plant roots aid in these processes by providing aerobic breakdown pathways. By using organic pollutants as a carbon source, chemoheterotrophic microbes produce stable byproducts including CO_2, NH_3, and water in aerobic environments, or CH_4 in anaerobic environments (Saeed and Sun 2017). A crucial source of oxygen for aerobic microbial activity is radial oxygen loss (ROL) from plant roots, which oxygenates the surrounding substrate (Sharma and Malaviya 2022). In addition to suspended bacteria, microbial communities form biofilms on the surface of bed materials, which are more effective in degrading organic matter due to higher microbial density and stability (Ahmed et al. 2008). Some microbial populations also have the ability to immobilise and transform heavy metals into less bioavailable forms, contributing to the overall detoxification process within CWs (Sharma and Malaviya 2022). Constructed wetlands, therefore, provide a diverse suite of biogeochemical processes for the removal of dissolved organic compounds.

14.7.4 Nitrogen

Nitrification and denitrification are the two main processes by which nitrogen is removed from SSCW. Nitrifying bacteria oxidise ammonia to nitrate in aerobic environments. In anoxic zones, denitrifying bacteria convert nitrates to nitrogen gas. Diffusion from the atmosphere and macrophyte roots provide the oxygen needed for nitrification. Additionally, the plants absorb nitrogen and incorporate it into the biomass. Additional removal processes consist of adsorption and volatilisation (Rehman et al. 2017). These processes are typically less significant than nitrification–denitrification though. Both the aerobic and anoxic zones of a CWs easily mineralise nitrogen-containing organics to ammonical nitrogen and this process is called ammonification.

$$\text{Organic nitrogen} = NH_4^+\text{-N}$$

Both organic and inorganic types of nitrogen can be found in wastewater. In SSF-CWs, when organic nitrogen is prevalent, ammonification becomes the initial transformation process. This microbial process converts organic nitrogen compounds—primarily amino acids—into ammonium (NH_4^+) through oxidative deamination, as shown in Eq. 14.1 (Saeed and Sun 2017). The process is exergonic and is influenced by environmental factors, particularly pH, with the optimal range for ammonification being 6.5–8.5 (Ahmed et al. 2008).

$$\text{Amino acids} \rightarrow \text{imino acids} \rightarrow \text{keto acids} \rightarrow NH_3 \tag{14.1}$$

Nitrification is a chemoautotrophic process. Nitrosomonas and Nitrobacter microbial genera, are responsible for biological nitrification. First, ammonical nitrogen is converted to nitrite; in the second step, nitrite is converted to nitrate. In unbalanced equation forms:

$$NH_4^+ + CO_2 + O_2 \xrightarrow{\text{Nitrosomonas}} NO_2^- + H_2O + \text{new cells}$$

$$NO_2^- + CO_2 + O_2 \xrightarrow{\text{Nitrobacter}} NO_3^- + H_2O + \text{new cells}$$

Nitrification constitutes the second phase of nitrogen transformation in SSF-CWs and is a strictly aerobic, two-step biochemical process. In the presence of oxygen, ammonium (NH_4^+) is initially oxidised to nitrite (NO_2^-) by chemolithotrophic bacteria such as Nitrosomonas, Nitrosospira, and Nitrosococcus. Subsequently, nitrite is converted to nitrate (NO_3^-) by facultative chemolithotrophic bacteria, primarily Nitrobacter and Nitrospira (Saeed and Sun 2017). This conversion requires an adequate oxygen supply and is highly dependent on redox potential and pH. One mg of the ammonical nitrogen must oxidise to nitrate with around 4.3 mg O_2 (Kurup 2018). Approximately 8.64 mg HCO_3 per mg of ammonical nitrogen oxidised and a considerable amount of alkalinity is consumed during this process. The conversion of ammonium to ammonia gas and subsequent volatilisation only occurs in systems when the pH value in the wastewater rises above the pKa of ammonium i.e. 9.3. Algal photosynthesis in wastewater often gives rise to high pH. Macrophytes in CW take up nitrogen in their mineralised form and turn it into their biomass. Wetlands cannot function well at removing nitrogen unless macrophytes are present (Herouvim et al. 2011), because they enhance nitrification by giving microorganisms in the rhizosphere surfaces and oxygen to feed on (Cui et al. 2013) and maximize the elimination of organic matter and the denitrification process by supplying carbon from root exudates (caused by photosynthetically fixed carbon, within a range of 5–25% C) (Wang et al. 2012). Depending on the supply of oxygen and the quality of soil, nitrogen conversion mainly takes place in soil depths upto 20 cm (Bhattarai et al. 2005). The results showed that generally, in an intermittently loaded vertical flow bed, there is a sufficient supply of oxygen that leads to a greater reduction (90%)

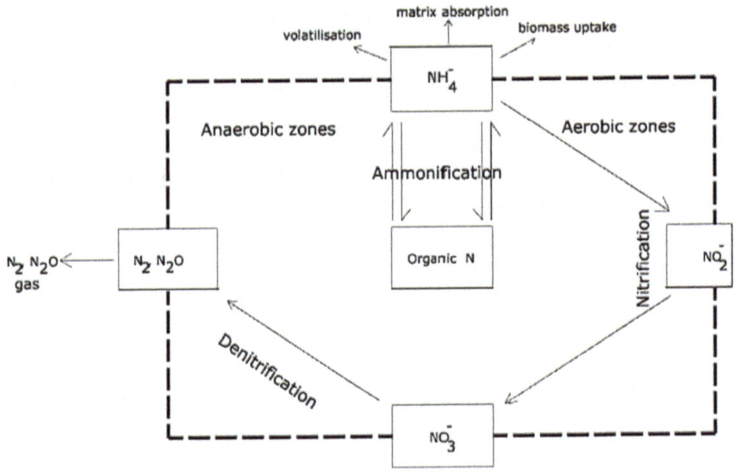

Fig. 14.2 Nitrogen transformations in SSCW

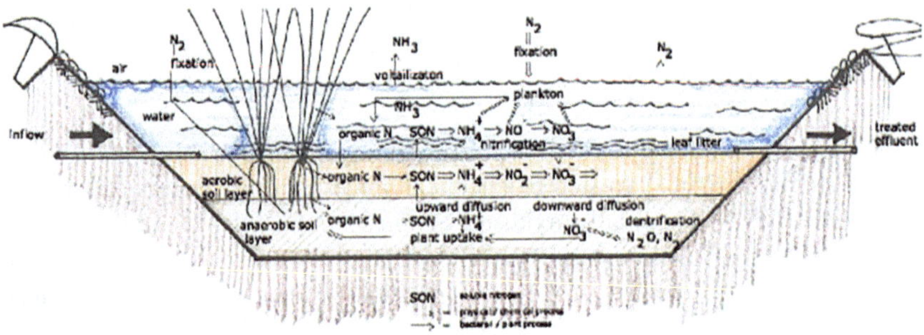

Fig. 14.3 Nitrogen cycle in constructed wetland

of NH4. Nitrification and nitrate reduction activities are maximum at a soil depth of 10 cm. Figures 14.2 and 14.3 depict nitrification and denitrification processes in a CW.

14.7.5 Phosphorus

Phosphorus in wastewater exists primarily as orthophosphate, condensed (dehydrated) phosphates, and organic phosphorus (Ahmed et al. 2008). The majority of phosphorus is changed into the orthophosphate forms (H_2PO_4, HPO_4^{-2}, PO_4^{-3}) by biological oxidation. These forms are primarily removed in SSCW as a result of adsorption, absorption, complexation and precipitation interactions with the minerals Al, Fe, Caaluminium, iron, calcium, and clay found in the filter media. Although there is some uptake of P into plant biomass, it is insignificant compared to the effects of sorption. Reduction of phosphorus under anaerobic conditions is also a possible mechanism of removal. Intermittent flow with wet and dry periods may improve the matrix's phosphorus fixing (Vymazal 2022). Phosphorus is mostly eliminated in SSCW by a confluence of physical, chemical, and biological processes. These include adsorption onto the wetland substrate and soil, uptake by plants and microbes, and chemical precipitation as insoluble compounds. Interactions with Al, Fe, Ca, and clay minerals within the filter media further facilitate removal through absorption and complexation (Ahmed et al. 2008). The efficiency of phosphorus removal in subsurface flow constructed wetlands is therefore strongly dependent on the filter material's physicochemical and hydrological properties, particularly its phosphorus-binding capacity, which is considered a critical design element (Vohla et al. 2011). In treatment wetlands, a medium with a high capacity to bind phosphorus is crucial.

Soluble inorganic phosphorus in wastewater migrates through soil pores and is primarily adsorbed onto mineral surfaces within the wetland substrate. According to Si et al. (2019), although metal cations in wastewater are present in substantial amounts, only a small proportion is absorbed into plant cells; the majority is adsorbed onto negatively charged functional groups (COO^-) on root cell walls. Furthermore, as Sharma and Malaviya in 2022 highlight, adsorbable pollutants (adsorbates) in the wastewater adhere to porous adsorbent surfaces such as plant roots, biochar, activated carbon, or certain plant-derived material provided sufficient contact time and surface availability (Sharma and Malaviya 2022). Maucieri et al. in 2020 demonstrated that plant uptake can enhance phosphorus (P) removal in constructed wetlands, with PO_4–P removal increasing by an average of 7.8% compared to unplanted controls. Among the species studied, *Typha latifolia* showed the highest efficiency, removing up to 1.5 times the PO_4–P input, followed by *Phalaris arundinacea*, which removed about 86%. In other species, over half of the phosphorus removal was primarily attributed to chemical precipitation rather than plant uptake. Reduction of phosphorus under anaerobic conditions is also a possible mechanism of removal (Maucieri et al. 2020). The removal mechanism of phosphorous in a CW is shown in Fig. 14.4. The main processes for removing phosphorus from CW are specific adsorptions by ligand exchange and precipitation reactions. An adsorption mechanism is shown by the influent's rapid phosphorus elimination. However, a significantly slower elimination rate frequently follows this quick removal. A gradual elimination process could involve the precipitation of insoluble phosphates, the transformation of monodentate into binuclear complexes or both.

14.7.6 Metals

There are two types of metals: soluble and particulate. When the metal exists as a weakly compound or ionic species, the former is the form that is bioavailable. Sorption, precipitation, complexation, sedimentation, erosion and diffusion are examples of physico-chemical processes that determine the distribution between particulate

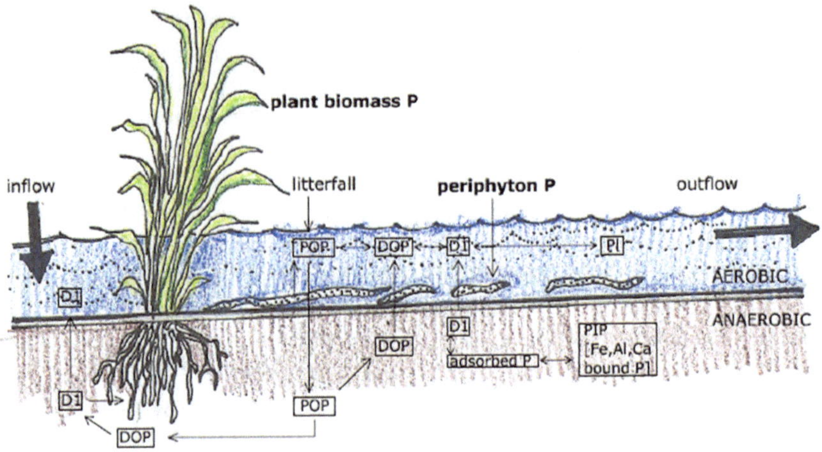

Fig. 14.4 Phosphorus cycle in constructed wetland

and dissolved phases. While lead tends to be mostly particulate-associated, some metals, like cadmium and zinc, have been discovered to have a larger affinity for the dissolved phase (Kumar et al. 2017). The flow suspended solids ratio, oxic/anoxic conditions, ionic strength, pH, dissolved and particulate organic carbon content, organic and inorganic ligand concentrations, and metal mobilisation by biochemically mediated reactions are some of the specific parameters that regulate the sediment–water partitioning of metals. Trace metals accumulate in the filter matrix due to their strong propensity for adsorption and complexation with organic substances (Al-Ghouti et al. 2023). In plants with high surface area to volume ratio, the process of metals adhering to the vegetation's surface is important. Similarly, the vegetation surface area available for the displacement of cations by the more strongly positively charged metal ions in the wastewater affects the cation exchange capacity (CEC). The carboxyl functional groups ($COOH^-$) in the humic acids of plant cellular tissue have been identified as the cause of cation exchange characteristic of SSCW vegetation. It has been discovered that the CEC value remains constant whether the plant is dead or alive (Vymazal 2022). Microbial mediated processes also contribute to metal removal. The precipitation of sulphates and metal oxides is caused by the presence of sulphate-reducing bacteria in the anaerobic zones and metal-oxidising bacteria in the aerobic zones (Yang et al. 2024). In SSCW treating metal-rich mine wastewaters, microbially-mediated iron oxidation by Thiobacillus ferroxidase, followed by iron hydroxide precipitation, is thought to be the most significant iron removal mechanism. This can be shown by an unbalanced equation as:

$$Fe_2^{2+} + O_2 + 4H_2O \rightarrow 2Fe(OH)_3 + 2H^+$$

Similarly for manganese:

$$Mn_2^{2+} + O_2 + 2H_2O \rightarrow 2MnO_2 + 2H^+$$

By introducing oxygen into the rhizosphere, wetland plants may promote the growth of microorganisms that oxidize metal. Sulphate ions are consumed via microbially-mediated sulphate reduction, which also yields hydrogen sulphide and alkalinity in the form of the bicarbonate ion (Mallick et al. 2022). In the unbalanced equation form, this can be shown as follows, with CH_2O standing in for a basic organic molecule:

$$(SO_2)^{4-} + CH_2O \rightarrow H_2S + HCO_3^-$$

After being discharged as a gas, the hydrogen sulphide ionises to produce sulphide ions (S^{2-}). Metal sulphide precipitates are the result of its rapid reaction with various metal ions. Sulphide precipitates form inside organic sediments and are denser than oxide precipitates. The extensive and irregular boundary layers between the aerobic and anaerobic zones in a CW, created by the root hairs in the rhizosphere, assist the precipitation of sulphide (Zheng et al. 2024). Additionally, sulphate-reducing bacteria use the carbohydrates and other organic compounds released by the macrophyte roots, together with decomposing plant matter, as a source of carbon. It is generally known that certain wetland species can directly absorb heavy metals. *Typha latifolia* does not accumulate heavy metals to dangerous amounts and has an excellent tolerance for them. The uptake of metals by the root hairs is decreased when an iron plaque is present in the plant root system. According to available data, some species, like phragmites have ecotypes (populations within a species) that are resistant to acid and metals and do not accumulate metals to a substantial degree (Morari et al. 2015). By facilitating the direct filtering of particulate matter, macrophytes can aid in the removal of metals. Metal hydroxide particles that have precipitated out of the solution can be effectively retained by macrophyte species with large plant surface areas. Pontier et al. (2001) observed that Zn and Cu are associated with organic impurities, carbonates and Fe oxyhydrates. Sediments become metal-enriched by diagenetic process. They also observed that carbonated form of Zn may be more common and abiotic factor can significantly affect the solubility of inorganic materials (Pontier et al. 2001).

14.7.7 Pathogens

A combination of physical, chemical and biological processes enable the wetlands to act as efficient biofilters. Seidel was the first to highlight the effect of halophytes on harmful bacteria and to apply this feature in the first CWs built in Germany (Ghosh et al. 2023). According to experiments, the effects differed according on the type of plants used. *Menthe aquatica, alisma plantago* and *J. effuses* were the most effective halophytes (Stottmeister et al. 2003). A CW's microbial ecology is just as intricate as that of any biological wastewater treatment system. For environmental engineers, eliminating harmful bacteria and viruses is crucial. In tropical and sub-tropical nations, helminth worms and protozoan pathogens are also very important. During wastewater flow in CW, pathogens are mostly eliminated via oxidation, adsorption, sedimentation, filtration, and exposure to biocides (Wang et al. 2021). Predation by zooplanktons and predators, along with antibiosis, are biological removal mechanisms. The mechanisms of natural die-off and predation cause these species numbers to rapidly decline once they are confined within the system.

M. aquatica, *P. australis* and *Sc. lacustris* are found to inhibit the growth of *E. coli* and the strongest among them is *Sc. lacustris*. Factors like adsorption and inactivation have a bearing on the microbial removal process, and these factors, in turn, depend on the design parameters, filter media characteristics and local climatic conditions (Wang et al. 2021). They found that systems using a combination of gravel and sand as filter media produced the best results. The Lemna pond is less effective at removing bacteria than vertical flow and a planted soil filter. Remarkably, the pond was able to eradicate the protozoa more effectively than the soil filter. The impact of phragmites and typha rhizospheres enhanced bacterial removal in comparison to unplanted controls in microcosm studies. CWs are effective in streptococci and coliforms removal, it has been observed that over 90% of the coliforms and over 80% of the faecal streptococci can be eradicated (Singh et al. 2024). According to Decamp and Warren (2000), the average *E. coli* removal rates were 96.6–98.9% for pilot-scale CW systems. For better efficiency, they recommended a vegetated soil bed, followed by a planted gravel bed. The parasite eggs, particularly Helminthes (IE, Ascaris, Trichuris and Hymenolepis), were reduced by 94% in aerobic and almost 99% in facultative conditions, respectively. Virus reduction was also found to be effectively attained up to 99% in 5.5 days of the retention period. They observed that removal efficiency was strongly dependent upon retention periods and temperature (Decamp and Warren 2000).

References

Ahmed S, Popov V, Trevedi RC (2008) Constructed wetland as tertiary treatment for municipal wastewater. Proc Inst Civ Eng Waste Resour Manag 161:77–84. https://doi.org/10.1680/warm.2008.161.2.77

Al-Ghouti MA, Ashfaq MY, Khan M, Al Disi Z, Da'na DA, Shoshaa R (2023) State-of-the-art adsorption and adsorptive filtration based technologies for the removal of trace elements: a critical review. Sci Total Environ 895:164854

Bhattarai SP, Su N, Midmore DJ (2005) Oxygation unlocks yield potentials of crops in oxygen-limited soil environments. In: Advances in agronomy. Elsevier, pp 313–377. https://doi.org/10.1016/S0065-2113(05)88008-3

Chandanshive VV, Rane NR, Tamboli AS, Gholave AR, Khandare RV, Govindwar SP (2017) Co-plantation of aquatic macrophytes *Typha angustifolia* and *Paspalum scrobiculatum* for effective treatment of textile industry effluent. J Hazard Mater 338:47–56. https://doi.org/10.1016/j.jhazmat.2017.05.021

Cui S, Shi Y, Groffman PM, Schlesinger WH, Zhu Y-G (2013) Centennial-scale analysis of the creation and fate of reactive nitrogen in China (1910–2010). Proc Natl Acad Sci 110:2052–2057. https://doi.org/10.1073/pnas.1221638110

Decamp O, Warren A (2000) Investigation of *Escherichia coli* removal in various designs of subsurface flow wetlands used for wastewater treatment. Ecol Eng 14:293–299. https://doi.org/10.1016/S0925-8574(99)00007-5

Du L, Trinh X, Chen Q, Wang C, Wang H, Xia X, Zhou Q, Xu D, Wu Z (2018) Enhancement of microbial nitrogen removal pathway by vegetation in integrated vertical-flow constructed wetlands (IVCWs) for treating reclaimed water. Bioresour Technol 249:644–651. https://doi.org/10.1016/j.biortech.2017.10.074

Galletti A, Verlicchi P, Ranieri E (2010) Removal and accumulation of Cu, Ni and Zn in horizontal subsurface flow constructed wetlands: Contribution of vegetation and filling medium. Sci Total Environ 408:5097–5105. https://doi.org/10.1016/j.scitotenv.2010.07.045

Ghosh A, Orasugh JT, Chattopadhyay D (2023) Constructed wetlands: the traditional system. In: Shah MP (ed) Recent trends in constructed wetlands for industrial wastewater treatment. Springer Nature Singapore, Singapore, pp 177–196. https://doi.org/10.1007/978-981-99-2564-3_9

Herouvim E, Akratos CS, Tekerlekopoulou A, Vayenas DV (2011) Treatment of olive mill wastewater in pilot-scale vertical flow constructed wetlands. Ecol Eng 37:931–939. https://doi.org/10.1016/j.ecoleng.2011.01.018

Hussein A, Scholz M (2017) Dye wastewater treatment by vertical-flow constructed wetlands. Ecol Eng 101:28–38. https://doi.org/10.1016/j.ecoleng.2017.01.016

Hussein A (2023) Textile wastewater treated by constructed wetlands–a critical review. J Ecol Eng 24

John Y, Langergraber G, Adyel TM, Emery David V (2020) Aeration intensity simulation in a saturated vertical up-flow constructed wetland. Sci Total Environ 708:134793. https://doi.org/10.1016/j.scitotenv.2019.134793

Khan NA, El Morabet R, Khan RA, Ahmed S, Dhingra A, Alsubih M, Khan AR (2020) Horizontal sub surface flow constructed wetlands coupled with tubesettler for hospital wastewater treatment. J Environ Manage 267:110627. https://doi.org/10.1016/j.jenvman.2020.110627

Kumar M, Gogoi A, Kumari D, Borah R, Das P, Mazumder P, Tyagi VK (2017) Review of perspective, problems, challenges, and future scenario of metal contamination in the urban environment. J Hazard Toxic Radioact Waste 21:04017007

Kurup R (2018) Implementation of EEI-anammox process for wastewater treatment. Water E-J 3:1–6

Lavrnić S, Alagna V, Iovino M, Anconelli S, Solimando D, Toscano A (2020) Hydrological and hydraulic behaviour of a surface flow constructed wetland treating agricultural drainage water in northern Italy. Sci Total Environ 702:134795

Li Y, Peng L, Li H, Liu D (2021) Clogging in subsurface wastewater infiltration beds: genesis, influencing factors, identification methods and remediation strategies. Water Sci Technol 83:2309–2326. https://doi.org/10.2166/wst.2021.155

Lu S, Gao X, Wu P, Li W, Bai X, Sun M, Wang A (2019) Assessment of the treatment of domestic sewage by a vertical-flow artificial wetland at different operating water levels. J Clean Prod 208:649–655

Lv Z, Ran X, Liu J, Feng Y, Zhong X, Jiao N (2024) Effectiveness of chemical oxygen demand as an indicator of organic pollution in aquatic environments. Ocean Land Atmos Res 3:0050. https://doi.org/10.34133/olar.0050

Machineni L (2019) Review on biological wastewater treatment and resources recovery: attached and suspended growth systems. Water Sci Technol 80:2013–2026

Malaviya P, Singh A (2012) Constructed wetlands for management of urban stormwater runoff. Crit Rev Environ Sci Technol 42:2153–2214. https://doi.org/10.1080/10643389.2011.574107

Mallick A, Dey S, Datta S, Barman M (2022) Bioremediation: an effective, significant and eco-friendly approach for sustainable management. In: Sustainable management of environmental contaminants: eco-friendly remediation approaches. Springer, pp 119–138

Maucieri C, Salvato M, Borin M (2020) Vegetation contribution on phosphorus removal in constructed wetlands. Ecol Eng 152:105853. https://doi.org/10.1016/j.ecoleng.2020.105853

Minakshi D, Sharma PK, Rani A, Malaviya P, Srivastava V, Kumar M (2022) Performance evaluation of vertical constructed wetland units with hydraulic retention time as a variable operating factor. Groundw Sustain Dev 19:100834. https://doi.org/10.1016/j.gsd.2022.100834

Morari F, Dal Ferro N, Cocco E (2015) Municipal wastewater treatment with *Phragmites australis* L. and *Typha latifolia* L. for irrigation reuse. Boron Heavy Metals Water Air Soil Pollut 226:56. https://doi.org/10.1007/s11270-015-2336-3

Nguyen T-P, Koyama M, Nakasaki K (2023) Effect of oxygen deficiency on organic matter decomposition during the early stage of composting. Waste Manag 160:43–50. https://doi.org/10.1016/j.wasman.2023.01.034

Nyberg ET, White SA, Jeffers SN, Bridges WC (2014) Removal of plant pathogen propagules from irrigation runoff using slow filtration systems: quantifying physical and biological components. Water Air Soil Pollut 225:1999. https://doi.org/10.1007/s11270-014-1999-5

Pant D, Adholeya A (2009) Nitrogen removal from biomethanated spentwash using hydroponic treatment followed by fungal decolorization. Environ Eng Sci 26:559–565. https://doi.org/10.1089/ees.2007.0328

Pontier H, Williams JB, May E (2001) Metals in combined conventional and vegetated road runoff control systems. Water Sci Technol 44:607–614. https://doi.org/10.2166/wst.2001.0887

Popov V, Ahmed S, Idris Qureshi M (2008) Simulation of constructed wetland performance using the advection diffusion equation. Presented at the WASTE MANAGEMENT 2008, Granada, Spain, pp 363–371. https://doi.org/10.2495/WM080381

Rehman F, Pervez A, Khattak BN, Ahmad R (2017) Constructed wetlands: perspectives of the oxygen released in the rhizosphere of macrophytes. CLEAN Soil Air Water 45

Reitsema RE, Meire P, Schoelynck J (2018) The future of freshwater macrophytes in a changing world: dissolved organic carbon quantity and quality and its interactions with macrophytes. Front Plant Sci 9:629. https://doi.org/10.3389/fpls.2018.00629

Reyes C, Apaz F, Niño Y, Barraza B, Arratia C, Ihle CF (2022) A review on steeply inclined settlers for water clarification. Miner Eng 184:107639. https://doi.org/10.1016/j.mineng.2022.107639

Romaní AM, Amalfitano S, Artigas J, Fazi S, Sabater S, Timoner X, Ylla I, Zoppini A (2013) Microbial biofilm structure and organic matter use in mediterranean streams. Hydrobiologia 719:43–58. https://doi.org/10.1007/s10750-012-1302-y

Saeed T, Sun G (2017) A comprehensive review on nutrients and organics removal from different wastewaters employing subsurface flow constructed wetlands. Crit Rev Environ Sci Technol 47:203–288. https://doi.org/10.1080/10643389.2017.1318615

Shahid MJ, Al-Surhanee AA, Kouadri F, Ali S, Nawaz N, Afzal M, Rizwan M, Ali B, Soliman MH (2020) Role of microorganisms in the remediation of wastewater in floating treatment wetlands: a review. Sustainability 12:5559

Sharma R, Malaviya P (2022) Constructed wetlands for textile wastewater remediation: a review on concept, pollutant removal mechanisms, and integrated technologies for efficiency enhancement. Chemosphere 290:133358. https://doi.org/10.1016/j.chemosphere.2021.133358

Sharma PK, Minakshi D, Rani A, Malaviya P (2018) Treatment efficiency of vertical flow constructed wetland systems operated under different recirculation rates. Ecol Eng 120:474–480. https://doi.org/10.1016/j.ecoleng.2018.07.004

Sharma PK, Rausa K, Rani A, Mukherjee S, Kumar M (2021) Biopurification of dairy farm wastewater through hybrid constructed wetland system: groundwater quality and health implications. Environ Res 200:111426. https://doi.org/10.1016/j.envres.2021.111426

Singh S, Raju NJ, Mehmood G, Gupta SK, Ahmed S (2024) A review of the current scenario and best possible solution for fecal sludge management (FSM) in India. Groundw Sustain Dev 27:101346. https://doi.org/10.1016/j.gsd.2024.101346

Soti A, Kulshreshtha NM, Singh S, Samaria A, Brighu U, Dontireddy G, Banda S, Gupta AB (2024) High rates of nitrogen removal in aerated VFCWs treating sewage through CNS cycle. Bioresour Technol 399:130620

Stefanakis AI, Seeger E, Dorer C, Sinke A, Thullner M (2016) Performance of pilot-scale horizontal subsurface flow constructed wetlands treating groundwater contaminated with phenols and petroleum derivatives. Ecol Eng 95:514–526. https://doi.org/10.1016/j.ecoleng.2016.06.105

Stottmeister U, Wießner A, Kuschk P, Kappelmeyer U, Kästner M, Bederski O, Müller RA, Moormann H (2003) Effects of plants and microorganisms in constructed wetlands for wastewater treatment. Biotechnol Adv 22:93–117. https://doi.org/10.1016/j.biotechadv.2003.08.010

Tomenko V, Ahmed S, Popov V (2007) Modelling constructed wetland treatment system performance. Ecol Model 205:355–364. https://doi.org/10.1016/j.ecolmodel.2007.02.030

Tomenko V, Ahmed S, Popov V (2006) Performance prediction of a constructed wetland wastewater treatment plant. In: Waste management and the environment III. Presented at the waste management 2006. WIT Press, Malta, pp 143–152. https://doi.org/10.2495/WM060161

Valipour A, Ahn Y-H (2016) Constructed wetlands as sustainable ecotechnologies in decentralization practices: a review. Environ Sci Pollut Res 23:180–197. https://doi.org/10.1007/s11356-015-5713-y

Vohla C, Kõiv M, Bavor HJ, Chazarenc F, Mander Ü (2011) Filter materials for phosphorus removal from wastewater in treatment wetlands—a review. Ecol Eng 37:70–89. https://doi.org/10.1016/j.ecoleng.2009.08.003

Vymazal J (2011) Long-term performance of constructed wetlands with horizontal sub-surface flow: ten case studies from the Czech Republic. Ecol Eng 37:54–63. https://doi.org/10.1016/j.ecoleng.2009.11.028

Vymazal J (2014) Constructed wetlands for treatment of industrial wastewaters: a review. Ecol Eng 73:724–751. https://doi.org/10.1016/j.ecoleng.2014.09.034

Vymazal J (2022) The historical development of constructed wetlands for wastewater treatment. Land 11:174. https://doi.org/10.3390/land11020174

Vymazal J, Kröpfelová L (2011) A three-stage experimental constructed wetland for treatment of domestic sewage: first 2 years of operation. Ecol Eng 37:90–98. https://doi.org/10.1016/j.ecoleng.2010.03.004

Vymazal J, Greenway M, Tonderski K, Brix H, Mander Ü (2006) Constructed wetlands for wastewater treatment. In: Verhoeven JTA, Beltman B, Bobbink R, Whigham DF (eds) Wetlands and natural resource management, ecological studies. Springer Berlin Heidelberg, pp 69–96. https://doi.org/10.1007/978-3-540-33187-2_5

Wang X, Yang G, Feng Y, Ren G, Han X (2012) Optimizing feeding composition and carbon–nitrogen ratios for improved methane yield during anaerobic co-digestion of dairy, chicken manure and wheat straw. Bioresour Technol 120:78–83. https://doi.org/10.1016/j.biortech.2012.06.058

Wang M, Zhu J, Mao X (2021) Removal of pathogens in onsite wastewater treatment systems: a review of design considerations and influencing factors. Water 13:1190

Yang L, Li W, Liu J, Zhu H, Mu H, Hu K, Li J, Dong S (2024) Nitrate-dependent ferrous oxidation: feasibility, mechanism, and application prospects for wastewater treatment. J. Water Process Eng. 60:105226

Zheng J, Guo D, Zhang J, Zhang T, Yang L, Li B, Lan J, Ren Y (2024) Construction of an ideotype root system architecture of subsurface flow constructed wetland macrophytes by vertical spatial stress: strengthening of rhizosphere effects and determination of appropriate substrate depth. Environ Res 259:119523

Chapter 15
Recent Advances in Sewage Treatment Technologies and Sustainable Sludge Management Practices

Jatin Kumar Choudhary, Mallikarjuna Prasad Polavarapu, Akshay Kumar Singh, Sushil Kumar Shukla, and Manoj Kumar

Abstract The treatment and management of sewerage and sludge are experiencing a revolutionary makeover, propelled by pressures of dealing with environmental degradation, health issues, and resource shortages. This chapter offers an in-depth appraisal of recent developments in sewerage treatment and sludge management, with special focus on the shortcomings of traditional systems and the onset of novel, greener alternatives. The traditional processes like Activated Sludge Process and Trickling Filters are plagued with recurring problems of high energy requirements, inefficient removal of contaminants, and complicated waste disposal. To surmount the challenges, the chapter addresses state-of-the-art technology that involves membrane bioreactor (MBR), forward osmosis (FO), novel biological processes such as Anammox and aerobic granular sludge, and physico-chemical methods like advanced oxidation processes (AOPs) and electrocoagulation. Nature-based and phytoremediation methods are also covered from the viewpoint of their cost-effectiveness and eco-compatibility. Key modern practices in managing sludge include resource recovery, in the form of biogas, biochar, and nutrient reclamation, and thermal treatment processes of pyrolysis and plasma treatment. The chapter also combines the advances in a circular economy context, focusing on energy-neutral STPs, carbon footprint minimization, and resource valorization. Policy frameworks, institutional capacity, and public–private partnerships are also attended to. The concluding part of the chapter is a far-sighted vision in the form of smart, modular, and regenerative sanitation systems fuelled by artificial intelligence that could lead us on the path of sustainable and decentralized approaches for wastewater treatment. In the long run, the chapter emphasizes innovation-led, circular, and climate-resilient approaches in developing future-proof sewage and sludge management.

J. K. Choudhary · M. P. Polavarapu · A. K. Singh · S. K. Shukla (✉) · M. Kumar
Department of Environmental Sciences, Central University of Jharkhand, Ranchi, Jharkhand, India
e-mail: sushil.shukla@cuj.ac.in

S. Madhav et al. (eds.), *Municipal Sewage and Sludge: Treatment and Disposal Strategies*, Environmental Science and Engineering,
https://doi.org/10.1007/978-3-032-07597-0_15

Keywords Advanced sewage treatment technologies · Sludge management · Circular economy · Wastewater treatment · Sustainable water management

15.1 Introduction

The management of sewage and sludge has undergone a remarkable evolution over the past century (Guven et al. 2023). Early sewage systems were primarily designed for the swift removal of wastewater from populated areas, with little concern for environmental impacts (Zhao et al. 2023a). Over time, the focus shifted towards treating sewage before discharge, leading to the development of conventional treatment technologies such as sedimentation tanks, trickling filters, and activated sludge processes (Sangamnere et al. 2023). Sludge, once regarded purely as waste, gradually attracted attention for its potential to be stabilised and disposed of safely (Guven et al. 2023). However, the modern era demands more than the basic treatment of sewage, owing to the rise of environmental consciousness and stricter regulations. It calls for innovative approaches prioritising sustainability, energy efficiency, and resource recovery (Awewomom et al. 2024).

Innovation in sewage treatment and sludge management is no longer optional, as it has become vital for safeguarding environmental and human health (Mishra et al., 2023). Untreated or poorly treated sewage can lead to widespread contamination of water bodies, soil degradation, loss of biodiversity, and serious public health crises (Wu et al. 2024). Moreover, traditional methods often contribute to greenhouse gas emissions and squander valuable resources contained within wastewater. New technologies and practices aim to treat sewage more effectively and reclaim water, energy, and nutrients, aligning wastewater management with principles of the circular economy and sustainable development (Mabrouki 2025).

Globally, sewage treatment has made significant progress, with many developed nations achieving high rates of wastewater treatment and resource recovery (Shehata et al. 2023). However, challenges persist, particularly in developing countries where infrastructure, technology adoption, and funding may lag (Mabrouki 2025). In India, while urban centres have made strides in establishing sewage treatment facilities, a considerable portion of wastewater still goes untreated, especially in rapidly growing peri-urban and rural areas (Singh et al., 2023a, 2023b, 2024, 2025). The need for advanced, decentralised, and sustainable treatment systems is urgent to address both environmental degradation and public health concerns (Mabrouki 2025).

The purpose of this chapter is to examine the latest developments in sludge management and sewage treatment, emphasizing how cutting-edge methods and technology can overcome the drawbacks of traditional systems. It looks at new trends, the importance of energy efficiency and lowering carbon emissions, resource recovery techniques, and successful implementations from the real world. With this research, the chapter aims to give readers a thorough grasp of how wastewater management is changing, with an emphasis on how it relates to India's expanding demands and the objectives of global sustainability.

15.2 Limitations of Conventional Sewage Treatment and Sludge Practices

Modern wastewater management procedures have their roots in traditional sewage treatment systems, which provide substantial advancements over crude disposal methods (Sathya et al. 2022). Trickling filters and the activated sludge process are two of the most used conventional techniques. While trickling filters employ a bed of media over which wastewater is trickled, allowing microorganisms to develop a biofilm that breaks down impurities, the activated sludge process depends on aerating wastewater to promote microbial breakdown of organic pollutants (Dai et al. 2022). Although these techniques have successfully lowered wastewater's suspended solids and biochemical oxygen demand (BOD). It should be mentioned that these were made for a very different world than the one we live in now, one with fewer pollutants and less consideration for the environment, thus they have several restrictions as shown in Fig. 15.1.

The shortcomings of these conventional methods have grown more apparent as wastewater properties have evolved and environmental regulations have tightened. The high energy consumption of these devices is one of the main shortcomings (Chen et al. 2025). One of the most energy-intensive phases of treatment is the activated sludge process, which requires constant aeration to sustain microbial activity. Concerns regarding these treatment facilities' environmental sustainability are raised by their significant reliance on mechanical systems, which also raises operating costs and significantly increases greenhouse gas emissions (Guo et al. 2024).

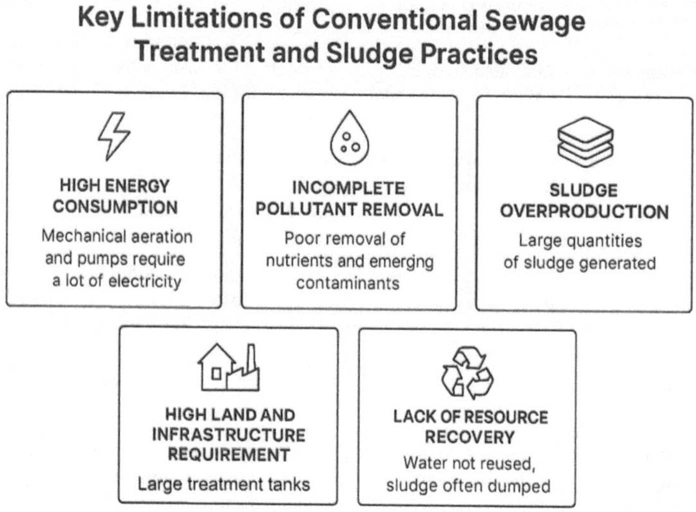

Fig. 15.1 Key limitations of conventional sewage treatment and sludge practices

Another important concern is the incomplete removal of contaminants. Conventional systems can effectively handle organic debris and pathogens, but they frequently struggle to handle contemporary contaminants such as microplastics, endocrine-disrupting substances, medications, and personal care items (Shukla and Ahammad 2023). These newly discovered pollutants can mostly evade conventional treatment methods before finding their way into natural water bodies and endangering human health and ecosystems (Mutegoa 2024).

Sludge disposal issues exacerbate the drawbacks of conventional systems. It is necessary to treat, dewater, and properly dispose of the sludge produced by primary and secondary treatment (Gupta et al. 2024a, b). Sludge management is complicated by the presence of heavy metals, viruses, and persistent organic pollutants, which makes land application or disposal more challenging under current regulatory frameworks Table 15.1. Sludge management becomes a major operational burden for treatment plants when the expenses and environmental hazards of managing sludge frequently exceed the advantages (Liu et al. 2022).

Furthermore, the land and operating needs of the typical treatment plants present considerable challenges, especially in rapidly urbanizing areas. For example, trickling filters require enormous land areas to accommodate biofilm beds, while activated sludge systems require big aeration tanks and settling basins (Zhang et al. 2023). Because of their large spatial footprint, conventional plants are not appropriate for areas with high population densities or where land costs are prohibitive. The efficiency and dependability of these conventional systems are further strained by their

Table 15.1 Limitations of conventional sewage treatment and sludge practices

Sl. No.	Aspect	Traditional method	Limitations	References
1	Treatment processes	Activated sludge process, trickling filters	Effective for basic pollutants, but not for emerging contaminants	Shukla and Ahammad (2023)
2	Contaminant removal	Biological degradation, basic sedimentation	Poor removal of micro-pollutants, pharmaceuticals, and microplastics	Gupta et al. (2024a, b)
3	Energy consumption	Mechanical aeration, pumping systems	High operational energy demand leading to an increased carbon footprint	Gupta et al. (2024a, b)
4	Land requirement	Large aeration tanks and biofilm beds	High land footprint, unsuitable for dense urban areas	Gupta et al. (2024a, b)
5	Sludge production	Biomass generation during biological treatment	Large volumes of sludge require costly treatment and safe disposal	Kurniawan et al. (2025)
6	Operational complexity	Continuous monitoring and requirement of skilled labour	Vulnerable to fluctuations in wastewater characteristics; costly to manage	Cairone et al. (2024)

operational complexity, reliance on skilled labour, and susceptibility to changes in influent characteristics (Gupta et al. 2024a, b).

All of these drawbacks point to the pressing need for sludge management and sewage treatment methods that are more resource-conscious, energy-efficient, and sustainable. In addition to addressing these issues, the new generation of technology aims to reframe wastewater as a useful resource as opposed to a waste product.

15.3 Recent Technological Advancements in Sewage Treatment

The increasing complexity of wastewater contaminants and the urgent need for sustainable management have spurred significant technological developments in sewage treatment. These advancements offer better treatment efficacy, lower operational costs, and a smaller environmental impact by resolving the shortcomings of conventional methods (Wang et al. 2024a, b). Innovations ranging from membrane technologies to biological processes, physico-chemical approaches, and nature-inspired solutions represent a shift toward integrated, flexible, and resource-recovering systems as shown in Fig. 15.2 (Wang et al. 2024a, b).

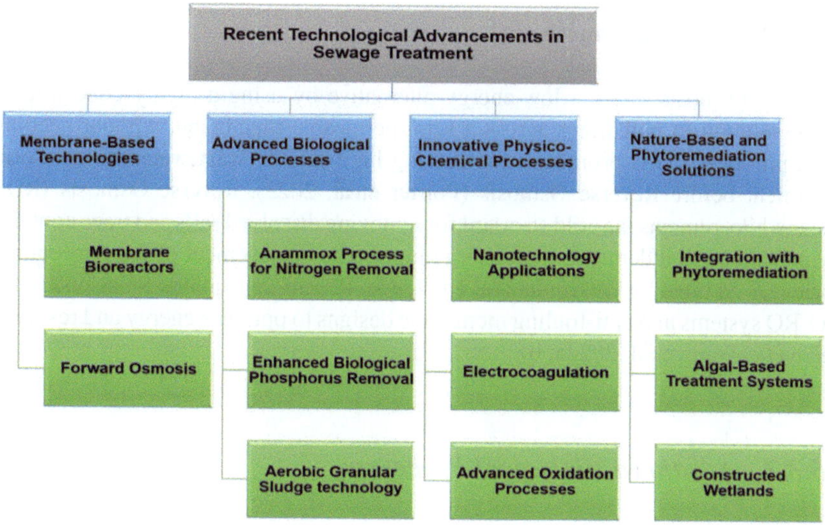

Fig. 15.2 Classification of recent sewage treatment technologies

15.3.1 Membrane-Based Technologies

As membrane technologies allow for the fine separation of contaminants at the molecular level, they have completely changed the potential of wastewater treatment. These systems are becoming more and more important for water reuse and recycling programs since they continuously produce high-quality effluents (Ang et al. 2024).

15.3.1.1 Membrane Bioreactors (MBRs)

MBRs integrate a biological degradation process with membrane filtration in a single compact system. The suspended growth reactor biologically treats wastewater, while membranes (typically ultrafiltration or microfiltration) physically separate biomass and solids from treated water (Fadillah et al. 2024). MBRs produce superior quality effluent suitable for direct non-potable reuse. Advantages include smaller space requirements, reduced sludge production, and greater operational flexibility. However, membrane fouling, due to microbial products and solids, remains a major operational bottleneck, often necessitating intensive maintenance protocols like backwashing and chemical cleaning (Rajendran et al. 2025).

15.3.1.2 Forward Osmosis (FO)

Forward Osmosis offers a low-energy alternative by using a highly concentrated draw solution to pull water across a semi-permeable membrane, without needing high pressure. FO is promising for treating high-salinity wastewaters or as a pre-treatment before Reverse Osmosis (Poirier et al. 2023). Reverse Osmosis (RO), meanwhile, remains the gold standard for removing dissolved salts and trace contaminants, although it demands high energy input. The management of RO concentrate (brine) is an ongoing environmental challenge. Recent innovations focus on hybrid FO-RO systems and anti-fouling membrane designs to optimise energy and resource recovery (Wang et al. 2024a, b).

15.3.2 Advanced Biological Processes

New-generation biological processes are making sewage treatment more efficient, targeted, and sustainable by leveraging specific microbial capabilities.

15.3.2.1 Anammox Process for Nitrogen Removal

The Anammox (Anaerobic Ammonium Oxidation) process is a groundbreaking development for nitrogen management. It bypasses the traditional nitrification–denitrification sequence by enabling specialized bacteria to directly convert ammonium and nitrite into nitrogen gas without the need for oxygen or organic carbon. This process can cut energy consumption by up to 60% compared to conventional nitrogen removal and significantly reduce greenhouse gas emissions from treatment plants (Liu et al. 2024a, b, c).

15.3.2.2 Aerobic Granular Sludge (AGS) Technology

AGS represents a shift from conventional floc-based systems to dense microbial aggregates. Granules offer improved settling characteristics and contain stratified micro-environments, allowing simultaneous aerobic, anoxic, and anaerobic reactions (Fadillah et al. 2024). This multi-functional structure enables integrated organic matter removal, nitrification–denitrification, and biological phosphorus removal within a single reactor. Additionally, AGS systems exhibit resilience to hydraulic and organic shock loads, making them ideal for varying influent qualities (Fadillah et al. 2024).

15.3.2.3 Enhanced Biological Phosphorus Removal (EBPR)

Another strategic advancement that biologically harnesses polyphosphate-accumulating organisms (PAOs) to achieve high phosphorus removal. PAOs uptake phosphorus beyond their metabolic needs, allowing its removal as part of the biomass (Li et al. 2024a, b). This not only reduces chemical usage but also facilitates phosphorus recovery for fertilizer production, promoting circular economy principles in wastewater management.

15.3.3 Innovative Physico-chemical Processes

Physico-chemical advancements are crucial for addressing contaminants that resist biological degradation, including pharmaceutical residues, industrial solvents, and heavy metals.

15.3.3.1 Advanced Oxidation Processes (AOPs)

AOPs generate highly reactive species such as hydroxyl radicals that non-selectively attack and degrade complex organic molecules into harmless end-products like carbon dioxide and water (Wang et al. 2022). Techniques include ozone-based AOPs, Fenton's reagent (Fe_{2+}/H_2O_2), UV-photolysis, and TiO_2 photocatalysis. AOPs are particularly effective for removing persistent organic pollutants (POPs) and are being increasingly employed for wastewater polishing before reuse or discharge into sensitive ecosystems (Wang et al. 2022).

15.3.3.2 Electrocoagulation

Electrocoagulation is a technique for treating wastewater that uses electric current to extract dissolved, emulsified, or suspended contaminants (Boinpally et al. 2023). It employs an electric current to sacrificial electrodes (usually iron or aluminium), producing coagulant ions in situ that destabilize suspended solids, emulsified oils, and heavy metals. The key advantages are minimal chemical usage, easy automation, and the production of compact, less-hazardous sludge. However, electrode maintenance and energy optimization remain areas of ongoing research (Boinpally et al. 2023).

15.3.3.3 Nanotechnology Applications

This method involves the use of nanoparticles, nanocomposites, and nano-membranes for advanced contaminant removal. Nanomaterials, due to their enormous surface area and unique physicochemical properties, excel at adsorbing heavy metals, degrading organic contaminants, and exhibiting antimicrobial properties. Yet, the long-term environmental fate of nanoparticles necessitates careful consideration to avoid secondary pollution (Shah et al. 2024).

15.3.4 Nature-Based and Phytoremediation Solutions

Nature-based solutions (NbS) are revolutionizing conventional sewage treatment paradigms by providing environmentally friendly, low-energy, and sustainable alternatives (Sandhwar et al. 2025). By utilizing organic biological, chemical, and physical processes, these systems effectively purify wastewater while also supporting biodiversity, sequestering carbon, and improving the landscape (Singh et al., 2025). Nature-based sewage treatment solutions are becoming more and more popular as the world turns its attention to green infrastructure and the circular economy (Sandhwar et al. 2025).

15.3.4.1 Constructed Wetlands

Constructed wetlands are engineered systems designed to replicate the water purification processes observed in natural wetlands. They utilize a synergistic combination of hydrological regimes, substrate filtration, microbial communities, and vegetation to treat municipal, industrial, and agricultural wastewater (Zhao et al. 2023a, b, c). The design may include subsurface flow (SSF) wetlands or free-water surface (FWS) wetlands. These systems achieve high removal efficiencies for biochemical oxygen demand (BOD), suspended solids, nutrients (nitrogen and phosphorus), pathogens, and even heavy metals. Beyond water quality improvement, constructed wetlands serve as carbon sinks, enhance local biodiversity, and provide critical green spaces within urban and peri-urban landscapes (Saeed et al. 2022). Their cost-effectiveness, low maintenance demands, resilience to hydraulic fluctuations, and minimal energy consumption make them particularly attractive for decentralized wastewater treatment in rural communities and developing regions (Zhao et al. 2023a, b, c).

15.3.4.2 Algal-Based Treatment Systems

Algal-based treatment systems utilize the natural nutrient uptake capabilities of microalgae to remove contaminants, particularly nitrogen and phosphorus, from wastewater (Li et al. 2022a, b). In photobioreactors or open pond systems, algae photosynthetically assimilate inorganic nutrients, thereby preventing eutrophication in downstream aquatic ecosystems. Moreover, algal systems produce substantial quantities of biomass, which can be valorized into biofuels, fertilizers, animal feed, or even bioplastics, supporting circular economy models (Li et al. 2022a, b). Recent innovations involve hybrid algae-bacterial consortia, where algae provide oxygen to bacterial partners, thus eliminating the need for mechanical aeration, which is a major energy sink in conventional treatment (Phyu et al. 2025). Algal systems are not only effective for nutrient polishing but also have potential applications in heavy metal removal, pharmaceutical degradation, and carbon dioxide sequestration, making them versatile platforms for integrated wastewater management (Phyu et al. 2025).

15.3.4.3 Integration with Phytoremediation

Phytoremediation strategies exploit the natural ability of certain plant species to absorb, accumulate, degrade, or immobilize a wide range of pollutants, including organic compounds, heavy metals, and emerging contaminants (Madhav et al. 2024). In wastewater management, phytoremediation is incorporated through systems such as floating treatment wetlands, vegetated drainage channels, bioswales, and vertical green walls (Maniam et al. 2022). Strategically selected plants such as *Typha*, *Phragmites*, *Vetiver*, and *Canna* species potentially enhance contaminant removal through mechanisms like rhizofiltration, phytodegradation, and phytostabilization (Madhav et al. 2024). Additionally, phytoremediation-based systems contribute secondary benefits such as urban heat island mitigation, aesthetic improvement, habitat creation, and improved microclimate regulation. These multi-functional landscapes align well with United Nations Sustainable Development Goals (SDGs), particularly those related to clean water, climate action, and sustainable cities (Maniam et al. 2022).

15.4 Modern Approaches to Sludge Management

Modern wastewater treatment systems no longer view sludge merely as a waste product (Khan et al. 2023). Instead, sludge is being reinterpreted as a useful source of nutrition, energy, and recyclable materials. This transition emphasizes waste reduction, circular resource use, and carbon–neutral operations, all of which align with global sustainability goals. Some of the newest developments in sludge management are examined in the parts that follow and in Table 15.2, with an emphasis on reducing sludge production, recovering embedded resources, and using cutting-edge treatment technologies (Tarpani and Azapagic 2023).

Table 15.2 Comparison of modern sludge management approaches

Sl. No.	Technique	Function	Main benefits	Best application	References
1	Oxic settling anaerobic (OSA) process	Reduce biomass during treatment	Less sludge, lower cost	Municipal WWTPs with conventional ASP	Talepour et al. (2025)
2	Biohydrogen production	Clean energy from fermentation	Hydrogen fuel	Experimental or pilot-scale renewable energy	Zhao et al. (2023a, b, c)
3	Anaerobic digestion (AD)	Convert organics to methane	Biogas, sludge stabilization	Centralized WWTPs with high organic loads	Talepour, et al. (2025)
4	Biochar production	Pyrolysis of dried sludge	Soil enhancer, carbon sink	Rural/agricultural zones	Khan et al. (2023)
5	Thermal plasma	Total decomposition of sludge	Inert slag, synthesis gas	Toxic, industrial, or pharmaceutical sludge	Silva et al. (2025)
6	Thermal hydrolysis	Enhance digestion	Increased gas yield, better dewatering	High-load urban WWTPs	Zhao et al. (2023a, b, c)
7	Gasification	Convert sludge to syngas	Heat and power generation	Energy-oriented treatment facilities	Li et al. (2024a, b)
8	Struvite recovery	Capture P and N from sludge	Fertilizer-grade crystals	Plants with nutrient-rich sludge streams	Li et al. (2024a, b)

15.4.1 Sludge Reduction at Source

Sludge production begins during the secondary treatment phase of sewage processing (Wang et al. 2023). Conventionally, excess biomass from microbial treatment requires significant handling and disposal. However, process-level innovations now enable a reduction in sludge generation at its origin, reducing downstream burdens and operational costs. One of the most promising methods is the Oxic Settling Anaerobic (OSA) process (Takdastan et al. 2025). This technique loops treated sludge between aerobic and anaerobic reactors. The alternating conditions encourage microbial decay, resulting in less excess biomass and significantly lower sludge yields, often reduced remarkably by 30–50% compared to standard activated sludge systems (Morello et al. 2023). The OSA process also improves settling characteristics and decreases the need for frequent sludge wasting (Morello et al. 2023).

Uncoupled metabolism techniques are another noteworthy strategy that prevents microbial production by interfering with the energy conversion process. Chemical or thermal pretreatment to promote endogenous respiration and cellular disintegration, such as light heat application or alkaline dosage (Morello et al. 2023). Microwave or ultrasound disintegration, which increases biodegradability by weakening microbial cell walls.

These approaches are particularly useful in space-limited or energy-sensitive wastewater treatment plants, offering a cost-effective path toward sustainable sludge handling.

15.4.2 Resource Recovery from Sludge

Organic carbon, nutrients, and trace elements are abundant in sewage sludge. Modern plants increasingly concentrate on obtaining value from sludge through material recovery, nitrogen cycling, and bioenergy production rather than disposal (Śniatała et al. 2024). Treatment plants can become more self-sufficient and low-impact systems by integrating resource recovery into sludge management. It creates economic value, lessens environmental impact, and advances more general sustainability objectives, such as circular agriculture, cleaner water, and renewable energy. It's a potent illustration of how, with the correct technology and mindset, environmental problems can be transformed into possibilities (Śniatała et al. 2024).

15.4.2.1 Biogas and Biohydrogen Production

Anaerobic digestion (AD) is one of the oldest methods for obtaining energy from sludge. Without oxygen, this biological process breaks down organic waste to produce biogas, primarily methane, which can be improved and delivered into natural gas systems or used to power treatment facilities. More sophisticated types, such as thermophilic digestion, which works at higher temperatures, are more efficient and more effective at eliminating germs (Ram et al. 2024). Co-digestion, which involves treating sludge and food waste together, increases process productivity by producing more gas and stabilizing operations. Meanwhile, the production of biohydrogen through dark fermentation is emerging as a clean, useful energy source that is especially relevant to infrastructure that is low-carbon and prepared for the future (Ram et al. 2024).

15.4.2.2 Biochar, Bioplastics and Fertilizers

Sludge is being tapped to recover materials. For instance, pyrolysis, the process of heating sludge without oxygen to create biochar, can improve soil health, assist retain nutrients, and even help lock carbon into the land, reducing emissions (Zhang et al. 2023). It is an excellent substance for enhancing soil properties and absorbing pollutants due to its high porosity. Another exciting breakthrough is the microbial manufacture of bioplastics, particularly polyhydroxyalkanoates (PHAs). These biodegradable polymers are created by bacteria that eat the volatile fatty acids created during sludge fermentation, turning waste into beneficial packaging or agricultural products (Khan et al. 2023).

15.4.2.3 Nutrient (N, P, K) Recovery

Nutrient recovery is another crucial area in which sludge can prove beneficial. Phosphorus and ammonium can be recovered by processes like struvite crystallization, which produces slow-release fertilizers that can be applied to crops without the dangers of raw sludge (Valchev et al. 2024). Other methods, like ion exchange, membrane separation, and ammonia stripping, aid in the recovery of nitrogen (N) and potassium (K), which are essential nutrients for crop production. These recovered nutrients aid in preventing nutrient pollution in water bodies and lessen the demand for the manufacture of synthetic fertilizers (Valchev et al. 2024).

15.4.3 Advanced Sludge Treatment Technologies

Effective and sustainable sludge management has emerged as a top priority as wastewater treatment plants transform into resource recovery facilities. These days, modern technologies are used not only to decrease the volume of sludge but also to improve pathogen elimination, increase energy yield, and allow for the safe reuse of material. The precision, effectiveness, and potential for the future of these techniques far surpass those of conventional thickening and drying (Taşeli et al. 2025).

15.4.3.1 Thermal Hydrolysis

Thermal hydrolysis is a pre-treatment process that subjects dewatered sludge to high-pressure saturated steam, typically at 160–180 °C (Sahu et al. 2022). This rapid thermal shock breaks down complex organic compounds, ruptures microbial cell walls, and makes the sludge more amenable to anaerobic digestion. As a result, plants using this method commonly observe a 30–50% increase in biogas production, better dewaterability, and a significant reduction in pathogens and odours. By enhancing biodegradability, thermal hydrolysis turns sludge into a more productive feedstock for energy recovery, while improving the safety and handling of the final biosolid (Giwa et al. 2023).

15.4.3.2 Pyrolysis and Gasification

These thermochemical conversion technologies offer a cleaner and more efficient route to energy and material recovery (Begum et al. 2024). Pyrolysis operates in an oxygen-free environment at temperatures between 300 and 700 °C, decomposing sludge into biochar, bio-oil, and non-condensable gases (Siwal et al. 2022). Gasification, in contrast, introduces a controlled amount of oxygen or steam and operates above 800 °C, favouring the production of syngas, a versatile fuel composed mainly of hydrogen and carbon monoxide (Gururani et al. 2022). Both processes achieve over 90% volume reduction, destroy harmful contaminants such as pharmaceutical residues and pathogens, and generate reusable outputs (Sharma et al. 2022a, b). Biochar can be used as a soil amendment and carbon sink, while syngas and bio-oil can be integrated into a local energy system. These technologies mark a significant shift toward low-waste, high-value sludge processing (Zhu et al. 2022).

15.4.3.3 Thermal Plasma Treatment

Thermal plasma treatment, a high-temperature procedure that uses plasma arcs (> 1500 °C) to dissolve organic and inorganic components in sludge, is at the forefront of sludge management (Silva et al. 2025). In contrast to other thermal processes, plasma produces almost complete contamination breakdown, resulting in vitrified slag, which is a stable, glass-like substance that can be used in construction and pure synthesis gas for energy applications (Silva et al. 2025). This method offers complete sterilization, destruction of persistent pollutants, and zero waste discharge, making it ideal for complex or hazardous industrial sludges. Though energy-intensive, it's well-suited for niche applications where environmental safety is non-negotiable, and traditional methods fall short (Zhou et al. 2024).

15.5 Challenges, Innovations, and the Future of Sewage and Sludge Management

As cities continue to grow and environmental laws become more stringent, the field of sewage and sludge management is undergoing significant change. Circular, resource-centric models that prioritize efficiency, sustainability, and resilience are replacing traditional linear approaches (Zhao et al. 2023a, b, c). However, several monetary, societal, and technological obstacles persist despite significant progress in recovery and treatment technology (Sharma et al. 2022a, b). This section provides a vision for the next generation of intelligent, networked, and regenerative wastewater systems in addition to examining these difficulties and current developments.

15.5.1 Technical Limitations and Process Complexity

Membrane bioreactors (MBRs), advanced oxidation processes (AOPs), and thermal plasma reactors are examples of advanced treatment methods. These methods frequently have significant energy consumption, complex operating requirements, and restricted scalability (Rao et al. 2022). The integration of these systems within aging infrastructure remains technically challenging, particularly where decentralized installations lack a consistent power supply or skilled operators. Furthermore, the recovery of nutrients and energy from sludge often suffers from suboptimal yield due to system inefficiencies and microbial instability under real-world conditions (Rao et al. 2022).

15.5.2 Societal Acceptance and Institutional Barriers

Despite proven efficacy, public resistance to the reuse of treated wastewater and biosolids stemming from perceived health risks and cultural norms undermines adoption. Institutional inertia, inadequate regulatory frameworks, and the absence of economic incentives further slow the transition from linear waste disposal to resource recovery paradigms (Chen et al. 2024). Behavioural economics and risk communication must now become integral to engineering solutions. Building trust and social acceptance requires bridging the gap between popular perception and scientific fact. Proactive policy interventions and community engagement can facilitate a smoother transition towards circular waste management practices (Chen et al. 2024).

15.5.3 Research Gaps and Need for Technological Integration

Contemporary systems lack holistic integration of artificial intelligence, machine learning, and sensor-driven automation, which are critical for predictive diagnostics and adaptive control. Additionally, few studies comprehensively address the coupling of low-energy biological treatments with high-efficiency recovery units. Fundamental gaps persist in scaling hybrid systems and assessing their life cycle impacts, particularly in rural and peri-urban contexts.

15.5.4 Future Vision: Smart, Modular, and Regenerative Sanitation Systems

The future of sewage treatment and sludge management is being rapidly transformed by the integration of smart technologies, modular system designs, and regenerative sanitation approaches. These advancements aim to address persistent challenges such as inadequate sanitation coverage, energy-intensive processes, and unsustainable waste disposal methods, particularly in the context of rapidly urbanizing and resource-stressed regions.

Smart sanitation systems incorporate real-time monitoring and automation technologies that enhance the efficiency, reliability, and responsiveness of wastewater treatment operations. The use of enabled sensors, artificial intelligence, and data analytics platforms allows for predictive maintenance, optimized energy consumption, and dynamic adjustment of treatment parameters based on fluctuating influent characteristics. For instance, machine learning models can predict BOD and nutrient loads, thereby enabling proactive operational decisions and reducing treatment costs (Mondal 2024).

Modular sanitation systems are gaining traction as scalable and adaptable solutions that can be rapidly deployed in decentralized settings. These systems are particularly beneficial for peri-urban and rural communities where centralized wastewater infrastructure is lacking or financially unfeasible. Modular units can be customized to treat various types of waste streams including domestic sewage, faecal sludge, and greywater using biological, physicochemical, or membrane-based treatment technologies (Mondal 2024). The plug-and-play design facilitates ease of transport, assembly, and expansion, enabling service providers to respond flexibly to changing demographic and environmental conditions.

Regenerative sanitation goes a step further by reimagining waste as a resource. This paradigm emphasizes nutrient recovery, energy generation, and water reuse within a circular economy framework. Technologies such as anaerobic digestion, struvite precipitation, and hydrothermal carbonization not only stabilize sewage and sludge but also yield valuable by-products including biogas, fertilizers, and biochar (Morello et al. 2022). For example, decentralized anaerobic digesters integrated with nutrient recovery systems have demonstrated the potential to meet up to 80% of the energy requirements of small-scale treatment facilities while recovering more than 60% of phosphorus and nitrogen content from sludge.

In addition, climate-resilient and off-grid solutions are being prioritized in the design of next-generation sanitation systems, particularly in low- and middle-income countries where infrastructure is vulnerable to extreme weather events. Solar-powered treatment modules, passive filtration systems, and waterless sanitation options are being developed to operate with minimal environmental impact while enhancing public health outcomes (Chang et al. 2023).

15.6 Conclusions

During the last decades, both sewage treatment and sludge management have undergone profound changes, shifting from legacy, energy-dependent, and linear schemes to efficient, sustainable, and circular ones. The employment of cutting-edge technologies ranging from membrane-based systems and innovative biological processes to physico-chemical practices has dramatically enhanced pollutant removal efficiency, cut operational footprints, and increased overall reliability of WWTPs.

Concurrently, the technology of sludge management has also seen paradigm shifts with new approaches in the form of resource-reduction strategies, resource recovery methods and high-temperature treatment methods like pyrolysis and plasma gasification. In addition to volume reduction, these technologies also facilitate the retrieval of valuable resources, supporting economic feasibility and sustainable treatment systems.

The chapter emphasizes the urgency for innovation-led management of sewage, especially with a focus on urbanization, climate variability, and clean water demands. Adopting modular, local, and intelligent treatment systems with artificial intelligence and real-time monitoring resources is necessary in order to attain operational effectiveness, adaptation management, and resilience in both metropolitan and rural environments. The principle of sustainable water futures is all about the implementation of circular economy principles in which wastewater is no longer viewed as waste but as a resource. Closing the loop is not merely a desirable but a necessary objective through reuse of water, energy neutrality, and recycling of nutrients. Future policies, infrastructure planning, and research are expected to align in developing carbon–neutral, resource-positive, and socially inclusive sanitation systems that promote environment, economic, and social well-being.

References

Ang WJ, Teow YH, Chang ZH, Mohammad AW, Wan TW (2024) Innovative ceramic membrane plate filtration system for sustainable semiconductor industry wastewater treatment: a pilot scale study. Chem Eng J 496:153767. https://doi.org/10.1016/j.cej.2024.153767

Awewomom J, Dzeble F, Takyi YD, Ashie WB, Ettey ENYO, Afua PE, Sackey LNA, Opoku F, Akoto O (2024) Addressing global environmental pollution using environmental control techniques: a focus on environmental policy and preventive environmental management. Discov Environ 2:8. https://doi.org/10.1007/s44274-024-00033-5

Begum YA, Kumari S, Jain SK, Garg MC (2024) A review on waste biomass-to-energy: integrated thermochemical and biochemical conversion for resource recovery. Environ Sci Adv 3:1197–1216. https://doi.org/10.1039/D4VA00109E

Boinpally S, Kolla A, Kainthola J, Kodali R, Vemuri J (2023) A state-of-the-art review of the electrocoagulation technology for wastewater treatment. Water Cycle 4:26–36. https://doi.org/10.1016/j.watcyc.2023.01.001

Cairone S, Hasan SW, Choo K-H, Lekkas DF, Fortunato L, Zorpas AA, Korshin G, Zarra T, Belgiorno V, Naddeo V (2024) Revolutionizing wastewater treatment toward circular economy and carbon neutrality goals: pioneering sustainable and efficient solutions for automation and advanced process control with smart and cutting-edge technologies. J Water Process Eng 63:105486. https://doi.org/10.1016/j.jwpe.2024.105486

Chang H, Zhao Y, Bisinella V, Damgaard A, Christensen TH (2023) Climate change impacts of conventional sewage sludge treatment and disposal. Water Res 240:120109. https://doi.org/10.1016/j.watres.2023.120109

Chen L, Chen Z, Liu Y, Lichtfouse E, Jiang Y, Hua J, Osman AI, Farghali M, Huang L, Zhang Y, Rooney DW, Yap P-S (2024) Benefits and limitations of recycled water systems in the building sector: a review. Environ Chem Lett 22:785–814. https://doi.org/10.1007/s10311-023-01683-2

Chen C, Zhang X, Webster C (2025) Truncation in the scaling of urban pollution control. NPJ Urban Sustain 5:9. https://doi.org/10.1038/s42949-025-00197-w

Dai H, Sun Y, Wan D, Abbasi HN, Guo Z, Geng H, Wang X, Chen Y (2022) Simultaneous denitrification and phosphorus removal: a review on the functional strains and activated sludge processes. Sci Total Environ 835:155409. https://doi.org/10.1016/j.scitotenv.2022.155409

Fadillah G, Alarifi NTS, Suryawan IWK, Saleh TA (2024) Advances in designed reactors for water treatment process: a review highlighting the designs and performance. J Water Process Eng 63:105417. https://doi.org/10.1016/j.jwpe.2024.105417

Giwa AS, Maurice NJ, Luoyan A, Liu X, Yunlong Y, Hong Z (2023) Advances in sewage sludge application and treatment: process integration of plasma pyrolysis and anaerobic digestion with the resource recovery. Heliyon 9:e19765. https://doi.org/10.1016/j.heliyon.2023.e19765

Guo D, Li B, Yu W, Baroutian S, Young BR (2024) A system engineering perspective for net zero carbon emission in wastewater and sludge treatment industry: a review. Sustain Prod Consump 46:369–381. https://doi.org/10.1016/j.spc.2024.02.033

Gupta A, Kumar S, Bajpai Y, Chaturvedi K, Johri P, Tiwari RK, Vivekanand V, Trivedi M (2024a) Pharmaceutically active micropollutants: origin, hazards and removal. Front Microbiol 15:1339469. https://doi.org/10.3389/fmicb.2024.1339469

Gupta V, Gonzales HB, Hungwe D, Jaldhari R, Xu H, Setyawan MIB, Wang B, Yang P, Takahashi F (2024) Behavioural insights into sewage sludge hydrochar acceptance through the theory of planned behaviour and cross table analysis in Jaipur City, India. https://doi.org/10.2139/ssrn.5038102

Gururani P, Bhatnagar P, Bisht B, Jaiswal KK, Kumar V, Kumar S, Vlaskin MS, Grigorenko AV, Rindin KG (2022) Recent advances and viability in sustainable thermochemical conversion of sludge to bio-fuel production. Fuel 316:123351. https://doi.org/10.1016/j.fuel.2022.123351

Guven H, Ersahin ME, Ozgun H, Ozturk I, Koyuncu I (2023) Energy and material refineries of future: wastewater treatment plants. J Environ Manage 329:117130. https://doi.org/10.1016/j.jenvman.2022.117130

Khan R, Shukla S, Kumar M, Zuorro A, Pandey A (2023) Sewage sludge derived biochar and its potential for sustainable environment in circular economy: advantages and challenges. Chem Eng J 471:144495. https://doi.org/10.1016/j.cej.2023.144495

Kurniawan TA, Yap P-S, Chen Z (2025) Techniques for pollutant removal, nutrient recovery, and energy production from landfill leachates: a review. Environ Chem Lett 23:517–577. https://doi.org/10.1007/s10311-024-01805-4

Li M, Zamyadi A, Zhang W, Dumée LF, Gao L (2022a) Algae-based water treatment: a promising and sustainable approach. J Water Process Eng 46:102630. https://doi.org/10.1016/j.jwpe.2022.102630

Li S, Show PL, Ngo HH, Ho S-H (2022b) Algae-mediated antibiotic wastewater treatment: a critical review. Environ Sci Ecotechnol 9:100145. https://doi.org/10.1016/j.ese.2022.100145

Li A, Han H, Zheng K, Zhu M, Xu K, Xu J, Jiang L, Wang Y, Su S, Hu S, Xiang J (2024a) Sludge pyrolysis integrated biomass gasification to promote syngas: comparison of different biomass. Sci Total Environ 908:168278. https://doi.org/10.1016/j.scitotenv.2023.168278

Li G, Srinivasan V, Tooker NB, Wang D, Yan Y, Onnis-Hayden A, Gu AZ (2024b) Distinct microdiversity of phosphate accumulating organisms (PAOs) between side-stream and conventional enhanced biological phosphorus removal (EBPR) systems with performance implications. Water Res 266:122280. https://doi.org/10.1016/j.watres.2024.122280

Liu X, Huang H, Iqbal A, Chen J, Zan F, Jiang C, Chen G (2022) Sustainability analysis of implementing sludge reduction in overall sludge management process: where do we stand? Waste Manage 152:80–93. https://doi.org/10.1016/j.wasman.2022.08.013

Liu L, Guo Z, Wang Y, Yin L, Zuo W, Tian Y, Zhang J (2024a) Low energy-consumption oriented membrane fouling control strategy in anaerobic fluidized membrane bioreactor. Chemosphere 359:142254. https://doi.org/10.1016/j.chemosphere.2024.142254

Liu S, Cai C, Sun F, Ma M, An T, Chen C (2024b) Advanced nitrogen removal of landfill leachate treatment with anammox process: a critical review. J Water Process Eng 58:104756. https://doi.org/10.1016/j.jwpe.2023.104756

Liu X, Sathishkumar K, Zhang H, Saxena KK, Zhang F, Naraginti S, Anbarasu K, Rajendiran R, Rajasekar A, Guo X (2024c) Frontiers in environmental cleanup: recent advances in remediation of emerging pollutants from soil and water. J Hazard Mater Adv 16:100461. https://doi.org/10.1016/j.hazadv.2024.100461

Mabrouki J (ed) (2025) Circular economy applications in energy policy. IGI Global Scientific Publishing, Hershey, Pennsylvania (701 E. Chocolate Avenue, Hershey, Pennsylvania, 17033, USA). https://doi.org/10.4018/979-8-3693-6680-6

Madhav S, Mishra R, Kumari A, Srivastav AL, Ahamad A, Singh P, Ahmed S, Mishra PK, Sillanpää M (2024) A review on sources identification of heavy metals in soil and remediation measures by phytoremediation-induced methods. Int J Environ Sci Technol 21:1099–1120. https://doi.org/10.1007/s13762-023-04950-5

Maniam G, Zakaria NA, Leo CP, Vassilev V, Blay KB, Behzadian K, Poh PE (2022) An assessment of technological development and applications of decentralized water reuse: a critical review and conceptual framework. Wires Water 9:e1588. https://doi.org/10.1002/wat2.1588

Mishra R, Singh E, Kumar A, Singh AK, Madhav S, Shukla SK, Kumar S (2023) Metal pollution in marine environment: sources and impact assessment, in: Metals in Water. Elsevier, pp. 175–193. https://doi.org/10.1016/B978-0-323-95919-3.00006-9

Mondal P (2024) AI and IoT in smart water management for urban sustainability. Uncertain Discour Appl 1:151–157. https://doi.org/10.48313/uda.v1i2.36

Morello R, Di Capua F, Esposito G, Pirozzi F, Fratino U, Spasiano D (2022) Sludge minimization in mainstream wastewater treatment: mechanisms, strategies, technologies, and current development. J Environ Manage 319:115756. https://doi.org/10.1016/j.jenvman.2022.115756

Morello R, Di Capua F, Sahinkaya E, Esposito G, Pirozzi F, Fratino U, Spasiano D (2023) Operational strategies enhancing sewage sludge minimization in a combined integrated fixed-film activated sludge—oxic settling anaerobic system. J Environ Manage 345:118808. https://doi.org/10.1016/j.jenvman.2023.118808

Mutegoa E (2024) Efficient techniques and practices for wastewater treatment: an update. Discov Water 4:69. https://doi.org/10.1007/s43832-024-00131-8
Phyu K, Zhi S, Graham DW, Cao Y, Xu X, Liu J, Wang H, Zhang K (2025) Impact of indigenous vs. cultivated microalgae strains on biomass accumulation, microbial community composition, and nutrient removal in algae-based dairy wastewater treatment. Bioresour Technol 426:132349. https://doi.org/10.1016/j.biortech.2025.132349
Poirier K, Lotfi M, Garg K, Patchigolla K, Anthony EJ, Faisal NH, Mulgundmath V, Sahith JK, Jadhawar P, Koh L, Morosuk T, Al Mhanna N (2023) A comprehensive review of pre- and post-treatment approaches to achieve sustainable desalination for different water streams. Desalination 566:116944. https://doi.org/10.1016/j.desal.2023.116944
Rajendran DS, Devi EG, Subikshaa VS, Sethi P, Patil A, Chakraborty A, Venkataraman S, Kumar VV (2025) Recent advances in various cleaning strategies to control membrane fouling: a comprehensive review. Clean Techn Environ Policy 27:649–664. https://doi.org/10.1007/s10098-024-03000-z
Ram C, Rani P, Kumar A (2024) Recent developments in biohydrogen production from wastewater: a review. Biocatal Biotransform 42:1–18. https://doi.org/10.1080/10242422.2023.2181046
Rao B, Wang G, Xu P (2022) Recent advances in sludge dewatering and drying technology. Drying Technol 40:3049–3063. https://doi.org/10.1080/07373937.2022.2043355
Saeed T, Majed N, Kumar Yadav A, Hasan A, Jihad Miah M (2022) Constructed wetlands for drained wastewater treatment and sludge stabilization: role of plants, microbial fuel cell and earthworm assistance. Chem Eng J 430:132907. https://doi.org/10.1016/j.cej.2021.132907
Sahu AK, Mitra I, Kleiven H, Holte HR, Svensson K (2022) Chapter 24—Cambi thermal hydrolysis process (CambiTHP) for sewage sludge treatment. In: An A, Tyagi V, Kumar M, Cetecioglu Z (eds) Clean energy and resource recovery. Elsevier, pp 405–422. https://doi.org/10.1016/B978-0-323-90178-9.00020-2
Sandhwar VK, Saxena S, Saxena D, Tiwari A, Parikh SM (2025). Chapter 13—Future trends and emerging technologies in water quality management. In: Dubey AK, Srivastav AL, Kumar A, Garcia Marquez FP, Giannakoudakis DA (eds) Computational automation for water security. Elsevier, pp 229–249. https://doi.org/10.1016/B978-0-443-33321-7.00012-3
Sangamnere R, Misra T, Bherwani H, Kapley A, Kumar R (2023) A critical review of conventional and emerging wastewater treatment technologies. Sustain Water Resour Manag 9:58. https://doi.org/10.1007/s40899-023-00829-y
Sathya K, Nagarajan K, Carlin Geor Malar G, Rajalakshmi S, Raja Lakshmi P (2022) A comprehensive review on comparison among effluent treatment methods and modern methods of treatment of industrial wastewater effluent from different sources. Appl Water Sci 12:70. https://doi.org/10.1007/s13201-022-01594-7
Shah AA, Walia S, Kazemian H (2024) Advancements in combined electrocoagulation processes for sustainable wastewater treatment: a comprehensive review of mechanisms, performance, and emerging applications. Water Res 252:121248. https://doi.org/10.1016/j.watres.2024.121248
Sharma J, Joshi M, Bhatnagar A, Chaurasia AK, Nigam S (2022a) Pharmaceutical residues: one of the significant problems in achieving 'clean water for all' and its solution. Environ Res 215:114219. https://doi.org/10.1016/j.envres.2022.114219
Sharma LD, Sarangthem I, Thangjam R, Sadhukhan R, Oinam N, Yanglem B, Banarjee L, Anal HJ, Jatav HS (2022) Sewage sludge and its health risk assessment: opportunities and challenges. In: Rajput VD, Yadav AN, Jatav HS, Singh SK, Minkina T (eds) Sustainable management and utilization of sewage sludge. Springer International Publishing, Cham, pp 205–225. https://doi.org/10.1007/978-3-030-85226-9_10
Shehata N, Egirani D, Olabi AG, Inayat A, Abdelkareem MA, Chae K-J, Sayed ET (2023) Membrane-based water and wastewater treatment technologies: Issues, current trends, challenges, and role in achieving sustainable development goals, and circular economy. Chemosphere 320:137993. https://doi.org/10.1016/j.chemosphere.2023.137993
Shukla R, Ahammad SZ (2023) Performance assessment of a modified trickling filter and conventional activated sludge process along with tertiary treatment in removing emerging pollutants

from urban sewage. Sci Total Environ 858:159833. https://doi.org/10.1016/j.scitotenv.2022.159833

Silva L, Miranda F, Moreira Júnior P, Ferreira A, Fuji M, Cruz A, Essiptchouk A, Petraconi Filho G (2025) Investigating chemical changes, material properties, and off-gas emissions in sewage sludge through thermal plasma treatment. J Environ Manage 375:124256. https://doi.org/10.1016/j.jenvman.2025.124256

Singh AK, Kumar M, Bauddh K, Singh A, Singh P, Madhav S, Shukla SK (2023a) Environmental impacts of air pollution and its abatement by plant species: A comprehensive review. Environ Sci Pollut Res 30:79587–79616. https://doi.org/10.1007/s11356-023-28164-x

Singh AK, Shukla SK, Singh P, Madhav S, Tripathi A (2023b) Assessment of air pollution tolerance and anticipated performance index of roadside trees in urban and semi-urban regions. Environ Monit Assess 195:1135. https://doi.org/10.1007/s10661-023-11759-9

Singh AK, Tripathi G, Choudhary JK, Shukla SK, Singh A, Kumar M (2024) Harnessing geospatial approaches in air quality mapping and monitoring. In: Sharma C, Shukla AK, Pathak S, Singh VP (Eds.), Sustainable development and geospatial technology. Springer Nature Switzerland, Cham, pp. 181–195. https://doi.org/10.1007/978-3-031-65703-0_12

Singh AK, Behera S, Choudhary JK, Shukla SK (2025) Evaluating the role of tree flora in air pollution mitigation and heavy metals accumulation near brick kilns. Atmos Pollut Res 16:102436. https://doi.org/10.1016/j.apr.2025.102436

Siwal SS, Sheoran K, Saini AK, Vo D-VN, Wang Q, Thakur VK (2022) Advanced thermochemical conversion technologies used for energy generation: advancement and prospects. Fuel 321:124107. https://doi.org/10.1016/j.fuel.2022.124107

Śniatała B, Al-Hazmi HE, Sobotka D, Zhai J, Mąkinia J (2024) Advancing sustainable wastewater management: a comprehensive review of nutrient recovery products and their applications. Sci Total Environ 937:173446. https://doi.org/10.1016/j.scitotenv.2024.173446

Takdastan A, Talepour N, Taherian M (2025) A review of the oxic-settling-anaerobic (OSA) process for sustainable sludge minimisation from biological treatment of wastewater. Environ Technol Rev 14:1–16. https://doi.org/10.1080/21622515.2024.2439068

Tarpani RRZ, Azapagic A (2023) Life cycle sustainability assessment of advanced treatment techniques for urban wastewater reuse and sewage sludge resource recovery. Sci Total Environ 869:161771. https://doi.org/10.1016/j.scitotenv.2023.161771

Taşeli BK, Jacob-Lopes E, Zepka LQ, Deprá M (2025) Wastewater treatment and sludge management systems—the gutter-to-good approaches: the gutter-to-good approaches. BoD—books on demand

Valchev D, Ribarova I, Borisov B, Radovanov V, Lyubomirova V, Kostova I, Dimova G, Karpuzova O, Lazarova S (2024) Valuable elements in sludge from eight municipal wastewater treatment plants in relation to their recovery potential. Environ Sci Eur 36:11. https://doi.org/10.1186/s12302-023-00837-x

Wang J, Deng J, Du E, Guo H (2022) Reevaluation of radical-induced differentiation in UV-based advanced oxidation processes (UV/hydrogen peroxide, UV/peroxydisulfate, and UV/chlorine) for metronidazole removal: kinetics, mechanism, toxicity variation, and DFT studies. Sep Purif Technol 301:121905. https://doi.org/10.1016/j.seppur.2022.121905

Wang Q, Li X, Liu W, Zhai S, Xu Q, Huan C, Nie S, Ouyang Q, Wang H, Wang A (2023) Carbon source recovery from waste sludge reduces greenhouse gas emissions in a pilot-scale industrial wastewater treatment plant. Environ Sci Ecotechnol 14:100235. https://doi.org/10.1016/j.ese.2022.100235

Wang C, Qi W-K, Zhang S-J, Liu L-F, Peng Y-Z (2024a) Innovation for continuous aerobic granular sludge process in actual municipal sewage treatment: self-circulating up-flow fluidized bed process. Water Res 260:121862. https://doi.org/10.1016/j.watres.2024.121862

Wang L, Sun X, Gao F, Yang Y, Song R (2024b) Solar membrane distillation: an emerging technology for reverse osmosis concentrated brine treatment. Desalination 592:118124. https://doi.org/10.1016/j.desal.2024.118124

Wu X, Nawaz S, Li Y, Zhang H (2024) Environmental health hazards of untreated livestock wastewater: potential risks and future perspectives. Environ Sci Pollut Res 31:24745–24767. https://doi.org/10.1007/s11356-024-32853-6

Zhang H, Liu W-H, Liu C-X, Sun P, Zeng Y-P, Gao Y-Y, Wang H-F, Zeng RJ (2023) Enhancing waste management and nutrient recovery: Preparation of adsorption-type sludge-biochar value-added fertilizer from sewage sludge and *Pistia stratiotes*. J Clean Prod 429:139642. https://doi.org/10.1016/j.jclepro.2023.139642

Zhao C, Liu H, Liu H, Zhan X, Zhang X, Fu B, Shi M, Lu S, Huang F, Cui M, Li Y (2023a) Influence of thermal hydrolysis on sludge anaerobic digestion: release of humic acid promotes electron transport of methanogen. Chem Eng J 451:138455. https://doi.org/10.1016/j.cej.2022.138455

Zhao L, Fu G, Zeng A, Cheng B, Song Z, Hu Z (2023b) Effects of different aeration strategies and ammonia-nitrogen loads on nitrification performance and microbial community succession of mangrove constructed wetlands for saline wastewater treatment. Chemosphere 339:139685. https://doi.org/10.1016/j.chemosphere.2023.139685

Zhao W, Liu Y, Guo D, Zou D (2023c) Development application of rural domestic sewage treatment project in cold areas of Northeast China: opportunities and challenges. J Water Process Eng 56:104326. https://doi.org/10.1016/j.jwpe.2023.104326

Zhou Y, Zhu L, Yang B, Fan L, Meng X, Chu R, Jiang X, Li P, Li W, Chen H (2024) Heavy metal migration regimes in the production of syngas from solid waste by thermal plasma treatment. J Hazard Mater 461:132698. https://doi.org/10.1016/j.jhazmat.2023.132698

Zhu X, Labianca C, He M, Luo Z, Wu C, You S, Tsang DCW (2022) Life-cycle assessment of pyrolysis processes for sustainable production of biochar from agro-residues. Biores Technol 360:127601. https://doi.org/10.1016/j.biortech.2022.127601

MIX
Papier aus verantwortungsvollen Quellen
Paper from responsible sources
FSC® C105338

If you have any concerns about our products,
you can contact us on
ProductSafety@springernature.com

In case Publisher is established outside the EU,
the EU authorized representative is:
Springer Nature Customer Service Center GmbH
Europaplatz 3, 69115 Heidelberg, Germany

Printed by Libri Plureos GmbH
in Hamburg, Germany